# Lecture Notes in Mathematics

2112



More information about this series at http://www.springer.com/series/304

Christian Bär • Christian Becker

# Differential Characters

Christian Bär
Institut für Mathematik
Universität Potsdam
Potsdam, Germany

Christian Becker
Institut für Mathematik
Universität Potsdam
Potsdam, Germany

ISBN 978-3-319-07033-9 ISBN 978-3-319-07034-6 (eBook)
DOI 10.1007/978-3-319-07034-6
Springer Cham Heidelberg New York Dordrecht London

Lecture Notes in Mathematics ISSN print edition: 0075-8434
ISSN electronic edition: 1617-9692

Library of Congress Control Number: 2014945637

Mathematics Subject Classification (2010): 53C08, 55N20



Printed on acid-free paper

Springer is part of Springer Science+Business Media (www.springer.com)

# Preface

This text provides a systematic introduction to differential characters, as introduced by Cheeger and Simons. Differential characters form a model of what is nowadays called differential cohomology. In degree 2, integral cohomology of a space $X$ classifies U(1)-bundles over $X$ via the first Chern class, while differential characters correspond to U(1)-bundles with a connection. Similarly, in degree 3, integral cohomology classes classify gerbes over $X$ while differential characters correspond to gerbes with additional geometric structure.

We construct the product which provides differential cohomology with a ring structure and we describe the fiber integration map. In both cases, we show uniqueness in the sense that these operations are determined by certain natural axioms. This shows in particular that the various very different descriptions in the literature are equivalent. We present natural and explicit geometric formulas for both the product and the fiber integration map.

The underlying space $X$ may be more general than a finite-dimensional manifold. We allow for "smooth spaces" which contains loop spaces of manifolds, for instance. This is important for applications like the transgression map.

Up to now, there does not exist much literature on the relative version of differential characters. We investigate them in detail. In degree 2, a relative differential character corresponds to a U(1)-bundle with connection and a section over a subspace. We derive long exact sequences which relate absolute and relative differential characters. Fiber integration for fibers with boundary is naturally considered in the relative framework. The module structure of relative differential cohomology over the ring of absolute differential characters is derived.

We discuss various applications including chain field theories and higher dimensional holonomies which occur as actions in string theory.

Potsdam,
June 2014

*Christian Bär*
*Christian Becker*

# Contents

# Differential Characters and Geometric Chains

Christian Bär and Christian Becker

**Abstract** We study Cheeger-Simons differential characters and provide geometric descriptions of the ring structure and of the fiber integration map. The uniqueness of differential cohomology (up to unique natural transformation) is proved by deriving an explicit formula for any natural transformation between a differential cohomology theory and the model given by differential characters. Fiber integration for fibers with boundary is treated in the context of relative differential characters. As applications we treat higher-dimensional holonomy, parallel transport, and transgression.

## 1 Introduction

Differential characters were introduced by Cheeger and Simons in [24]. Let $X$ be a differentiable manifold. A differential character of degree $k$ on $X$ is a homomorphism $h : Z_{k-1}(X;\mathbb{Z}) \to \mathrm{U}(1)$. Here $Z_{k-1}(X;\mathbb{Z})$ denotes the group of smooth integral-valued singular cycles of degree $k-1$. It is supposed that the evaluation on boundaries is given by integration of a form, more precisely, there exists a differential form $\mathrm{curv}(h) \in \Omega^k(X)$ such that $h(\partial c) = \exp\left(2\pi i \int_c \mathrm{curv}(h)\right)$. The form $\mathrm{curv}(h)$ is uniquely determined by $h$ and is called its *curvature*. We denote the set of all differential characters on $X$ of degree $k$ by $\widehat{H}^k(X;\mathbb{Z})$.

In degree $k = 1$ a differential character is essentially a smooth $\mathrm{U}(1)$-valued function on $X$. If one is given a $\mathrm{U}(1)$-bundle over $X$ with connection, then one

Christian Bär
Universität Potsdam, Institut für Mathematik, Am Neuen Palais 10, 14469 Potsdam, Germany, e-mail: baer@math.uni-potsdam.de

Christian Becker
Universität Potsdam, Institut für Mathematik, Am Neuen Palais 10, 14469 Potsdam, Germany, e-mail: becker@math.uni-potsdam.de

can associate a differential character by mapping any 1-cycle to the holonomy of the bundle along this cycle. This sets up a bijection between isomorphism classes of U(1)-bundles with connection to the set of differential characters of degree $k = 2$. In a similar way, differential characters of higher degree correspond to "higher U(1)-gauge theories" like Hitchin gerbes in degree $k = 3$.

The Chern class provides a bijection between $H^2(X;\mathbb{Z})$ and the set of isomorphism classes of U(1)-bundles (without connection). Hence $\widehat{H}^2(X;\mathbb{Z})$ may be considered as a geometric enrichment of the singular cohomology group $H^2(X;\mathbb{Z})$. In fact, in any degree there is an analogous map $c : \widehat{H}^k(X;\mathbb{Z}) \to H^k(X;\mathbb{Z})$ associating to a differential character its *characteristic class*. This observation can be axiomatized and leads to the concept of *differential cohomology theory*. Differential characters form a model for differential cohomology. We give a constructive proof of the uniqueness of differential cohomology up to unique natural transformations by deriving an explicit formula for any natural transformation between a differential cohomology theory and differential characters.

Pointwise multiplication provides $\widehat{H}^k(X;\mathbb{Z})$ with an obvious abelian group structure. This is the *addition* on $\widehat{H}^k(X;\mathbb{Z})$. There is a less obvious *multiplication* $\widehat{H}^k(X;\mathbb{Z}) \times \widehat{H}^l(X;\mathbb{Z}) \to \widehat{H}^{k+l}(X;\mathbb{Z})$ which turns $\widehat{H}^*(X;\mathbb{Z})$ into a ring. We show that a set of natural axioms uniquely determines the ring structure. Again, the proof is constructive and gives us an explicit geometric description of the ring structure, quite different from the original definition in [24].

Like for singular cohomology and for differential forms there is a concept of *fiber integration* for differential characters. We show that naturality and two compatibility conditions uniquely determine the fiber integration map. Let $\pi : E \to X$ be a fiber bundle with closed oriented fibers $F$. For the fiber integration map $\widehat{\pi}_! : \widehat{H}^{k+\dim(F)}(E;\mathbb{Z}) \to \widehat{H}^k(X;\mathbb{Z})$ we obtain the geometric formula

$$(\widehat{\pi}_! h)(z) = h(\lambda(z)) \cdot \exp\Big(2\pi i \int_{a(z)} \fint_F \operatorname{curv}(h)\Big).$$

Here $\lambda$ is a *transfer map* and essentially does the following: given a cycle $z$ in $X$ look at the homology class represented by $z$ and choose a closed manifold whose fundamental class also represents this homology class. Then pull back the bundle $E$ to this manifold and take a representing cycle of the fundamental class of the resulting total space. This is then a cycle in $E$ which can be inserted into $h$. The "correction factor" $\exp\big(2\pi i \int_{a(z)} \fint_F \operatorname{curv}(h)\big)$ involves the fiber integration $\fint$ of differential forms and a chain $a(z)$ associated with $z$. It ensures that the construction is independent of the choices.

The uniqueness results for fiber integration and for differential cohomology together show that the various fiber integration maps for different models of differential cohomology in the literature are all equivalent.

There is the technical problem that not every homology class can be represented by a manifold. For this reason we have to allow for certain "manifolds" with singularities, called *stratifolds*. We use stratifolds to define *geometric chains* in order to provide a geometric description of singular homology theory.

There is a second reason to consider differential characters on more general "smooth spaces", rather than manifolds only. Certain infinite-dimensional manifolds have to be allowed because we want to apply the theory to the loop space of a manifold, for instance.

The multiplication $*$ and the fiber integration map are compatible: Given $h \in \widehat{H}^k(X;\mathbb{Z})$ and $f \in \widehat{H}^l(E;\mathbb{Z})$, we show that the *up-down formula* holds:

$$\widehat{\pi}_!(\pi^* h * f) = h * (\widehat{\pi}_! f) \in \widehat{H}^{k+l-\dim F}(X;\mathbb{Z}).$$

If the fibers of the bundle bound, then the fiber integrated differential character turns out to be topologically trivial. This means that its characteristic class vanishes. One finds an explicit topological trivialization involving the curvature. A special case of this situation is the well-known *homotopy formula*. Let $f : [0,1] \times X \to Y$ be a homotopy between smooth maps $f_0, f_1 : X \to Y$ and $h \in \widehat{H}^k(Y;\mathbb{Z})$. Then we find

$$f_1^* h - f_0^* h = \iota\Big( \int_0^1 f_s^* \mathrm{curv}(h) ds \Big) .$$

We also consider the groups of *relative differential characters*, denoted $\widehat{H}^k(\varphi;\mathbb{Z})$, where $\varphi : A \to X$ is a smooth map. In degree $k = 1$ they correspond to smooth U(1)-valued functions on $X$ with a lift to an $\mathbb{R}$-valued function over $A$. In degree $k = 2$ they correspond to U(1)-bundles with connection over $X$ with a section over $A$. We derive long exact sequences relating absolute and relative differential characters. Since differential cohomology theories are not cohomology theories in the usual sense, these exact sequences are more subtle than those in singular cohomology theory, for instance. Our sequences provide criteria for a differential character to be topologically trivial over $A$. *Fiber integration for fibers with boundary* can now be defined. It is a map $\widehat{\pi}_!^E : \widehat{H}^{k+\dim(F)}(E;\mathbb{Z}) \to \widehat{H}^{k+1}(\mathrm{id}_X;\mathbb{Z})$.

We apply fiber integration to construct transgression maps to the loop space $\mathcal{L}(X)$ of a smooth manifolds $X$ and more general mapping spaces. Transgression along $S^1$ is a homomorphism $\widehat{H}^k(X;\mathbb{Z}) \to \widehat{H}^{k-1}(\mathcal{L}(X);\mathbb{Z})$. It is constructed by pull-back of differential characters from $X$ to $\mathcal{L}(X) \times S^1$ using the evaluation map followed by integration over the fiber of the trivial bundle. Analogously, we define transgression along any oriented closed manifold $\Sigma$. Using fiber integration for fibers with boundary we also define transgression along a compact oriented manifold with boundary.

Differential characters are *thin invariant*: A smooth singular chain $c \in C_k(X;\mathbb{Z})$ is called *thin* if the integral of any $k$-form over $c$ vanishes. For instance this happens if $c$ is supported on a $(k-1)$-dimensional submanifold.

Differential characters of degree $k$ vanish on boundaries of thin $k$-chains. In particular, they are invariant under barycentric subdivision.

We apply the notion of thin invariance to chain field theories, a modification of topological quantum field theories in the sense of Atiyah. Generalizing work of Bunke and others, we show that chain field theories are invariant under thin 2-morphisms.

## *Related Concepts and Literature*

Differential characters were introduced by Cheeger and Simons in [24]. The ring of differential characters is a particular model of what is nowadays called *differential cohomology*. Meanwhile there are various different models for differential cohomology. We briefly mention those models and hint to the literature.

The particular case of degree 3 differential cohomology is known as the isomorphism group of U(1)-gerbes with connections. Gerbes may be described either as sheaves of groupoids [9] or as cycles of the total complex of a truncated Čech-de Rham double complex [19]. The latter goes under the name of the smooth Deligne complex.

More geometric models of gerbes have been introduced by Hitchin [44] and Murray [51, 52]. The latter are called bundle gerbes and are frequently used in various applications. The notion of bundle 2-gerbes provides a generalization to represent degree 4 differential cohomology [63].

Besides the model of differential characters, differential cohomology in arbitrary degree can also be constructed in terms of the hypercohomology of sequences of sheaves [9], by the smooth Deligne complex [19], by differential forms with singularities [23], by de Rham-Federer currents [40, 43, 41, 42], by differential cocycles [45] or by stratifolds [12]. An axiomatic treatment of differential cohomology has been developed by Simons and Sullivan in [60].

Differential refinements for a large class of generalized cohomology theories have been constructed by Hopkins-Singer in [45]. Different constructions were obtained more recently in [10] and [13]. A particularly interesting example of differential generalized cohomology is differential $K$-theory. An axiomatic treatment of differential $K$-theory has been developed by Bunke and Schick in [14]. Geometric models of differential $K$-theory appear in the work of Freed and Lott [32] and Simons and Sullivan [61]. Differential algebraic $K$-theory is discussed in [11, 16].

Differential cohomology groups also appear in mathematical physics in various contexts. For instance, they are used to describe the anomaly bundle on loop space [35, 9, 20]. The classical Chern-Simons action is the holonomy of a certain degree 4 differential cohomology class [24, 30]. Analogously, the Wess-Zumino-Witten term is the holonomy of a degree 3 differential cohomology class [22, 36, 19, 21, 58]. In mathematical physics, the local data that describe connections on (bundle) gerbes go under the name of $B$-fields. Their field strengths are 3-forms associated with the gerbe with connection. Bundle 2-gerbes and there trivializations (both with connections) are used to describe geometric String structures [53, 21, 68]. $D$-brane charges in String theory are certain classes in differential $K$-theory [31].

**Acknowledgements** It is a great pleasure to thank Matthias Kreck for very helpful discussion. Moreover, the authors thank *Sonderforschungsbereich 647* funded by *Deutsche Forschungsgemeinschaft* for financial support.

## 2 Smooth Spaces

Differential characters were introduced by Cheeger and Simons in [24] on finite-dimensional smooth manifolds. We will need to consider differential characters on more general spaces $X$. First of all, $X$ may be a manifold with a nonempty boundary. Secondly, we have to allow certain infinite-dimensional spaces because we want to include examples such as the loop space $X = \mathcal{L}(M) = C^\infty(S^1, M)$ of a finite-dimensional manifold $M$. Thirdly, $X$ may also be any oriented compact regular $p$-stratifold as in [47]. Stratifolds will be needed to represent homology classes.

One convenient class of spaces to work with is that of differential spaces in the sense of Sikorski [62]. Recall their definition:

**Definition 1.** A *differential space* is a pair $(X, C^\infty(X))$ where $X$ is a topological space and $C^\infty(X)$ is a subset of the set $C^0(X)$ of all continuous real-valued functions such that the following holds:

- *Initial topology:* $X$ carries the weakest topology for which all functions in $C^\infty(X)$ are continuous;
- *Locality:* If $f \in C^0(X)$ is such that for every point in $X$ there is a function $g \in C^\infty(X)$ coinciding with $f$ on a neighborhood of that point, then $f \in C^\infty(X)$;
- *Composition with smooth functions:* If $f_1, \dots, f_k \in C^\infty(X)$ and $g$ is a smooth function defined on an open neighborhood of $f_1(X) \times f_k(X) \subset \mathbb{R}^k$, then $g \circ (f_1, \dots, f_k) \in C^\infty(X)$.

The functions in $C^\infty(X)$ are called *smooth functions.* A map $f : X \to Y$ between differential spaces is called *smooth* if smooth functions on $Y$ pull back to smooth functions on $X$ along $f$. This way we obtain the category of differential spaces.

On differential spaces one can define tangent vectors, $k$-forms, their exterior differential and one can pull back forms. The usual rules such as the

Stokes theorem apply [50]. In addition to that we will need that certain definitions of homology and cohomology which are equivalent in the case of manifolds remain equivalent.

**Definition 2.** A differential space is called a *smooth space* if the following holds:

- *Continuous versus smooth singular (co-)homology:* The inclusion of the complex of smooth singular chains (with integral coefficients) into that of continuous singular chains induces isomorphisms for the corresponding homology and cohomology theories;
- *de Rham theorem:* Integration of differential forms induces an isomorphism from de Rham cohomology to smooth singular cohomology with real coefficients;
- *Stratifold- versus singular homology:* Pushing forward fundamental cycles induces an isomorphism from the bordism theory of oriented $p$-stratifolds to smooth singular homology theory with integral coefficients.

Finite-dimensional manifolds (possibly with boundary), stratifolds and also infinite-dimensional Fréchet manifolds such as the loop space of a compact manifold are all examples for smooth spaces, see [48, Ch. 7] for infinite-dimensional manifolds and [47, 28] for stratifolds.

*Remark 3.* Instead of differential spaces one could also use diffeological spaces as in [46] to define smooth spaces in Definition 2. A smooth space would then be defined as a diffeological space satisfying the properties in Definition 2. These properties are not automatic; by [46, p. 272] there are diffeological spaces for which the de Rham map fails to be an isomorphism.

# 3 Refined Smooth Singular Homology

Let $X$ be a smooth space in the sense explained above. For $n \in \mathbb{N}_0$, we denote by $C_n(X;\mathbb{Z})$ the abelian group of smooth singular $n$-chains in $X$ with integral coefficients. The spaces of $n$-cycles and $n$-boundaries of the complex $(C_n(X;\mathbb{Z}), \partial)$ are denoted by $Z_n(X;\mathbb{Z})$ and $B_n(X;\mathbb{Z})$, respectively. Denote the space of smooth $n$-forms on $X$ by $\Omega^n(X)$.

**Definition 4 (Thin Chains).** A smooth singular chain $y \in C_n(X;\mathbb{Z})$ is called *thin* if

$$\int_y \omega = 0$$

for all $\omega \in \Omega^n(X)$. We denote by $S_n(X;\mathbb{Z}) \subset C_n(X;\mathbb{Z})$ the subgroup of thin $n$-chains in $X$.

This definition of thin chains is similar to that of *thin homotopies* in the literature, see e.g. [2, 18]. Thin homotopies will not occur in this paper, however.

If $X$ and $Y$ are smooth spaces and if $f : X \to Y$ is a smooth map, then if $c \in C_n(X;\mathbb{Z})$ is thin, so is $f_*c \in C_n(Y;\mathbb{Z})$. Namely, for any $\omega \in \Omega^n(Y)$ we have

$$\int_{f_*c} \omega = \int_c f^*\omega = 0.$$

Hence $f_*(S_n(X;\mathbb{Z})) \subset S_n(Y;\mathbb{Z})$ and thus $f_*$ induces a homomorphism $f_* : C_n(X;\mathbb{Z})/S_n(X;\mathbb{Z}) \to C_n(Y;\mathbb{Z})/S_n(Y;\mathbb{Z})$.

Denote the equivalence class of $c \in C_n(X;\mathbb{Z})$ in $C_n(X;\mathbb{Z})/S_n(X;\mathbb{Z})$ by $[c]_{S_n}$. By definition, integration of an $n$-form $\omega \in \Omega^n(X)$ descends to a linear map $C_n(X;\mathbb{Z})/S_n(X;\mathbb{Z}) \to \mathbb{R}$, $[c]_{S_n} \mapsto \int_c \omega$.

Moreover, thin chains are preserved by the boundary operator. Namely, for $c \in S_{n+1}(X;\mathbb{Z})$ and any $\eta \in \Omega^n$ we have by the Stokes theorem

$$\int_{\partial c} \eta = \int_c d\eta = 0.$$

Thus $\partial S_{n+1}(X;\mathbb{Z}) \subset S_n(X;\mathbb{Z})$. The boundary operator induces a homomorphism

$$C_{n+1}(X;\mathbb{Z})/S_{n+1}(X;\mathbb{Z}) \xrightarrow{\partial} B_n(X;\mathbb{Z})/\partial S_{n+1}(X;\mathbb{Z})\,.$$

Since $Z_n(X;\mathbb{Z}) \subset C_n(X;\mathbb{Z})$ and $\partial S_{n+1}(X;\mathbb{Z}) \subset S_n(X;\mathbb{Z})$ we have a natural homomorphism

$$Z_n(X;\mathbb{Z})/\partial S_{n+1}(X;\mathbb{Z}) \longrightarrow C_n(X;\mathbb{Z})/S_n(X;\mathbb{Z})\,. \tag{1}$$

Denote the equivalence class of $z \in Z_n(X;\mathbb{Z})$ in $Z_n(X;\mathbb{Z})/\partial S_{n+1}(X;\mathbb{Z})$ by $[z]_{\partial S_{n+1}}$. Integration of differential forms induces well-defined maps

$$\Omega^n(X) \times C_n(X;\mathbb{Z})/S_n(X;\mathbb{Z}) \to \mathbb{R}, \quad (\eta, [c]_{S_n}) \mapsto \int_{[c]_{S_n}} \eta := \int_c \eta,$$

and

$$\Omega^n(X) \times Z_n(X;\mathbb{Z})/\partial S_{n+1}(X;\mathbb{Z}) \to \mathbb{R}, \quad (\eta, [z]_{\partial S_{n+1}}) \mapsto \int_{[z]_{\partial S_{n+1}}} \eta := \int_z \eta\,.$$

The Stokes theorem says

$$\int_{[c]_{S_n}} d\eta = \int_{\partial [c]_{S_n}} \eta.$$

Recall that for a closed form $\omega \in \Omega^n(X)$, integration over a smooth singular cycle $z \in Z_n(X;\mathbb{Z})$ corresponds to evaluation of the de Rham class $[\omega]_{\mathrm{dR}} \in$

$H^n_{\mathrm{dR}}(X)$ on the homology class $[z] \in H_n(X;\mathbb{Z})$, i.e.,

$$\int_z \omega = \langle [\omega]_{\mathrm{dR}}, [z] \rangle .$$

*Remark 5.* The quotients $C_n(X;\mathbb{Z})/S_n(X;\mathbb{Z})$ and $Z_n(X;\mathbb{Z})/\partial S_{n+1}(X;\mathbb{Z})$ are geometrically very natural and appear in elementary constructions: for instance, if $X$ is a closed smooth oriented $n$-manifold (or, more generally, an oriented compact $n$-dimensional regular $p$-stratifold without boundary) and if $c, c' \in Z_n(X;\mathbb{Z})$ represent the fundamental class of $X$, then they are homologous, i.e., there exists $y \in C_{n+1}(X;\mathbb{Z})$ with $c - c' = \partial y$. For dimensional reasons $C_{n+1}(X;\mathbb{Z}) = S_{n+1}(X;\mathbb{Z})$, hence $[c]_{\partial S_{n+1}} = [c']_{\partial S_{n+1}}$. In fact, in this case $H_n(X;\mathbb{Z}) = Z_n(X;\mathbb{Z})/B_n(X;\mathbb{Z}) = Z_n(X;\mathbb{Z})/\partial S_{n+1}(X;\mathbb{Z})$.

If $X$ has a boundary and $c, c' \in C_n(X;\mathbb{Z})$ represent the fundamental class of $X$ in $H_n(X, \partial X;\mathbb{Z})$, then we can find $y \in C_{n+1}(X;\mathbb{Z}) = S_{n+1}(X;\mathbb{Z})$ such that $c - c' - \partial y$ is supported in the boundary of $X$ and is hence thin. Therefore $[c]_{S_n} = [c']_{S_n}$ in this case.

Generalizations of these elementary observations are crucial for the construction of geometric chains in the next section.

## 4 Geometric Chains

We now define our notion of geometric chains. The idea is to represent singular homology classes in $X$ by manifolds because this geometric description is well adapted for a geometric definition of fiber integration for Cheeger-Simons differential characters as we shall see. There is the problem however, that not all homology classes are representable by smooth manifolds. Fortunately, Kreck's stratifolds [47] provide a suitable generalization of manifolds which repairs this defect.

For $n \in \mathbb{N}_0$ let $\mathcal{C}_n(X)$ be the set of diffeomorphism classes of smooth maps $f : M \to X$ where $M$ is an oriented compact $n$-dimensional regular $p$-stratifold with boundary $\partial M$, and the boundary itself is a closed stratifold, compare [47, pp. 35 and 43]. Here two maps $f : M \to X$ and $f' : M' \to X$ are called diffeomorphic if there is an orientation preserving diffeomorphism $F : M \to M'$ such that

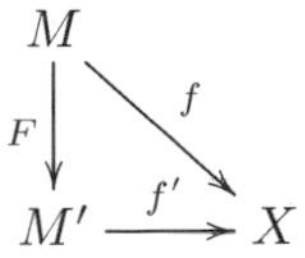

$$\begin{array}{ccc} M & & \\ {\scriptstyle F}\downarrow & \searrow^{f} & \\ M' & \xrightarrow{f'} & X \end{array}$$

commutes. The equivalence class of $f : M \to X$ is denoted by $[M \xrightarrow{f} X]$. For $n < 0$ put $\mathcal{C}_n(X) := \{0\}$. If $f : X \to Y$ is a smooth map, then we define $f_* : \mathcal{C}_n(X) \to \mathcal{C}_n(Y)$ by $f_*([M \xrightarrow{g} X]) := [M \xrightarrow{f \circ g} Y]$.

Disjoint union defines a structure of abelian semigroup on $\mathcal{C}_n(X)$. The boundary operator $\partial : \mathcal{C}_n(X) \to \mathcal{C}_{n-1}(X)$ is given by restriction to the geometric boundary. For the boundary orientation we use the convention that an outward pointing tangent vector of $M$ at a regular point $p$ of $\partial M$ followed by an oriented basis of $T_p(\partial M)$ yields an oriented basis of $T_pM$.

We define a homomorphism $\varphi_n : \mathcal{C}_n(X) \to C_n(X;\mathbb{Z})/S_n(X;\mathbb{Z})$ as follows: For $f : M \to X$ choose a representing $n$-chain $c$ of the fundamental class of $M$ in $H_n(M, \partial M;\mathbb{Z})$. Then the equivalence class of $c$ in $C_n(M;\mathbb{Z})/S_n(M;\mathbb{Z})$ is independent of the particular choice of $c$ and we put $\varphi_n([M \xrightarrow{f} X]) := [f_*(c)]_{S_n}$.

Similarly, if $\partial M = \emptyset$, then the equivalence class in $Z_n(M;\mathbb{Z})/\partial S_{n+1}(M;\mathbb{Z})$ of an $n$-cycle $c$ representing the fundamental class of $M$ in $H_n(M;\mathbb{Z})$ does not depend on the particular choice of $c$ and we can define $\psi_n : \mathcal{Z}_n(X) \to Z_n(X;\mathbb{Z})/\partial S_{n+1}(X;\mathbb{Z})$ by $\psi_n([M \xrightarrow{f} X]) := [f_*(c)]_{\partial S_{n+1}}$.

We call elements of $\mathcal{C}_n(X)$ *geometric chains* and elements of

$$\begin{aligned} \mathcal{Z}_n(X) &:= \{\zeta \in \mathcal{C}_n(X) \mid \partial\zeta = 0\} \\ \text{and} \quad \mathcal{B}_n(X) &:= \{\zeta \in \mathcal{C}_n(X) \mid \exists \beta \in \mathcal{C}_{n+1}(X) : \partial\beta = \zeta\} \end{aligned}$$

*geometric cycles* and *geometric boundaries*, respectively. We obtain the following commutative diagram:

$$\begin{array}{ccccccccc} \cdots \longrightarrow & \mathcal{C}_{n+1}(X) & \xrightarrow{\partial} & \mathcal{B}_n(X) & \xrightarrow{\text{inclusion}} & \mathcal{Z}_n(X) & \xrightarrow{\text{inclusion}} & \mathcal{C}_n(X) & \longrightarrow \cdots \\ & \downarrow{\scriptstyle \varphi_{n+1}} & & \downarrow{\scriptstyle \psi_n|_{\mathcal{B}_n(X)}} & & \downarrow{\scriptstyle \psi_n} & & \downarrow{\scriptstyle \varphi_n} & \\ \cdots \longrightarrow & \frac{C_{n+1}(X;\mathbb{Z})}{S_{n+1}(X;\mathbb{Z})} & \xrightarrow{\partial} & \frac{B_n(X;\mathbb{Z})}{\partial S_{n+1}(X;\mathbb{Z})} & \xrightarrow{\text{inclusion}} & \frac{Z_n(X;\mathbb{Z})}{\partial S_{n+1}(X;\mathbb{Z})} & \longrightarrow & \frac{C_n(X;\mathbb{Z})}{S_n(X;\mathbb{Z})} & \longrightarrow \cdots \end{array} \tag{2}$$

The map $Z_n(X;\mathbb{Z})/\partial S_{n+1}(X;\mathbb{Z}) \to C_n(X;\mathbb{Z})/S_n(X;\mathbb{Z})$ is the one from (1). Diagram (2) is natural. In particular, for any smooth map $f : X \to Y$ the diagram

$$\begin{array}{ccc} \mathcal{C}_n(X) & \xrightarrow{f_*} & \mathcal{C}_n(Y) \\ \downarrow{\scriptstyle \varphi_n} & & \downarrow{\scriptstyle \varphi_n} \\ \frac{C_n(X;\mathbb{Z})}{S_n(X;\mathbb{Z})} & \xrightarrow{f_*} & \frac{C_n(Y;\mathbb{Z})}{S_n(Y;\mathbb{Z})} \end{array}$$

commutes and similarly for $\psi_n$.

From now on, we will, by slight abuse of notation, write $[\zeta]_{\partial S_{n+1}}$ instead of $\psi_n(\zeta)$ for $\zeta \in \mathcal{Z}_n(X)$ and $[\beta]_{S_n}$ instead of $\varphi_n(\beta)$ for $\beta \in \mathcal{C}_n(X)$.

For an oriented stratifold $M$ we denote by $\overline{M}$ the same stratifold with reversed orientation. Then $[M \xrightarrow{f} X] \mapsto [\overline{M} \xrightarrow{f} X]$ is an involution on $\mathcal{C}_n(X)$ which commutes with $\partial$. Furthermore, $\zeta + \overline{\zeta} \in \mathcal{B}_n(X)$ for any $\zeta \in \mathcal{Z}_n(X)$ because $f \sqcup f : M \sqcup \overline{M} \to X$ is bounded by $f : [0,1] \times M \to X$. In other words, the involution $\overline{\phantom{x}} : \mathcal{Z}_n(X) \to \mathcal{Z}_n(X)$ induces $-\mathrm{id}$ on homology,

$$[\bar{\zeta}] = -[\zeta] \text{ in } \mathcal{H}_n(X) := \mathcal{Z}_n(X)/\mathcal{B}_n(X).$$

In particular, the geometric homology $\mathcal{H}_n(X) := \mathcal{Z}_n(X)/\mathcal{B}_n(X)$ is an abelian group, not just a semigroup.

The reason for using stratifolds instead of manifolds is the fact that the homomorphisms $\psi_n : \mathcal{Z}_n(X) \to Z_n(X;\mathbb{Z})/\partial S_{n+1}(X;\mathbb{Z})$ induce isomorphisms on homology (see [47, Thm. 20.1]) :

$$\mathcal{H}_n(X) := \frac{\mathcal{Z}_n(X)}{\mathcal{B}_n(X)} \longrightarrow \frac{Z_n(X;\mathbb{Z})/\partial S_{n+1}(X;\mathbb{Z})}{B_n(X;\mathbb{Z})/\partial S_{n+1}(X;\mathbb{Z})} = \frac{Z_n(X;\mathbb{Z})}{B_n(X;\mathbb{Z})} = H_n(X;\mathbb{Z})\,.$$

The *cross product* of geometric chains is defined by

$$\times : \mathcal{C}_k(X) \otimes \mathcal{C}_{k'}(X') \to \mathcal{C}_{k+k'}(X \times X'),$$
$$[M \xrightarrow{g} X] \otimes [M' \xrightarrow{g'} X'] \mapsto [M \times M' \xrightarrow{g\times g'} X \times X'].$$

By [47, Thm. 20.1] this cross product in $\mathcal{H}_*$ is compatible with the usual cross product in $H_*$.

*Remark 6.* At various occasions we will have to extend homomorphisms $Z_n(X;\mathbb{Z}) \to G$ to homomorphisms $C_n(X;\mathbb{Z}) \to G$ where $G$ is an abelian group. Since $B_{n-1}(X;\mathbb{Z})$ is free, the exact sequence

$$0 \to Z_n(X;\mathbb{Z}) \xrightarrow{i} C_n(X;\mathbb{Z}) \xrightarrow{\partial} B_{n-1}(X;\mathbb{Z}) \to 0$$

splits, though not canonically. In particular, any basis of $Z_n(X;\mathbb{Z})$ can be extended to a basis of $C_n(X;\mathbb{Z})$. Therefore, any group homomorphism $Z_n(X;\mathbb{Z}) \to G$ can be extended as a group homomorphism to $C_n(X;\mathbb{Z}) \to G$ by defining it in an arbitrary manner on the complementary basis elements.

**Lemma 7 (Representation by Geometric Chains).** *There are homomorphisms $\zeta : C_{n+1}(X;\mathbb{Z}) \to \mathcal{C}_{n+1}(X)$, $a : C_n(X;\mathbb{Z}) \to C_{n+1}(X;\mathbb{Z})$, and $y : C_{n+1}(X;\mathbb{Z}) \to Z_{n+1}(X;\mathbb{Z})$ such that*

$$\partial\zeta(c) = \zeta(\partial c) \qquad \text{for all } c \in C_{n+1}(X;\mathbb{Z}); \quad (3)$$
$$[\zeta(c)]_{S_{n+1}} = [c - a(\partial c) - \partial a(c + y(c))]_{S_{n+1}} \qquad \text{for all } c \in C_{n+1}(X;\mathbb{Z}); \quad (4)$$
$$[\zeta(z)]_{\partial S_{n+1}} = [z - \partial a(z)]_{\partial S_{n+1}} \qquad \text{for all } z \in Z_{n+1}(X;\mathbb{Z}). \quad (5)$$

*Proof.* a) For any $z \in Z_n(X;\mathbb{Z})$ the singular homology class represented by $z$ lies in the image of the map induced by $\psi_n$. Hence we may choose a geometric cycle $\zeta(z) \in \mathcal{Z}_n(X)$ such that $[z]_{\partial S_{n+1}} - [\zeta(z)]_{\partial S_{n+1}} \in B_n(X;\mathbb{Z})/\partial S_{n+1}(X;\mathbb{Z})$. We may thus choose a smooth singular chain $a(z) \in C_{n+1}(X;\mathbb{Z})$ such that (5) holds. In particular, if $z = \partial c \in B_n(X;\mathbb{Z})$ is a smooth singular boundary, then $\zeta(z) = \zeta(\partial c) \in \mathcal{B}_n(X)$ is a geometric boundary.

Since $Z_n(X;\mathbb{Z})$ is free, the choices in $z \mapsto \zeta(z)$ and $z \mapsto a(z)$ can be made such that $\zeta : Z_n(X;\mathbb{Z}) \to \mathcal{Z}_n(X)$ and $a : Z_n(X;\mathbb{Z}) \to C_{n+1}(X;\mathbb{Z})$ are homomorphisms. One simply makes choices on elements of a basis of $Z_n(X;\mathbb{Z})$ and extends as a homomorphism. In particular, we then have $\zeta(0) = 0$. We perform this construction in all degrees $n \in \mathbb{N}_0$. By Remark 6 we can extend $a$ to a homomorphism $a : C_n(X;\mathbb{Z}) \to C_{n+1}(X;\mathbb{Z})$.

b) We construct an extension of the homomorphism $\zeta$ to a homomorphism from singular chains to geometric chains such that it commutes with the boundary operations. As an auxiliary tool, we first define a group homomorphism $\alpha : C_{n+1}(X;\mathbb{Z}) \to \mathcal{C}_{n+1}(X)$ by choosing $\alpha(c)$ on basis elements and extending as a homomorphism. On the basis elements of $Z_{n+1}(X;Z)$ we set $\alpha(c) = \zeta(c)$. On the complementary basis elements we choose $\alpha(c)$ such that $\partial\alpha(c) = \zeta(\partial c)$. This can be done since $\zeta(\partial c)$ is a geometric boundary. We then have

$$[\partial(c - a(\partial c) - \partial a(c))]_{\partial S_{n+1}} = [\partial c - \partial a(\partial c)]_{\partial S_{n+1}} \overset{(5)}{=} [\zeta(\partial c)]_{\partial S_{n+1}} = \partial[\alpha(c)]_{S_{n+1}}. \tag{6}$$

Hence there exists a smooth singular cycle $y(c) \in Z_{n+1}(X;\mathbb{Z})$ such that

$$[c - a(\partial c) - \partial a(c) - y(c)]_{S_{n+1}} = [\alpha(c)]_{S_{n+1}}. \tag{7}$$

We can choose $c \mapsto y(c)$ as a group homomorphism $y : C_{n+1}(X;\mathbb{Z}) \to Z_{n+1}(X;\mathbb{Z})$ by defining it on basis elements, as explained above. On the basis elements of $Z_{n+1}(X;\mathbb{Z})$ we set $y(c) = 0$. Condition (5) implies that (7) holds in this case. On the complementary basis elements, we choose $y(c) \in Z_{n+1}(X;\mathbb{Z})$ such that (7) holds.

We have $\zeta(y(c)) \in \mathcal{Z}_{n+1}(X)$ and $a(y(c)) \in C_{n+2}(X;\mathbb{Z})$ with

$$[y(c) - \partial a(y(c))]_{\partial S_{n+2}} = [\zeta(y(c))]_{\partial S_{n+2}}.$$

If $c \in C_{n+1}(X;\mathbb{Z})$ is a cycle we have $\alpha(c) + \zeta(y(c)) = \zeta(c) + \zeta(0) = \zeta(c)$. We may thus extend the homomorphism $\zeta : Z_{n+1}(X;\mathbb{Z}) \to \mathcal{Z}_{n+1}(X)$ constructed above to a homomorphism $\zeta : C_{n+1}(X;\mathbb{Z}) \to \mathcal{C}_{n+1}(X)$ by setting $\zeta(c) := \alpha(c) + \zeta(y(c)) \in \mathcal{C}_{n+1}(X)$. We perform this construction in all degrees $n \in \mathbb{N}_0$.

c) We have constructed a group homomorphism $\zeta : C_{n+1}(X;\mathbb{Z}) \to \mathcal{C}_{n+1}(X)$ such that in addition to (5) we have for all $c \in C_{n+1}(X;\mathbb{Z})$:

$$\partial\zeta(c) = \partial\alpha(c) + \partial\zeta(y(c)) = \zeta(\partial c) + 0 = \zeta(\partial c)$$

which is (3) and

$$\begin{aligned}
[c - a(\partial c) - &\partial a(c + y(c))]_{S_{n+1}} \\
&= [c - a(\partial c) - \partial a(c) - y(c)]_{S_{n+1}} + [y(c) - \partial a(y(c))]_{S_{n+1}} \\
&\overset{(7),(5)}{=} [\alpha(c)]_{S_{n+1}} + [\zeta(y(c))]_{S_{n+1}} \\
&= [\zeta(c)]_{S_{n+1}}.
\end{aligned} \tag{8}$$

which is (4). □

Now we turn to fiber bundles. Let $F \hookrightarrow E \twoheadrightarrow X$ be a fiber bundle whose fibers are compact oriented manifolds possibly with boundary. For a geometric chain $\zeta = [M \xrightarrow{g} X] \in \mathcal{C}_k(X)$ (or a geometric cycle $\zeta \in \mathcal{Z}_k(X)$ if $F$ has a boundary) let

$$\begin{array}{ccc} g^*E & \xrightarrow{\widetilde{g}} & E \\ \downarrow & & \downarrow \\ M & \xrightarrow{g} & X \end{array}$$

be the pull-back of the fiber bundle to $M$. Since $M$ and $F$ do not both have a boundary, $g^*E$ is an $(k+\dim F)$-dimensional compact oriented stratifold with boundary. The orientation of $g^*E$ is chosen such that an oriented horizontal tangent basis (defined by the orientation of $M$) followed by an oriented tangent basis along the fiber yields an oriented tangent basis of the total space. Put

$$\mathrm{PB}_E(\zeta) := [g^*E \xrightarrow{\widetilde{g}} E] \in \mathcal{C}_{k+\dim F}(E).$$

This defines homomorphisms $\mathrm{PB}_E : \mathcal{Z}_k(X) \to \mathcal{C}_{k+\dim F}(E)$ and also $\mathrm{PB}_E : \mathcal{C}_k(X) \to \mathcal{C}_{k+\dim F}(E)$ if $\partial F = \emptyset$. The following holds:

- For each $\zeta \in \mathcal{Z}_k(X)$ we have

$$\partial(\mathrm{PB}_E\zeta) = \begin{cases} \overline{\mathrm{PB}_{\partial E}(\zeta)}, & \text{if } k \text{ is odd,} \\ \mathrm{PB}_{\partial E}(\zeta), & \text{if } k \text{ is even.} \end{cases} \tag{9}$$

- If $\partial F = \emptyset$, then we have for all $\zeta \in \mathcal{C}_k(X)$

$$\partial(\mathrm{PB}_E\zeta) = \mathrm{PB}_E(\partial\zeta). \tag{10}$$

- $\mathrm{PB}_\bullet$ is natural in the following sense: Whenever we have a commutative diagram

$$\begin{array}{ccc} E & \xrightarrow{H} & E' \\ \downarrow & & \downarrow \\ X & \xrightarrow{h} & X' \end{array}$$

  where $h$ is smooth and $H$ restricts to an orientation preserving diffeomorphism $E_x \to E'_{h(x)}$ for any $x \in X$, then

$$\begin{array}{ccc} \mathcal{C}_{k+\dim F}(E) & \xrightarrow{H_*} & \mathcal{C}_{k+\dim F}(E') \\ \uparrow{\scriptstyle \mathrm{PB}_E} & & \uparrow{\scriptstyle \mathrm{PB}_{E'}} \\ \mathcal{Z}_k(X) & \xrightarrow{h_*} & \mathcal{Z}_k(X') \end{array} \tag{11}$$

commutes (replace $\mathcal{Z}_k$ by $\mathcal{C}_k$ if $\partial F = \emptyset$).

- $\mathrm{PB}_\bullet$ is compatible with integration of differential forms in the following sense: For all differential forms $\omega \in \Omega^{k+\dim F}(E)$ and all $\zeta \in \mathcal{Z}_k(X)$ we have
$$\int_{[\mathrm{PB}_E \zeta]_{S_{k+\dim F}}} \omega = \int_{[\zeta]_{\partial S_{k+1}}} \fint_F \omega\,. \tag{12}$$
Here $\fint$ denotes the ordinary fiber integration of differential forms. If $\partial F = \emptyset$ replace $[\zeta]_{\partial S_{k+1}}$ by $[\zeta]_{S_k}$ and demand (12) for all $\zeta \in \mathcal{C}_k(X)$.
- $\mathrm{PB}_\bullet$ is functorial with respect to composition of fiber bundle projections: For a fiber bundle $\kappa : N \to E$ with compact oriented fibers over a fiber bundle $\pi : E \to X$ with compact oriented fibers, we have the composite fiber bundle $\pi \circ \kappa : N \to X$ with the composite orientation. In this case, we have
$$\mathrm{PB}_{\pi\circ\kappa} = \mathrm{PB}_\kappa \circ \mathrm{PB}_\pi. \tag{13}$$
- $\mathrm{PB}_\bullet$ is compatible with the fiber product of bundles: For fiber bundles $E \to X$ and $E' \to X'$ with compact oriented fibers and geometric chains $\zeta = [M \xrightarrow{g} X] \in \mathcal{C}_k(X)$ and $\zeta' = [M' \xrightarrow{g'} X'] \in \mathcal{C}_{k'}(X')$, we have:
$$\begin{aligned}&\mathrm{PB}_{E\times E'}(\zeta \times \zeta')\\ &= (-1)^{k'\cdot \dim F}\mathrm{PB}_E(\zeta) \times \mathrm{PB}_{E'}(\zeta') \in \mathcal{C}_{k+k'+\dim F\times F'}(E \times E').\end{aligned} \tag{14}$$

Properties (9), (10), (12), and (13) are readily checked. The sign in (14) is caused by the conventions on orientations. To verify (11) we observe that there is an orientation preserving diffeomorphism $J : E \to h^*E'$ such that

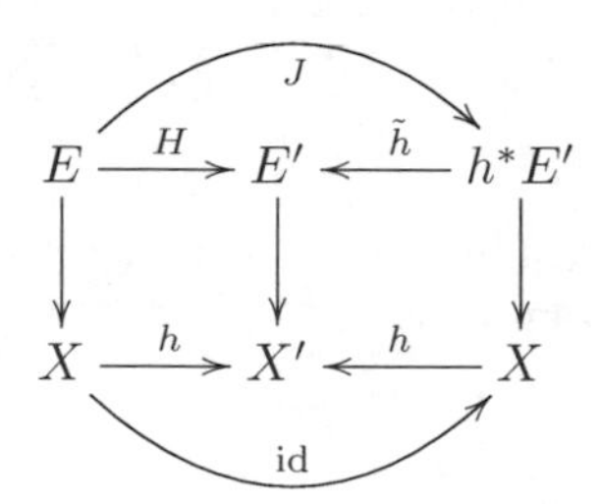


commutes. Now for any $\zeta = [M \xrightarrow{g} X] \in \mathcal{C}_k(X)$ we get an induced orientation preserving diffeomorphism $g^*J : g^*E \to g^*h^*E'$ such that

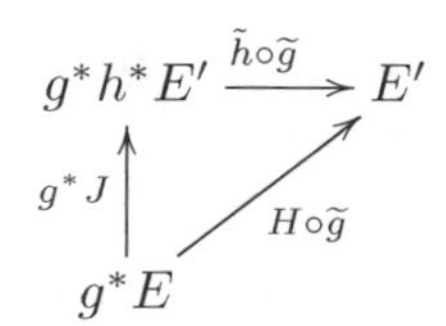

commutes. Thus $[g^*E \xrightarrow{H\circ\widetilde{g}} E'] = [g^*h^*E' \xrightarrow{\widetilde{h}\circ\widetilde{g}} E'] \in \mathcal{C}_{k+\dim F}(E')$. We compute

$$\begin{aligned}
\mathrm{PB}_{E'}(h_*(\zeta)) &= \mathrm{PB}_{E'}([M \xrightarrow{h\circ g} X']) \\
&= [g^*h^*E' \xrightarrow{\widetilde{h}\circ\widetilde{g}} E'] \\
&= [g^*E \xrightarrow{H\circ\widetilde{g}} E'] \\
&= H_*([g^*E \xrightarrow{\widetilde{g}} E]) \\
&= H_*(\mathrm{PB}_E(\zeta))
\end{aligned}$$

and (11) is shown.

*Remark 8. Transfer map on cycles.* We construct a transfer map on the level of singular cycles. Let $\zeta : C_{k-\dim F}(X;\mathbb{Z}) \to \mathcal{C}_{k-\dim F}(X)$ be the homomorphism from Lemma 7. We construct a homomorphism $\lambda : Z_{k-\dim F}(X;\mathbb{Z}) \to Z_k(E;\mathbb{Z})$ such that

$$[\lambda(z)]_{\partial S_{k+1}} = [\mathrm{PB}_E(\zeta(z))]_{\partial S_{k+1}} \tag{15}$$

for all cycles $z \in Z_{k-\dim F}(X;\mathbb{Z})$. For any $z$ in a basis of $Z_{k-\dim F}(X;\mathbb{Z})$ we choose a cycle $\lambda(z) \in Z_k(E;\mathbb{Z})$ representing $[\mathrm{PB}_E(\zeta(z))]_{\partial S_{k+1}}$ and extend $\lambda$ as a homomorphism. In particular, $\lambda$ maps $B_{k-\dim F}(X;\mathbb{Z})$ to $B_k(E;\mathbb{Z})$. We perform this construction in all degrees $k \geq \dim F$.
*Extension to chains.* We extend the transfer map $\lambda : Z_{k-\dim F}(X;\mathbb{Z}) \to Z_k(E;\mathbb{Z})$ to a homomorphism $\lambda : C_{k-\dim F}(X;\mathbb{Z}) \to C_k(E;\mathbb{Z})$ in an appropriate manner. First, we extend $\lambda : Z_{k-\dim F}(X;\mathbb{Z}) \to Z_k(E;\mathbb{Z})$ to a homomorphism $\gamma : C_{k-\dim F}(X;\mathbb{Z}) \to C_k(E;\mathbb{Z})$ as described in Remark 6. On the basis elements of $Z_{k-\dim F}(X;\mathbb{Z})$ we set $\gamma(c) := \lambda(c)$. On the complementary basis elements we choose $k$-chains $\gamma(c)$ such that $\partial\gamma(c) = \lambda(\partial c)$. This is possible since $\lambda(\partial c)$ is a boundary.

We then have:

$$\begin{aligned}
\partial[\gamma(c)]_{S_k} &= [\partial\gamma(c)]_{\partial S_k} \\
&= [\lambda(\partial c)]_{\partial S_k} \\
&\overset{(15)}{=} [\mathrm{PB}_E(\zeta(\partial c))]_{\partial S_k} \\
&\overset{(3)}{=} [\mathrm{PB}_E(\partial\zeta(c))]_{\partial S_k} \\
&\overset{(10),(2)}{=} \partial[\mathrm{PB}_E(\zeta(c))]_{S_k}\,.
\end{aligned}$$

Hence there exists a cycle $w(c) \in Z_k(E;\mathbb{Z})$ such that

$$[\gamma(c) - w(c)]_{S_k} = [\mathrm{PB}_E(\zeta(c))]_{S_k}. \tag{16}$$

We can choose $c \mapsto w(c)$ as a group homomorphism $w : C_{k-\dim F}(X;\mathbb{Z}) \to Z_k(X;\mathbb{Z})$ by defining it on basis elements, as explained above. On the basis elements of $Z_{k-\dim F}(X;\mathbb{Z})$ we set $w(c) = 0$. Condition (15) implies that (16) holds in this case. On the complementary basis elements, we choose $w(c) \in Z_k(E;\mathbb{Z})$ such that (16) holds.

If $c \in C_{k-\dim F}(X;\mathbb{Z})$ is a cycle, we have $\gamma(c) - w(c) = \lambda(c) + 0 = \lambda(c)$. We set $\lambda(c) := \gamma(c) - w(c)$ for general $c \in C_{k-\dim F}(X;\mathbb{Z})$.

*Transfer map on chains.* We have extended the transfer map on cycles to a group homomorphism $\lambda : C_{k-\dim F}(X;\mathbb{Z}) \to C_k(E;\mathbb{Z})$ with

$$\partial\lambda(c) = \lambda(\partial c) \tag{17}$$

and

$$[\lambda(c)]_{S_k} = [\gamma(c) - w(c)]_{S_k} \overset{(16)}{=} [\mathrm{PB}_E(\zeta(c))]_{S_k} \tag{18}$$

The transfer map $\lambda$ should be thought of as the pull-back mapping on the level of chains.

*Remark 9. Transfer map and fiber integration of differential forms.* From (12), we conclude that for any differential form $\omega \in \Omega^k(E)$ and any smooth singular chain $c \in C_{k-\dim F}(X;\mathbb{Z})$, we have:

$$\int_{\lambda(c)} \omega = \int_{[\zeta(c)]_{S_{k-\dim F}}} \fint_F \omega\,. \tag{19}$$

In particular, if $\omega$ is a closed form, (4) yields:

$$\int_{\lambda(c)} \omega = \int_{c-a(\partial c)} \fint_F \omega\,. \tag{20}$$

For a cycle $z \in Z_{k-\dim F}(X;\mathbb{Z})$ and $\omega \in \Omega^k(E)$, we also have:

$$\int_{\lambda(z)} \omega = \int_{[\zeta(z)]_{S_{k-\dim F}}} \fint_F \omega = \int_{[\zeta(z)]_{\partial S_{k-\dim F+1}}} \fint_F \omega\,. \tag{21}$$

Finally, for a cycle $z \in Z_{k-\dim F}(X;\mathbb{Z})$ and a closed form $\omega \in \Omega^k(E)$, we have:

$$\int_{\lambda(z)} \omega = \int_{[\zeta(z)]_{\partial S_{k-\dim F+1}}} \fint_F \omega = \int_{z-\partial a(z)} \fint_F \omega = \int_z \fint_F \omega\,. \tag{22}$$

*Remark 10. Transfer map and fiber integration on singular cohomology.* Let $F \hookrightarrow E \to X$ be a fiber bundle with compact oriented fibers without boundary. The construction of the Leray-Serre spectral sequence in [59] involves the construction of Eilenberg-Zilber type maps $EZ : C_p(X;\mathbb{Z}) \otimes C_q(F;\mathbb{Z}) \to E^0_{p,q}$ for all $p, q \in \mathbb{N}_0$. These maps induce a map of bigraded chain complexes

$$(C_\bullet(X;\mathbb{Z})\otimes C_\bullet(F;\mathbb{Z}), \mathbb{1}\otimes\partial_F) \xrightarrow{EZ} (E^0_{\bullet,\bullet}, d_0)\,.$$

The induced maps on homology yield identifications $C_p(X;H_q(F_x;\mathbb{Z})) \xrightarrow{\overline{EZ}} E^1_{p,q}$. Here $F_x$ denotes the fiber of the bundle over $x\in X$ and $\{H_q(F_x;\mathbb{Z})\}_{x\in X}$ the corresponding local coefficient system.

We consider the special case $q=\dim(F)$. Since the bundle $F\hookrightarrow E\to X$ has compact oriented fibers the local coefficient system $\{H_q(F_x;\mathbb{Z})\}_{x\in X}$ has a canonical section $x\mapsto [F_x]$ where $[F_x]\in H_{\dim F}(F_x;\mathbb{Z})$ is the fundamental class.

The maps $\mathbb{Z}\to H_{\dim(F)}(F_x;\mathbb{Z})$, $k\mapsto k\cdot[F_x]$, induce a homomorphism of chain complexes

$$(C_\bullet(X;\mathbb{Z}),\partial)\to (C_\bullet(X;H_{\dim(F)}(F_x;\mathbb{Z})),\partial) \xrightarrow{\overline{EZ}} (E^1_{\bullet,\bullet},d_1)\,.$$

On the homology of the last two chain complexes we get the well-known identification $H_p(X;H_q(F_x;\mathbb{Z}))\xrightarrow{\cong} E^2_{p,q}$ for the case $q=\dim F$.

Let $c\in C_{k-\dim F}(X;\mathbb{Z})$ be a smooth singular chain in the base $X$. Let $[\mu]\in H^k(E;\mathbb{Z})$ be a cohomology class on the total space and $\mu\in C^k(E;\mathbb{Z})$ a cocycle representing it. Fiber integration for singular cohomology as constructed in [6] maps the class $[\mu]\in H^k(E;\mathbb{Z})$ to

$$\pi_![\mu] := \big[c\mapsto \mu(\overline{EZ}(c\otimes[F_x]))\big]\in H^{k-\dim F}(X;\mathbb{Z})\,.$$

By the constructions of the pull-back operation $\mathrm{PB}_E$ on smooth chains and the transfer map $\lambda$ on singular chains, the chain $\lambda(c)\in C_k(E;\mathbb{Z})$ represents the equivalence class $\overline{EZ}((c-a(\partial c)-\partial a(c+y(c)))\otimes[F_x])\in E^1_{k-\dim F,\dim F}$ of smooth singular $k$-chains in $E$. Combining this observation with the definition of the map $\pi_!: H^k(E;\mathbb{Z})\to H^{k-\dim F}(X;\mathbb{Z})$ we obtain:

$$\begin{aligned}
\pi_![\mu] &= \big[c\mapsto \mu(\overline{EZ}(c\otimes[F_x]))\big]\\
&= \big[c\mapsto \mu(\overline{EZ}(c\otimes[F_x]))\big] + \big[\delta(c\mapsto \mu(\overline{EZ}(a(c)\otimes[F_x])))\big]\\
&= \big[c\mapsto \mu(\overline{EZ}(c\otimes[F_x]))\big] + \big[c\mapsto \mu(\overline{EZ}(a(\partial c))\otimes[F_x])\big]\\
&\qquad + \big[c\mapsto \underbrace{\mu(\overline{EZ}(\partial a(c+y(c))\otimes[F_x]))}_{=0}\big]\\
&= \big[c\mapsto \mu(\overline{EZ}(c-a(\partial c)-\partial a(c+y(c)))\otimes[F_x])\big]\\
&= \big[c\mapsto \mu(\lambda(c))\big]\\
&= [\mu\circ\lambda].
\end{aligned}\tag{23}$$

Thus pre-composition of cochains with the transfer map on chains yields the fiber integration on singular cohomology.

*Remark 11. Transfer map on homology.* As for fiber integration on singular cohomology, the Eilenberg-Zilber map from the Leray-Serre spectral sequence induces the so-called homology transfer

$$H_*(X;\mathbb{Z}) \to E^2_{*,\dim F} \twoheadrightarrow H_{*+\dim F}(E;\mathbb{Z}), \quad [z] \mapsto [\overline{EZ}(z \otimes [F_x])].$$

By construction, homology transfer is represented on the level of cycles by the transfer map $\lambda : Z_*(X;\mathbb{Z}) \to Z_{*+\dim F}(E;\mathbb{Z})$ constructed in Remark 8. Hence the name.

*Remark 12. Fiber integration, transfer and push-forward.* In the literature, fiber integration is sometimes referred to as cohomology transfer. Both homology and cohomology transfer can be defined for any smooth map between compact oriented smooth manifolds by conjugating the pull-back and push-forward maps with Poincaré duality, see e.g. [26, Ch. VIII, § 10]. Therefore, fiber integration is also referred to as push-forward.

## 5 Differential Characters

Differential characters were introduced by Cheeger and Simons in [24]. The group $\widehat{H}^k(X;\mathbb{Z})$ of differential characters in a smooth space has various equivalent descriptions. For instance, it is isomorphic to the smooth Deligne cohomology group $H^{k-1}_{\mathcal{D}}(X;\mathrm{U}(1))$, see e.g. [19]. Differential characters can also be described by differential forms with singularities as in [23] or as de Rham-Federer currrents as in [41, 42, 43]. The groups of differential characters are often referred to as differential cohomology. We use the original definition of differential characters due to Cheeger and Simons.

We first recall the definition and some elementary properties of Cheeger-Simons differential characters. Then we give a new proof of a result of Simons and Sullivan saying that for any differential cohomology theory there is a unique natural transformation to the model given by differential characters. Our proof yields an explicit formula for this natural transformation. Similarly, we reprove the abstract uniqueness result for the ring structure due to Simons and Sullivan by deriving an explicit formula from the axioms.

Stratifolds enter the game because they can be used to represent *homology classes*. However, we do not modify the definition of differential characters as in [12]. The usage of stratifolds in [12] to represent *cohomology classes* is responsible for the limitation to finite-dimensional manifolds. Instead of stratifolds one could also use Baas-Sullivan pseudomanifolds. It was proposed in [34] to use them to describe differential characters.

## 5.1 Definition and Examples

Let $X$ be a smooth space. We denote by $\widehat{H}^k(X;\mathbb{Z})$ the abelian group of degree $k \geq 1$ differential characters, i.e.[1],

$$\widehat{H}^k(X;\mathbb{Z}) := \left\{ h \in \mathrm{Hom}(Z_{k-1}(X;\mathbb{Z}), \mathrm{U}(1)) \,\middle|\, h \circ \partial \in \Omega^k(X) \right\}. \tag{24}$$

The notation $h \circ \partial \in \Omega^k(X)$ means that there exists a differential form $\omega \in \Omega^k(X)$ such that for every smooth singular chain $c \in C_k(X;\mathbb{Z})$, we have:

$$h(\partial c) = \exp\left(2\pi i \int_c \omega\right). \tag{25}$$

The differential form $\omega$ is uniquely determined by the differential character $h \in \widehat{H}^k(X;\mathbb{Z})$. Moreover, it is closed and has integral periods. This form $\omega =: \mathrm{curv}(h)$ is called the *curvature* of $h$. If $\mathrm{curv}(h) = 0$, then $h$ is called a *flat* differential character.

Moreover, a differential character $h$ determines a class $c(h) \in H^k(X;\mathbb{Z})$, constructed as follows: Since $Z_{k-1}(X;\mathbb{Z})$ is a free $\mathbb{Z}$-module, there exists a real lift $\tilde{h}$ of the differential character $h$, i.e., $\tilde{h} \in \mathrm{Hom}(Z_{k-1}(X;\mathbb{Z}),\mathbb{R})$ such that $h(z) = \exp(2\pi i \tilde{h}(z))$ for all $z \in Z_{k-1}(X;\mathbb{Z})$. Then set

$$\mu^{\tilde{h}} : C_k(X;\mathbb{Z}) \to \mathbb{Z}, \quad c \mapsto \int_c \mathrm{curv}(h) - \tilde{h}(\partial c). \tag{26}$$

Since curv is closed $\mu^{\tilde{h}}$ is a cocycle, and it follows from equation (25) that it takes integral values. The cohomology class $[\mu^{\tilde{h}}] \in H^k(X;\mathbb{Z})$ does not depend on the choice of the lift $\tilde{h}$. Now $c(h) := [\mu^{\tilde{h}}] \in H^k(X;\mathbb{Z})$ is called the *characteristic class* of $h$. If $c(h) = 0$, then $h$ is called a *topologically trivial* differential character.

By definition, any real lift $\tilde{h}$ of a differential character $h$ yields a cocycle for the characteristic class $c(h)$. Conversely, if $\mu \in C^k(X;\mathbb{Z})$ is a cocycle representing the cohomology class $c(h) \in H^k(X;\mathbb{Z})$, then we can find a real lift $\tilde{h}'$ such that $\mu = \mu^{\tilde{h}'} := \mathrm{curv}(h) - \delta\tilde{h}'$. For if $\tilde{h}$ is any real lift of $h$, then $\mu$ and $\mu^{\tilde{h}}$ are cohomologous, i.e. there exists a cochain $t \in C^{k-1}(X;\mathbb{Z})$ such that $\delta t = \mu^{\tilde{h}} - \mu$. Setting $\tilde{h}' := \tilde{h} + t$ yields a real lift of $h$ with $\mu^{\tilde{h}'} = \mathrm{curv}(h) - \delta\tilde{h} - \delta t = \mu^{\tilde{h}} - \delta t = \mu$.

Note that by (26), the image of $c(h)$ in $H^k(X;\mathbb{R})$ coincides with the image of the de Rham cohomology class $[\mathrm{curv}(h)]_{\mathrm{dR}}$ of $\mathrm{curv}(h)$ under the de Rham isomorphism.

[1] It is convenient to shift the degree of the differential characters by $+1$ as compared to the original definition from [24]. Thus a degree $k$ differential character has curvature and characteristic class of degree $k$.

*Remark 13.* Even though the abelian group U(1) is written multiplicatively, we write $\widehat{H}^k(X;\mathbb{Z})$ additively, i.e., for $h, h' \in \widehat{H}^k(X;\mathbb{Z})$ and $z \in Z_{k-1}(X;\mathbb{Z})$ we have

$$(h + h')(z) = h(z) \cdot h'(z).$$

The neutral element $0 \in \widehat{H}^k(X;\mathbb{Z})$ is the constant map

$$0(z) = 1.$$

The reason for this convention is that there is an additional multiplicative structure on $\widehat{H}^*(X;\mathbb{Z})$ analogous to the cup product turning it into a ring. The ring structure will be discussed in Sect. 6.

Let $\eta \in \Omega^{k-1}(X)$ be a differential form on $X$. We define a differential character $\iota(\eta) \in \widehat{H}^k(X;\mathbb{Z})$ by setting

$$\iota(\eta)(z) := \exp\left(2\pi i \int_z \eta\right). \tag{27}$$

Evaluating on boundaries, we see that in this case,

$$\operatorname{curv}(\iota(\eta)) = d\eta. \tag{28}$$

Taking $\widetilde{\iota(\eta)}(z) := \int_z \eta$ as real lift, we have by the Stokes theorem

$$\mu^{\widetilde{\iota(\eta)}}(x) = \int_x d\eta - \widetilde{\iota(\eta)}(\partial x) = \int_x d\eta - \int_{\partial x} \eta = 0$$

so that $h$ is topologically trivial. If also $d\eta = 0$, then $\operatorname{curv}(\iota(\eta)) = 0$, thus $h$ is flat.

We thus obtain a homomorphism $\iota : \Omega^{k-1}(X) \to \widehat{H}^k(X;\mathbb{Z})$. If the closed form $\eta$ has integral periods, then $\iota(\eta)(z) = 1$ for every $z$, thus $\iota(\eta) = 0$. A form $\eta \in \Omega^{k-1}(X)$ such that $\iota(\eta) = h \in \widehat{H}^k(X;\mathbb{Z})$ is called a *topological trivialization* of $h$.

Let $u \in H^{k-1}(X;\mathrm{U}(1))$. We define a differential character $j(u) \in \widehat{H}^k(X;\mathbb{Z})$ by setting

$$j(u)(z) := \langle u, [z] \rangle. \tag{29}$$

Thus we obtain an injective map $j : H^{k-1}(X;\mathrm{U}(1)) \to \widehat{H}^k(X;\mathbb{Z})$.

By $\Omega^k_{cl}(X)$ we denote the space of closed $k$-forms and by $\Omega^k_0(X) \subset \Omega^k_{cl}(X)$ the set of closed $k$-forms with integral periods. We identify the quotients

$$\frac{H^k(X;\mathbb{R})}{H^k(X;\mathbb{Z})_{\mathbb{R}}} \cong \frac{\Omega^k_{cl}(X)}{\Omega^k_0(X)}$$

using the de Rham isomorphism. Here $H^k(X;\mathbb{Z})_{\mathbb{R}} \subset H^k(X;\mathbb{R})$ denotes the image of $H^k(X;\mathbb{Z})$ in $H^k(X;\mathbb{R})$ under the natural map induced by the change of coefficients. Recall that $\iota : \Omega^{k-1}(X) \to \widehat{H}^k(X;\mathbb{Z})$ induces a homomorphism $\frac{\Omega^{k-1}(X)}{\Omega_0^{k-1}(X)} \to \widehat{H}^k(X;\mathbb{Z})$, again denoted $\iota$.

We obtain the following commutative diagram with exact rows and columns:

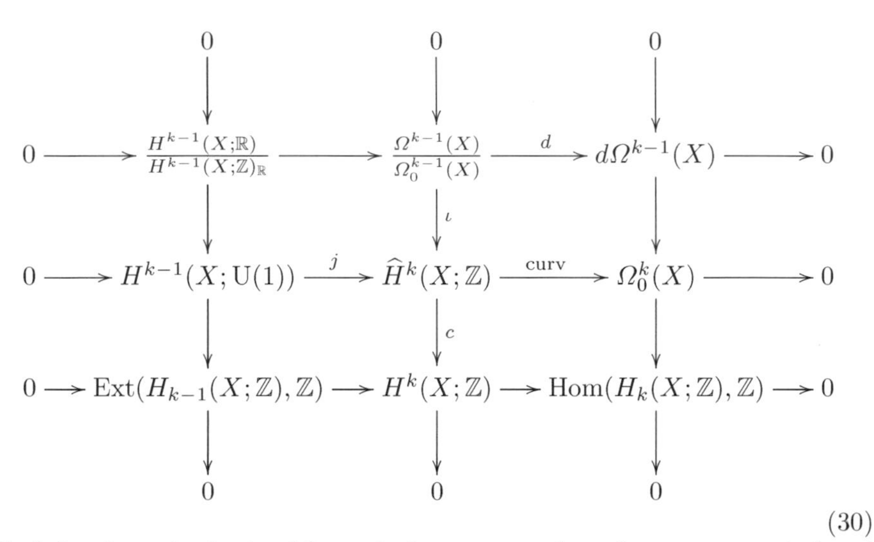

(30)

The left column is obtained from the long exact cohomology sequence induced by the coefficient sequence $0 \to \mathbb{Z} \to \mathbb{R} \to \mathrm{U}(1) \to 0$ together with the canonical identification of $\mathrm{Ext}(H_{k-1}(X;\mathbb{Z}),\mathbb{Z})$ with the torsion subgroup of $H^k(X;\mathbb{Z})$. The middle column says that a differential character admits a topological trivialization if and only if it is topologically trivial.

For reasons that will become apparent later, we extend the definition of the group $\widehat{H}^k(X;\mathbb{Z})$ by setting

$$\widehat{H}^k(X;\mathbb{Z}) := H^k(X;\mathbb{Z}) \qquad \text{for } k \leq 0. \tag{31}$$

This is the only possible choice compatible with the diagram (30). In particular, we have $\widehat{H}^k(X;\mathbb{Z}) = \{0\}$ for $k < 0$. For $k \leq 0$, we define the characteristic class $c : \widehat{H}^k(X;\mathbb{Z}) \to H^k(X;\mathbb{Z})$ to be the identity.

*Remark 14. Thin invariance.* By construction, the evaluation of differential characters is well defined on $Z_{k-1}(X;\mathbb{Z})/\partial S_k(X;\mathbb{Z})$: If $z \in Z_{k-1}(X;\mathbb{Z})$ with $z = \partial y$ and $\int_y \eta = 0$ for all $\eta \in \Omega^k(X)$, then we find:

$$h(z) = h(\partial y) = \exp\Big(2\pi i \underbrace{\int_y \mathrm{curv}(h)}_{=0}\Big) = 1\,.$$

We refer to this property of differential characters as *thin invariance*.

In particular, differential characters are invariant under barycentric subdivision of smooth singular cycles. This was already observed in [24, p. 55].

*Remark 15. Naturality.* If $f : X \to Y$ is a smooth map, then one can pull back differential characters $h \in \widehat{H}^k(Y;\mathbb{Z})$ on $Y$ to $X$ by

$$f^*h := h \circ f_*$$

where $f_* : Z_{k-1}(X;\mathbb{Z}) \to Z_{k-1}(Y;\mathbb{Z})$ is the induced map on cycles. This defines a homomorphism $f^* : \widehat{H}^k(Y;\mathbb{Z}) \to \widehat{H}^k(X;\mathbb{Z})$. One easily checks that $\operatorname{curv}(f^*h) = f^*\operatorname{curv}(h)$ and $c(f^*h) = f^*c(h)$.

*Remark 16. Evaluation on torsion cycles.* Let $h \in \widehat{H}^k(X;\mathbb{Z})$ and let $z \in Z_{k-1}(X;\mathbb{Z})$ be a cycle that represents a torsion class in $H_{k-1}(X;\mathbb{Z})$. Hence there exists an $N \in \mathbb{N}$ such that $N \cdot [z] = 0 \in H_{k-1}(X;\mathbb{Z})$. Choose $x \in C_k(X;\mathbb{Z})$ such that $N \cdot z = \partial x$. In particular, $z = \frac{1}{N} \cdot \partial x$ as real cycles. Then we have:

$$\begin{aligned}
h(z) &= \exp\Big(2\pi i \cdot \tilde{h}(z)\Big) \\
&= \exp\Big(2\pi i \cdot \tilde{h}(\frac{1}{N} \cdot \partial x)\Big) \\
&= \exp\Big(\frac{2\pi i}{N}\tilde{h}(\partial x)\Big) \\
&= \exp\Big(\frac{2\pi i}{N}\delta\tilde{h}(x)\Big) \\
&= \exp\frac{2\pi i}{N}\Big(\int_x \operatorname{curv}(h) - \mu^{\tilde{h}}(x)\Big).
\end{aligned}$$

If $\mu \in Z^k(X;\mathbb{Z})$ is another cocycle representing the characteristic class $c(h)$, then we have $\mu^{\tilde{h}} - \mu = \delta t$ for some $t \in C^{k+1}(X;\mathbb{Z})$. This yields

$$\frac{1}{N} \cdot (\mu^{\tilde{h}} - \mu)(x) = \delta t(\frac{1}{N} \cdot x) = t(\frac{1}{N} \cdot \partial x) = t(z) \in \mathbb{Z}.$$

Thus although the evaluation of $c(h)$ on $x$ is not well defined, by abuse of notation we may write

$$h(z) = \exp\frac{2\pi i}{N}\Big(\int_x \operatorname{curv}(h) - \langle c(h), x\rangle\Big). \tag{32}$$

In particular, if $h$ is topologically trivial and flat, then it vanishes on torsion cycles.

The latter fact can also be deduced from the commutative diagram (30): if $h$ is in the image of the map $\frac{H^{k-1}(X;\mathbb{R})}{H^{k-1}(X;\mathbb{Z})_{\mathbb{R}}} \to \widehat{H}^k(X;\mathbb{Z})$, then the real lift $\tilde{h}$ can be chosen to be a real cocycle. Thus $\tilde{h}$ vanishes on torsion cycles, and so does $h$.

*Remark 17.* Let $h \in \widehat{H}^k(X;\mathbb{Z})$ be a differential character on a smooth space $X$, and let $z \in Z_{k-1}(X;\mathbb{Z})$ be a smooth singular cycle. According to Lemma 7, we get a geometric cycle $\zeta(z) = [M \xrightarrow{g} X] \in \mathcal{Z}_{k-1}(X)$ and a smooth singular chain $a(z) \in C_k(X;\mathbb{Z})$ such that $[z - \partial a(z)]_{\partial S_k} = [\zeta(z)]_{\partial S_k}$. Since differential characters are thin invariant, we have

$$h(z) = h([\zeta(z)]_{\partial S_k}) \cdot h(\partial a(z)) = h([\zeta(z)]_{\partial S_k}) \cdot \exp\Big(2\pi i \int_{a(z)} \mathrm{curv}(h)\Big).$$

We may also pull back the differential character along the smooth map $g$ to the stratifold $M$. For dimensional reasons, $g^*h$ is topologically trivial and flat, hence $g^*h = \iota(\varrho)$ for a closed differential form $\varrho \in \Omega^{k-1}_{cl}(M)$. By definition, the evaluation of $h$ on $[\zeta(z)]_{\partial S_k}$ is the same as the evaluation of $g^*h$ on any representing chain of the fundamental class $[M] \in H_{k-1}(M;\mathbb{Z}) = Z_{k-1}(M;\mathbb{Z})/\partial S_k(M;\mathbb{Z})$ of the stratifold $M$. So we may write:

$$h(z) = h([\zeta(z)]_{\partial S_k}) \cdot \exp\Big(2\pi i \int_{a(z)} \mathrm{curv}(h)\Big) \tag{33}$$

$$= g^*h([M]) \cdot \exp\Big(2\pi i \int_{a(z)} \mathrm{curv}(h)\Big) \tag{34}$$

$$= \exp\Big(2\pi i \int_M \varrho\Big) \cdot \exp\Big(2\pi i \int_{a(z)} \mathrm{curv}(h)\Big). \tag{35}$$

We check that (34) is consistent with the property (25) that defines differential characters: for a boundary $z = \partial c \in B_{k-1}(X;\mathbb{Z})$ we choose $\zeta(c) \in \mathcal{C}_k(X)$ and $a(c) \in C_k(X;\mathbb{Z})$ as in Lemma 7. This yields:

$$\begin{aligned} h(\partial c) &= h(\partial[\zeta(c)]_{S_k}) \cdot \exp\Big(2\pi i \int_{a(\partial c)} \mathrm{curv}(h)\Big) \\ &= \exp\Big[2\pi i\Big(\int_{[\zeta(c)]_{S_k}} \mathrm{curv}(h) + \int_{a(\partial c)} \mathrm{curv}(h)\Big)\Big] \\ &\overset{(4)}{=} \exp\Big[2\pi i\Big(\int_c \mathrm{curv}(h) - \underbrace{\int_{\partial(a(c+y(c))} \mathrm{curv}(h)}_{=0}\Big)\Big] \\ &= \exp\Big(2\pi i \int_c \mathrm{curv}(h)\Big). \end{aligned}$$

We identify differential characters in low degrees as mentioned in [24, p. 54].

*Example 18.* $\mathrm{U}(1)$*-valued smooth functions.* Let $X$ be a differentiable manifold and let $k = 1$. We show $\widehat{H}^1(X;\mathbb{Z}) = C^\infty(X, \mathrm{U}(1))$. Any homomorphism $h : Z_0(X;\mathbb{Z}) \to \mathrm{U}(1)$ corresponds to a map $\bar{h} : X \to \mathrm{U}(1)$. For a fixed point $x_0 \in M$ we identify a neighborhood of $x_0$ with a ball such that $x_0$ corresponds to its center. For $x$ in this neighborhood we let $y(x)$ be the straight line from $x_0$ to $x$. By (25) we have

$$\bar{h}(x) = \bar{h}(x_0) \cdot \exp\Big(2\pi i \cdot \int_{y(x)} \mathrm{curv}(h)\Big).$$

This shows that $\bar{h}$ is smooth.

Conversely, given a smooth function $\bar{h} : X \to \mathrm{U}(1)$, we choose a smooth local lift $\tilde{h} : U \subset X \to \mathbb{R}$, i.e., $\exp(2\pi i \tilde{h}(x)) = \bar{h}(x)$, and put $\omega := d\tilde{h}$. This form $\omega$ does not depend on the choice of lift and is therefore a globally defined 1-form on $X$. Now $h : Z_0(X;\mathbb{Z}) \to \mathrm{U}(1)$ given by $h(\Sigma_j \alpha_j x_j) = \prod_j \bar{h}(x_j)^{\alpha_j}$ is a differential character with curvature $\omega$. Hence $h$ is flat if and only if $\bar{h}$ is locally constant. Moreover, $h$ is topologically trivial if and only if $\bar{h}$ has a *global* lift $\tilde{h} : X \to \mathbb{R}$.

For the characteristic class one can check that

$$c(h) = \bar{h}^* \theta$$

where $\theta \in H^1(\mathrm{U}(1);\mathbb{Z})$ is the fundamental class. From now on we will identify $\widehat{H}^1(X;\mathbb{Z}) = C^\infty(X, \mathrm{U}(1))$ and not distinguish between $h \in \widehat{H}^1(X;\mathbb{Z})$ and $\bar{h} \in C^\infty(X, \mathrm{U}(1))$.

*Example 19.* U(1)*-bundles with connection.* Let $X$ be a differentiable manifold and let $k = 2$. For a U(1)-bundle with connection $(P, \nabla)$ on $X$, the holonomy map associates to each smooth 1-cycle $z$ an element $h(z) \in \mathrm{U}(1)$. Let $\mathcal{P}^\nabla_c$ denote parallel transport along an oriented curve $c$ with respect to the connection $\nabla$. If $c$ is closed and $z$ is the cycle represented by $c$, then $h(z)$ is characterized by $\mathcal{P}^\nabla_c(p) = p \cdot h(z)$. Here $p \in P$ lies in the fiber over the initial point of $c$ and $h(z)$ does not depend on its choice.

This defines a differential character $h \in \widehat{H}^2(X;\mathbb{Z})$ whose curvature is $\mathrm{curv}(h) = \frac{-1}{2\pi i} R^\nabla$, where $R^\nabla$ is the curvature of $\nabla$. The characteristic class $c(h)$ is the first Chern class of $P$.

Conversely, any $h \in \widehat{H}^2(X;\mathbb{Z})$ is the holonomy map of a U(1)-bundle with connection and determines the bundle up to connection-preserving isomorphism. Hence differential characters in $\widehat{H}^2(X;\mathbb{Z})$ are in 1-1 correspondence with isomorphism classes of U(1)-bundles with connection.

*Change of connections.* Given a U(1)-bundle with connection $(P, \nabla)$ and a 1-form $\rho \in \Omega^1(X)$, we get a new connection $\nabla' = \nabla + i\rho$ on $P$. The differential character corresponding to $(P, \nabla')$ is obtained by adding $\iota(\frac{-1}{2\pi}\rho)$ to the character corresponding to $(P, \nabla)$.

*Topological trivializations.* If the U(1)-bundle $P \to X$ is topologically trivial, any trivialization $T : P \to X \times \mathrm{U}(1)$ yields a 1-1 correspondence of connections $\nabla$ on $P$ and differential forms $\vartheta(\nabla, T) \in \Omega^1(X)$.Under this correspondence, the connection 1-form of $\nabla$ is given as $(T \circ \mathrm{pr}_1)^*(-2\pi i \vartheta(\nabla, T))$. Parallel transport along a curve $c$ in $X$ with respect to a connection $\nabla$ on

$P$ corresponds to multiplication with $\exp\left(2\pi i \int_c \vartheta(\nabla, T)\right)$. In particular, the holonomy map of $(P, \nabla)$ is given as $c \mapsto \exp\left(2\pi i \int_c \vartheta(\nabla, T)\right)$, hence $h = \iota(\vartheta)$.

Conversely, given a 1-form $\varrho \in \Omega^1(X)$ such that $h = \iota(\varrho)$, then the first Chern class of the correpondìng U(1)-bundle $P$ vanishes, hence $P$ is topologically trivial. One can directly construct global sections and hence trivializations of the bundle $P$ from the 1-form $\varrho$. This is explained in detail in Example 65 below.

*Flat bundles.* If $P \to X$ is a U(1)-bundle which admits a flat connection $\nabla$, then $c_1(P)$ is a torsion class. The holonomy of $\nabla$ along a closed curve now only depends on the homotopy class of the curve and thus yields an element in $\mathrm{Hom}(\pi_1(X), \mathrm{U}(1)) \cong \mathrm{Hom}(H_1(X); \mathrm{U}(1)) \cong H^1(X; \mathrm{U}(1))$.

Conversely, for any homomorphism $\chi : \pi_1(X) \to \mathrm{U}(1)$, the U(1)-bundle $P := \widetilde{X} \times_\chi \mathrm{U}(1)$ associated to the universal cover via the representation $\chi$ has Chern class $c_1(P) = \chi \in \mathrm{Hom}(\pi_1(X), \mathrm{U}(1)) \cong H^1(X; \mathrm{U}(1))$. The canonical flat connection on the trivial bundle $\widetilde{X} \times \mathrm{U}(1)$ descends to a flat connection on $P$ with holonomy map $\chi$.

The 1-1 correspondence between isomorphism classes of flat bundles and homomorphisms $\pi_1(X) \to \mathrm{U}(1)$ thus obtained corresponds to the isomorphism $j : H^1(X, \mathrm{U}(1)) \xrightarrow{\cong} \widehat{H}^2_{\mathrm{flat}}(X; \mathbb{Z})$ of diagram (30).

*Example 20. Hitchin gerbes with connection.* Let $X$ be a differentiable manifold and let $k = 3$. Similar to the case $k = 2$ and U(1)-bundles with connection, there is a 1-1 correspondence between differential characters in $\widehat{H}^3(X; \mathbb{Z})$ and isomorphism classes of Hitchin gerbes with connection [44].

## 5.2 *Differential Cohomology*

There are several ways to define differential cohomology axiomatically as a functor $\widetilde{H}^*(\,\cdot\,; \mathbb{Z})$ from the category of smooth spaces to the category of $\mathbb{Z}$-graded abelian groups, together with natural transformations $\widetilde{\mathrm{curv}} : \widetilde{H}^*(\,\cdot\,; \mathbb{Z}) \to \Omega^*_0(\,\cdot\,)$ (curvature), $\widetilde{c} : \widetilde{H}^*(\,\cdot\,; \mathbb{Z}) \to H^*(\,\cdot\,; \mathbb{Z})$ (characteristic class), $\widetilde{\iota} : \Omega^{*-1}(\,\cdot\,)/\Omega^{*-1}_0(\,\cdot\,) \to \widetilde{H}^*(\,\cdot\,; \mathbb{Z})$ (topological trivialization) and $\widetilde{j} : H^{*-1}(\,\cdot\,; \mathrm{U}(1)) \to \widetilde{H}^*(\,\cdot\,; \mathbb{Z})$ (inclusion of flat classes). One difference of our definition from those used in [60] and [15] is that we require the functor to be defined on a class of spaces also containing stratifolds.

**Definition 21 (Differential Cohomology Theory).** A *differential cohomology* theory is a functor $\widetilde{H}^*(\,\cdot\,; \mathbb{Z})$ from the category of smooth spaces to the categoy of $\mathbb{Z}$-graded abelian groups, together with four natural transformations

- $\widetilde{\mathrm{curv}} : \widetilde{H}^*(\,\cdot\,; \mathbb{Z}) \to \Omega^*_0(\,\cdot\,)$, called *curvature*,
- $\widetilde{c} : \widetilde{H}^*(\,\cdot\,; \mathbb{Z}) \to H^*(\,\cdot\,; \mathbb{Z})$, called *characteristic class*,

- $\widetilde{\iota} : \Omega^{*-1}(\,\cdot\,)/\Omega_0^{*-1}(\,\cdot\,) \to \widetilde{H}^*(\,\cdot\,;\mathbb{Z})$, called *topological trivialization*, and
- $\widetilde{j} : H^{*-1}(\,\cdot\,;\mathrm{U}(1)) \to \widetilde{H}^*(\,\cdot\,;\mathbb{Z})$, called *inclusion of flat classes*,

such that for any smooth space $X$ the following diagram commutes and has exact rows and columns:

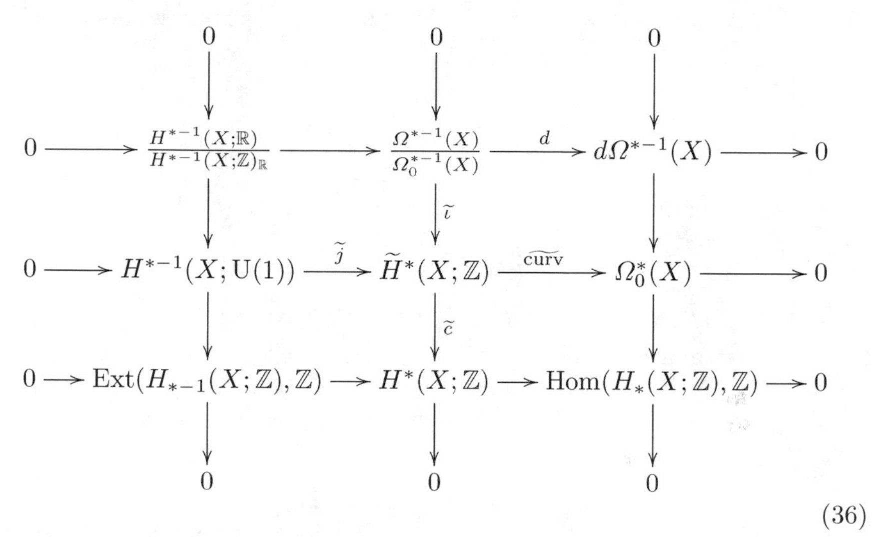

(36)

*Remark 22.* Note that the upper and lower rows as well as the left and right columns of (36) are exact sequence, independently of the differential cohomology theory $\widetilde{H}^*(\,\cdot\,;\mathbb{Z})$. Thus the requirement is that the middle row and column are exact sequences and the whole diagram commutes. Commutativity of the right upper quadrant means that $\widetilde{\mathrm{curv}} \circ \widetilde{\iota}$ is the exterior differential. Commutativity of the left lower quadrant means that $\widetilde{c} \circ \widetilde{j}$ is the connecting homomorphism in cohomology for the coefficient sequence $0 \to \mathbb{Z} \to \mathbb{R} \to \mathrm{U}(1) \to 0$. Hence our definition of differential cohomology coincides with that of *character functors* in [60, p. 46].

In this section, we show uniqueness of differential cohomology theories up to unique natural transformations. More precisely, for any differential cohomology theory $\widetilde{H}^*(\,\cdot\,;\mathbb{Z})$, there exists a unique natural transformation $\Xi : \widetilde{H}^*(\,\cdot\,;\mathbb{Z}) \to \widehat{H}^*(X;\mathbb{Z})$ that commutes with the identity on the other functors in diagram 36. Equivalent statements were proved in [60, Thm. 1.1] and in [15, Thm. 3.1]. Our proof differs from both in that for any fixed smooth space $X$ we obtain an explicit formula for $\Xi : \widetilde{H}^*(X;\mathbb{Z}) \to \widehat{H}^*(X;\mathbb{Z})$. However, we rely on [60, Lemma 1.1] to conclude that $\Xi$ commutes with the characteristic class.

The proof of uniqueness of differential cohomology up to unique natural transformation is done in two steps: We first show that if there exists a natural transformation, then it is uniquely determined.

**Theorem 23 (Uniqueness of Differential Cohomology I).** *Let $\widetilde{H}^*(\,\cdot\,;\mathbb{Z})$ be a differential cohomology theory in the sense of Definition 21. Suppose* *there exists a natural transformation* $\Xi : \widetilde{H}^*(\,\cdot\,;\mathbb{Z}) \to \widehat{H}^*(\,\cdot\,;\mathbb{Z})$ *that commutes with curvature and topological trivializations. Then* $\Xi$ *is uniquely determined by these requirements.*

*Proof.* Let $X$ be a smooth space. By assumption, we have a homomorphism $\Xi : \widetilde{H}^*(X;\mathbb{Z}) \to \widehat{H}^*(X;\mathbb{Z})$ satisfying

$$\Xi \circ \widetilde{\iota} = \iota\,, \tag{37}$$

$$\operatorname{curv} \circ \Xi = \widetilde{\operatorname{curv}}\,. \tag{38}$$

Moreover, naturality means that for any smooth map $f : Y \to X$ and any $x \in \widetilde{H}^*(X)$, we have:

$$f^*(\Xi(x)) = \Xi(f^*x)\,. \tag{39}$$

Now let $x \in \widetilde{H}^k(X;\mathbb{Z})$, and let $z \in Z_{k-1}(X;\mathbb{Z})$. We show that $\Xi(x)(z)$ is uniquely determined: Choose homomorphisms $\zeta : Z_{k-1}(X;\mathbb{Z}) \to \mathcal{Z}_{k-1}(X)$ and $a : Z_{k-1}(X;\mathbb{Z}) \to C_k(X;\mathbb{Z})$ as in Lemma 7 such that $[z - \partial a(z)]_{\partial S_k} = [\zeta(z)]_{\partial S_k}$. By Remark 14, differential characters are thin invariant. Thus we have

$$\begin{aligned}
\Xi(x)(z) &= \Xi(x)([\zeta(z)]_{\partial S_k}) \cdot \Xi(x)(\partial a(z)) \\
&\overset{(25)}{=} \Xi(x)([\zeta(z)]_{\partial S_k}) \cdot \exp\Big(2\pi i \int_{a(z)} \operatorname{curv}(\Xi(x))\Big)\,.
\end{aligned}$$

Write $\zeta(z) = [M \xrightarrow{g} X]$. For dimensional reasons, we have $\widetilde{c}(g^*x) = 0$. Thus by (36), we find $\varrho \in \Omega^{k-1}(X)$ such that $g^*x = \widetilde{\iota}([\varrho])$. This yields:

$$\begin{aligned}
\Xi(x)(z) &\overset{(38)}{=} g^*\Xi(x)([M]) \cdot \exp\Big(2\pi i \int_{a(z)} \widetilde{\operatorname{curv}}(x)\Big) \\
&\overset{(39)}{=} \Xi(g^*x)([M]) \cdot \exp\Big(2\pi i \int_{a(z)} \widetilde{\operatorname{curv}}(x)\Big) \\
&= \Xi(\widetilde{\iota}([\varrho]))([M]) \cdot \exp\Big(2\pi i \int_{a(z)} \widetilde{\operatorname{curv}}(x)\Big) \\
&\overset{(37)}{=} \iota(\varrho)([M]) \cdot \exp\Big(2\pi i \int_{a(z)} \widetilde{\operatorname{curv}}(x)\Big) \\
&\overset{(27)}{=} \exp\Big[2\pi i\Big(\int_M \varrho + \int_{a(z)} \widetilde{\operatorname{curv}}(x)\Big)\Big]\,.
\end{aligned} \tag{40}$$

We have derived an explicit formula for $\Xi$ and, in particular, proved its uniqueness. □

Now we take (40) to define a natural transformation $\Xi : \widetilde{H}^*(\,\cdot\,;\mathbb{Z}) \to \widehat{H}^*(\,\cdot\,;\mathbb{Z})$:

**Definition 24.** Let $\widetilde{H}^*(\,\cdot\,;\mathbb{Z})$ be a differential cohomology theory. We define a natural transformation $\Xi : \widetilde{H}^*(\,\cdot\,;\mathbb{Z}) \to \widehat{H}^*(\,\cdot\,;\mathbb{Z})$ as follows: Let $X$ be a smooth space and $x \in \widetilde{H}^k(X;\mathbb{Z})$. Choose homomorphisms $\zeta : Z_{k-1}(X;\mathbb{Z}) \to \mathcal{Z}_{k-1}(X)$ and $a : Z_{k-1}(X;\mathbb{Z}) \to C_k(X;\mathbb{Z})$ as in Lemma 7 such that $[z - \partial a(z)]_{\partial S_k} = [\zeta(z)]_{\partial S_k}$ for all $z \in Z_{k-1}(X;\mathbb{Z})$. Write $\zeta(z) = [M \xrightarrow{g} X]$. For dimensional reasons, we have $\widetilde{c}(g^*x) = 0$. Thus by (36), we find $\varrho \in \Omega^{k-1}(M)$ such that $g^*x = \widetilde{\iota}([\varrho])$. Now we set:

$$\Xi(x)(z) := \exp\Big[2\pi i\Big(\int_M \varrho + \int_{a(z)} \widetilde{\mathrm{curv}}(x)\Big)\Big]. \tag{41}$$

The following Lemma shows that $\Xi$ is well defined. The fact that $\zeta$ and $a$ are homomorphisms will be convenient for the proof of Theorem 26 but for formula (41) this is not relevant.

**Lemma 25.** *Let $X$ be a smooth space and $x \in \widetilde{H}^k(X;\mathbb{Z})$. Let $z \in Z_{k-1}(X;\mathbb{Z})$. Let $\zeta'(z) = [M' \xrightarrow{g'} X] \in \mathcal{Z}_{k-1}(X)$ and $a'(z) \in C_k(X;\mathbb{Z})$ be any choice of geometric cycle and singular chain such that $[z - \partial a'(z)]_{\partial S_k} = [\zeta'(z)]_{\partial S_k}$. Let $\varrho' \in \Omega^{k-1}(M')$ be any differential form such that ${g'}^*x = \widetilde{\iota}([\varrho'])$. Then we have*

$$\Xi(x)(z) = \exp\Big[2\pi i\Big(\int_{M'} \varrho' + \int_{a'(z)} \widetilde{\mathrm{curv}}(x)\Big)\Big]. \tag{42}$$

*Proof.* Since $\zeta(z)$ and $\zeta'(z)$ both represent the homology class of $z$, we find a geometric boundary $\partial\beta(z) \in \mathcal{B}_{k-1}(X)$ such that $\partial\beta(z) = \zeta'(z) - \zeta(z)$. Since

$$[\partial a(z) - \partial a'(z)]_{\partial S_k} = [\partial\beta(z)]_{\partial S_k} = \partial[\beta(z)]_{S_k},$$

we find a smooth singular cycle $w(z) \in Z_k(X;\mathbb{Z})$ such that

$$[a(z) - a'(z) - w(z)]_{S_k} = [\beta(z)]_{S_k}. \tag{43}$$

Write $\beta(z) = [N \xrightarrow{G} X]$, where $N$ is a $k$-dimensional oriented compact $p$-stratifold with boundary $\partial N = M' \sqcup \overline{M}$ and $g = G|_M$, $g' = G|_{M'}$. Since $H^k(N;\mathbb{Z}) = \{0\}$, we have $\widetilde{c}(G^*x) = 0$. By (36), we find a differential form $\eta \in \Omega^{k-1}(N)$ such that $G^*x = \widetilde{\iota}([\eta])$. Then we have

$$\widetilde{\iota}([\varrho']) - \widetilde{\iota}([\varrho]) = g^*x - {g'}^*x = G|_{\partial N}^*x = (G^*x)|_{\partial N} = \widetilde{\iota}([\eta])|_{\partial N}.$$

In particular, we have $\eta|_{\partial N} - (\varrho' - \varrho) \in \Omega_0^{k-1}(\partial N)$. Inserting this into (41) and (42), we find:

$$\begin{aligned}\Xi(x)(z) \cdot \exp\Big[2\pi i\Big(\int_{M'} \varrho' + \int_{a'(z)} \widetilde{\mathrm{curv}}(x)\Big)\Big]^{-1} \\ = \exp\Big[2\pi i\Big(\int_{M'} \varrho' - \int_M \varrho + \int_{a'(z)-a(z)} \widetilde{\mathrm{curv}}(x)\Big)\Big]\end{aligned}$$

$$\begin{aligned}
&= \exp\Big[2\pi i\Big(\int_{\partial N}\eta + \int_{a'(z)-a(z)}\widetilde{\mathrm{curv}}(x)\Big)\Big] \\
&= \exp\Big[2\pi i\Big(\int_N d\eta + \underbrace{\int_{-w(z)}\widetilde{\mathrm{curv}}(x)}_{\in\mathbb{Z}} + \int_{-[\beta(z)]_{S_k}}\widetilde{\mathrm{curv}}(x)\Big)\Big] \\
&= \exp\Big[2\pi i\Big(\int_N G^*\widetilde{\mathrm{curv}}(x) + \int_{-[\beta(z)]_{S_k}}\widetilde{\mathrm{curv}}(x)\Big)\Big] \\
&= \exp\Big(2\pi i\int_{\underbrace{G_*[N]_{S_k} - [\beta(z)]_{S_k}}_{=0}}\widetilde{\mathrm{curv}}(x)\Big) \\
&= 1\,.
\end{aligned}$$

This yields (42). □

Now we complete the proof of uniqueness of differential cohomology up to unique natural transformation by establishing existence of a natural transformation.

**Theorem 26 (Uniqueness of Differential Cohomology II).** *The map* $\Xi : \widetilde{H}^*(\,\cdot\,;\mathbb{Z}) \to \widehat{H}^*(\,\cdot\,;\mathbb{Z})$ *defined in* (41) *is a natural transformation and commutes with curvature, topological trivializations and inclusion of flat classes. More explicitly, we have*

$$\Xi \circ \widetilde{\iota} = \iota\,, \tag{44}$$

$$\Xi \circ \widetilde{j} = j\,, \tag{45}$$

$$\mathrm{curv} \circ \Xi = \widetilde{\mathrm{curv}}\,. \tag{46}$$

*For any smooth map* $f : Y \to X$*, and any* $x \in \widetilde{H}^k(X)$*, we have:*

$$f^*\Xi(x) = \Xi(f^*x)\,. \tag{47}$$

*Remark 27.* It follows from [60, Lemma 1.1], that $\Xi$ also satisfies

$$c \circ \Xi = \widetilde{c}\,.$$

*Proof (Proof of Theorem 26).* a) We first show that $\Xi$ takes values in $\widehat{H}^*(\,\cdot\,;\mathbb{Z})$. Let $X$ be a fixed smooth space and $x \in \widetilde{H}^k(X;\mathbb{Z})$. By construction, the maps $\zeta : Z_{k-1}(X;\mathbb{Z}) \to \mathcal{Z}_{k-1}(X)$ and $a : Z_{k-1}(X;\mathbb{Z}) \to C_k(X;\mathbb{Z})$ are group homomorphisms, first defined on basis elements and then extended linearly. Similarly, the choice of differential forms $\varrho \in \Omega^{k-1}(M)$ for $\zeta(z) = [M \xrightarrow{g} X]$ is made on a basis of $Z_{k-1}(X;\mathbb{Z})$. Extending linearly, the map $z \mapsto \exp\Big[2\pi i\Big(\int_M \varrho + \int_{a(z)}\widetilde{\mathrm{curv}}(x)\Big)\Big]$ defines a group homomorphism $\Xi(x) : Z_{k-1}(X;\mathbb{Z}) \to \mathrm{U}(1)$.

It remains to show that $\Xi(x)$ satisfies condition (25) for the homomorphism $z \mapsto \Xi(x)(z)$ to be a differential character. The argument is almost the same as in the proof of Lemma 25. Let $z = \partial c$ for some $c \in C_k(X;\mathbb{Z})$. By Lemma 7, the homomorphism $\zeta : C_k(X;\mathbb{Z}) \to \mathcal{C}_k(X;\mathbb{Z})$ satisfies $\zeta(\partial c) = \partial\zeta(c)$. We write $\zeta(c) = [N \xrightarrow{f} X]$, where $M = \partial N$ and $g = f|_M$.

If $N$ were an oriented smooth manifold with boundary, we would have $H^k(N;\mathbb{Z}) = \{0\}$. By the following argument (suggested to us by M. Kreck), we may also choose the stratifold $N$ such that its top dimensional cohomology vanishes: Replacing the top dimensional strata of $N$ and $M$ by the connected sum of their components if necessary, we may assume the top dimensional strata of $N$ and $M$ to be connected. This yields $H^k(N,M;\mathbb{Z}) \cong H^{k-1}(M;\mathbb{Z}) \cong \mathbb{Z}$, the first isomorphism being the boundary map. Now the long exact sequence of the pair $(N,M)$ yields $H^k(N;\mathbb{Z}) = \{0\}$.

Since $H^k(N;\mathbb{Z}) = \{0\}$, we have $f^*x = \widetilde{\iota}([\eta])$ for some $\eta \in \Omega^{k-1}(N)$. Since $\widetilde{\iota}$ is natural with respect to smooth maps, we have:

$$\widetilde{\iota}([\varrho]) = g^*x = (f^*x)|_{\partial N} = \widetilde{\iota}([\eta])|_{\partial N} = \widetilde{\iota}([\eta|_{\partial N}]) .$$

In particular, $\varrho - \eta|_M \in \Omega^{k-1}_0(M)$. This yields:

$$\begin{aligned}
\exp\Big(2\pi i \int_M \varrho\Big) &= \exp\Big(2\pi i \int_{\partial N} \eta\Big) \\
&= \exp\Big(2\pi i \int_N d\eta\Big) \\
&= \exp\Big(2\pi i \int_N \widetilde{\operatorname{curv}}(f^*x)\Big) .
\end{aligned}$$

Inserting this into (41), we obtain:

$$\begin{aligned}
\Xi(x)(\partial c) &= \exp\Big[2\pi i\Big(\int_{\partial N} \varrho + \int_{a(\partial c)} \widetilde{\operatorname{curv}}(x)\Big)\Big] \\
&= \exp\Big[2\pi i\Big(\int_N \widetilde{\operatorname{curv}}(f^*x) + \int_{a(\partial c)} \widetilde{\operatorname{curv}}(x)\Big)\Big] \\
&= \exp\Big[2\pi i\Big(\int_{[\zeta(c)]_{S_k}} \widetilde{\operatorname{curv}}(x) + \int_{a(\partial c)} \widetilde{\operatorname{curv}}(x)\Big)\Big] \\
&\overset{(4)}{=} \exp\Big[2\pi i\Big(\int_c \widetilde{\operatorname{curv}}(x) + \underbrace{\int_{\partial a(c+y(c))} \widetilde{\operatorname{curv}}(x)}_{=0}\Big)\Big] \\
&= \exp\Big(2\pi i \int_c \widetilde{\operatorname{curv}}(x)\Big) .
\end{aligned}$$

Thus $\Xi(x)$ is a differential character in $\widehat{H}^k(X;\mathbb{Z})$ with $\operatorname{curv}(\Xi(x)) = \widetilde{\operatorname{curv}}(x)$.

b) For any smooth space $X$, the map $\Xi : \widetilde{H}^*(X;\mathbb{Z}) \to \widehat{H}^*(X;\mathbb{Z})$ defined by (41) is additive. Thus $\Xi : \widetilde{H}^*(X;\mathbb{Z}) \to \widehat{H}^*(X;\mathbb{Z})$ is a degree 0 homomorphism of graded groups.

c) We show that $\Xi$ is natural with respect to smooth maps. Let $f : Y \to X$ be a smooth map. Let $x \in \widetilde{H}^k(X)$ and $z \in Z_{k-1}(Y;\mathbb{Z})$. We need to show that $\Xi(f^*x)(z) = f^*(\Xi(x))(z)$. Choose $\zeta(z) \in \mathcal{Z}_{k-1}(Y)$ and $a(z) \in C_k(Y;\mathbb{Z})$ such that $[z - \partial a(z)]_{\partial S_k} = [\zeta(z)]_{\partial S_k}$.

Write $\zeta(z) = [M \xrightarrow{g} Y]$. Setting $\zeta(f_*z) := f_*\zeta(z) = [M \xrightarrow{f\circ g} X]$ and $a(f_*z) := f_*a(z)$, we obtain

$$[f_*z - \partial a(f_*z)]_{\partial S_k} = f_*[z - \partial a(z)]_{\partial S_k} = f_*[\zeta(z)]_{\partial S_k} = [f_*\zeta(z)]_{\partial S_k}.$$

Now choose $\varrho \in \Omega^{k-1}(M)$ such that $(f \circ g)^*x = g^*(f^*x) = \widetilde{\iota}([\varrho])$. By Remark 15 and Lemma 25, we find:

$$\begin{aligned} f^*(\Xi(x))(z) &:= \Xi(x)(f_*z) \\ &= \exp\Big[2\pi i\Big(\int_M \varrho + \int_{a(f_*z)} \widetilde{\mathrm{curv}}(x)\Big)\Big] \\ &= \exp\Big[2\pi i\Big(\int_M \varrho + \int_{a(z)} \widetilde{\mathrm{curv}}(f^*x)\Big)\Big] \\ &= \Xi(f^*x)(z)\,. \end{aligned}$$

d) We show that $\Xi$ commutes with inclusions of flat classes. Let $u \in H^{k-1}(X;\mathrm{U}(1))$ and $z \in Z_{k-1}(X;\mathbb{Z})$. We choose $\zeta(z) = [M \xrightarrow{g} X]$ and $a(z) \in C_k(X;\mathbb{Z})$ as above. Note that $\widetilde{c}(g^*\widetilde{j}(u)) = 0$ for dimensional reasons. Thus $g^*u$ is the reduction mod $\mathbb{Z}$ of a class in $H^{k-1}(M;\mathbb{R})$. Let $\varrho \in \Omega^{k-1}(M)$ such that $g^*(\widetilde{j}(u)) = \widetilde{\iota}([\varrho])$. Since the upper left quadrant of diagram (36) commutes, the reduction mod $\mathbb{Z}$ of $[\varrho]_{\mathrm{dR}} \in H^{k-1}(M;\mathbb{R})$ coincides with $g^*u$. Moreover, the diagram (36) yields $\widetilde{\mathrm{curv}}(\widetilde{j}(u)) = 0$. Thus we have:

$$\begin{aligned} \Xi(\widetilde{j}(u))(z) &= \exp\Big[2\pi i\Big(\int_M \varrho + \int_{a(z)} \underbrace{\widetilde{\mathrm{curv}}(\widetilde{j}(u))}_{=0}\Big)\Big] \\ &= \exp\Big(2\pi i \int_M [\varrho]_{\mathrm{dR}}\Big) \\ &= \langle g^*u, [M]\rangle \\ &= \langle u, g_*[M]\rangle \\ &\overset{(5)}{=} \langle u, [z]\rangle \\ &\overset{(29)}{=} j(u)(z)\,. \end{aligned}$$

e) We show that $\Xi$ commutes with topological trivializations. Let $\varrho \in \Omega^{k-1}(X)$. Then we have:

$$
\begin{aligned}
\Xi(\widetilde{\iota}([\varrho]))(z) &= \exp\Big[2\pi i\Big(\int_M g^*\varrho + \int_{a(z)} \widetilde{\operatorname{curv}}(\widetilde{\iota}([\varrho]))\Big)\Big] \\
&\overset{(36)}{=} \exp\Big[2\pi i\Big(\int_{g_*[M]_{\partial S_k}} \varrho + \int_{a(z)} d\varrho\Big)\Big] \\
&= \exp\Big[2\pi i\Big(\int_{[\zeta(z)]_{\partial S_k}} \varrho + \int_{\partial a(z)} \varrho\Big)\Big] \\
&\overset{(5)}{=} \exp\Big(2\pi i \int_z \varrho\Big) \\
&\overset{(27)}{=} \iota(\varrho)(z)\,.
\end{aligned}
$$

□

# 6 The Ring Structure

In this section we discuss the ring structure on differential cohomology. Existence of a natural ring structure on $\widehat{H}^*(X;\mathbb{Z})$ compatible with curvature, characteristic class and topological trivializations was established in [24, Thm. 1.11] by an explicit formula using barycentric subdivision of singular chains and the chain homotopy from the subdivision to the identity. Simple formulas for the product are obtained for differential characters represented by differential forms with singularities as in [23] or by de Rham-Federer currents as in [43, Sec. 3].

An axiomatic definition of a ring structure on differential cohomology was established in [60], together with a proof that the ring structure is uniquely determined by these axioms (see [60, Thm. 1.2]). We use an axiomatic definition of the ring structure similar to the one in [60]. The sign convention for topological trivializations differs from the one in [60, p. 51] but coincides with the one in [14, Def. 1.2]. We give a corresponding axiomatic definition of an external or cross product and prove that this product is uniqely determined by the axioms. Uniquess of the external product has also been discussed in [49, Ch. 6]. Our proof has the advantage of giving an explicit geometric formula for the product.

**Definition 28.** An *internal product* of differential characters yields for any smooth space $X$ and any $(k,l) \in \mathbb{Z}\times\mathbb{Z}$ a map

$$
* : \widehat{H}^k(X;\mathbb{Z}) \times \widehat{H}^l(X;\mathbb{Z}) \to \widehat{H}^{k+l}(X;\mathbb{Z})\,, \quad (h,f) \mapsto h * f\,, \tag{48}
$$

such that the following holds:

1. *Ring structure.* The internal product $*$ is associative and $\mathbb{Z}$-bilinear, i.e. $(\widehat{H}^*(X;\mathbb{Z}),+,*)$ is a ring.

2. *Graded commutativity.* The product $*$ is graded commutative, i.e. for $h \in \widehat{H}^k(X;\mathbb{Z})$ and $f \in \widehat{H}^l(X;\mathbb{Z})$, we have $f * h = (-1)^{kl} h * f$.
3. *Naturality.* For any smooth map $g : Y \to X$ and $h, f \in \widehat{H}^*(X;\mathbb{Z})$, we have $g^*(h * f) = g^*h * g^*f$.
4. *Compatibility with curvature.* The curvature $\operatorname{curv} : \widehat{H}^*(X;\mathbb{Z}) \to \Omega^*_0(X)$ is a ring homomorphism, i.e. for $h, f \in \widehat{H}^*(X;\mathbb{Z})$, we have $\operatorname{curv}(h * f) = \operatorname{curv}(h) \wedge \operatorname{curv}(f)$.
5. *Compatibility with characteristic class.* The characteristic class is a ring homomorphism $c : \widehat{H}^*(X;\mathbb{Z}) \to H^*(X;\mathbb{Z})$, i.e. for $h, f \in \widehat{H}^*(X;\mathbb{Z})$, we have $c(h * f) = c(h) \cup c(f)$.
6. *Compatibility with topological trivialization.* For $\varrho \in \Omega^*(X)$ and $f \in \widehat{H}^l(X;\mathbb{Z})$, we have $\iota(\varrho) * f = \iota(\varrho \wedge \operatorname{curv}(f))$.

An internal product on differential cohomology induces an *external product* or *differential cohomology cross product*

$$\times : \widehat{H}^k(X;\mathbb{Z}) \times \widehat{H}^{k'}(X';\mathbb{Z}) \to \widehat{H}^{k+l}(X \times X';\mathbb{Z}), \quad h \times h' := \operatorname{pr}_1^* h * \operatorname{pr}_2^* h' .$$

Here $\operatorname{pr}_1, \operatorname{pr}_2$ denotes the projection on the first and second factor of $X \times X'$, respectively.

We may also define an *external product* or *differential cohomology cross product* axiomatically:

**Definition 29.** An *external product* of differential characters yields for any smooth spaces $X$ and $X'$ and any $(k, k') \in \mathbb{Z} \times \mathbb{Z}$ a map

$$\times : \widehat{H}^k(X;\mathbb{Z}) \times \widehat{H}^{k'}(X';\mathbb{Z}) \to \widehat{H}^{k+k'}(X \times X';\mathbb{Z}) , \quad (h, h') \mapsto h \times h' , \tag{49}$$

such that the following holds:

1. *Associativity, bilinearity.* The product $\times$ is associative and $\mathbb{Z}$-bilinear.
2. *Graded commutativity.* The product $\times$ is graded commutative, i.e. for $h \in \widehat{H}^k(X;\mathbb{Z})$ and $h' \in \widehat{H}^{k'}(X';\mathbb{Z})$, we have:
$$h' \times h = (-1)^{kk'} h \times h' . \tag{50}$$
3. *Naturality.* For any smooth maps $g : Y \to X$ and $g' : Y' \to X'$ and for $h \in \widehat{H}^*(X;\mathbb{Z})$ and $h' \in \widehat{H}^*(X';\mathbb{Z})$, we have:
$$(g \times g')^*(h \times h') = g^*h \times {g'}^* h' . \tag{51}$$
4. *Compatibility with curvature.* The curvature $\operatorname{curv} : \widehat{H}^*(X;\mathbb{Z}) \to \Omega^*_0(X)$ commutes with external products, in other words, for $h \in \widehat{H}^*(X;\mathbb{Z})$ and $h' \in \widehat{H}^*(X';\mathbb{Z})$, we have:
$$\operatorname{curv}(h \times h') = \operatorname{curv}(h) \times \operatorname{curv}(h') . \tag{52}$$

5. *Compatibility with characteristic class.* The characteristic class commutes with external products, i.e. for $h \in \widehat{H}^*(X;\mathbb{Z})$ and $h' \in \widehat{H}^*(X';\mathbb{Z})$, we have:
$$c(h \times h') = c(h) \times c(h'). \tag{53}$$

6. *Compatibility with topological trivialization.* For any form $\varrho \in \Omega^*(X)$ and any character $h' \in \widehat{H}^k(X';\mathbb{Z})$, we have:
$$\iota(\varrho) \times h' = \iota(\varrho \times \mathrm{curv}(h')). \tag{54}$$

An external product yields an internal product by setting $h * f := \Delta_X^*(h \times f)$ for any $h, f \in \widehat{H}^*(X;\mathbb{Z})$. Here $\Delta_X : X \to X \times X$ denotes the diagonal map.

Internal and external products are equivalent in the sense that any one determines the other. Starting with an internal product $*$, the induced external product recovers the original internal product: for any $h, f \in \widehat{H}^*(X;\mathbb{Z})$, we have

$$\Delta_X^*(h \times f) = \Delta_X^*(\mathrm{pr}_1^* h * \mathrm{pr}_2^* f) = (\mathrm{pr}_1 \circ \Delta_X)^* h * (\mathrm{pr}_2 \circ \Delta_X)^* f = h * f. \tag{55}$$

Conversely, starting with an external product $\times$, the induced internal product recovers the original external product: for $h \in \widehat{H}^*(X;\mathbb{Z})$ and $h' \in \widehat{H}^*(X';\mathbb{Z})$, we have

$$\begin{aligned}
\mathrm{pr}_1^* h * \mathrm{pr}_2^* h' &= \Delta_{X \times X'}(\mathrm{pr}_1^* h \times \mathrm{pr}_2^* h') \\
&= \Delta_{X \times X'}^*(\mathrm{pr}_1 \times \mathrm{pr}_2)^*(h \times h') \\
&= (\underbrace{(\mathrm{pr}_1 \times \mathrm{pr}_2) \circ \Delta_{X \times X'}}_{=\mathrm{id}_{X \times X'}})^*(h \times h') \\
&= h \times h'.
\end{aligned}$$

Internal products are useful, since they provide differential cohomology with a ring structure. On the other hand, external products are sometimes more useful for explicit calculations, as we shall see below.

In the following, we show that the ring structure on differential cohomology is uniquely determined by the axioms in Definition 28. By the discussion above, this is equivalent to the fact that the induced external product is uniquely determined by the axioms in Definition 29. To prove the latter, we start with the following special case:

**Lemma 30 (Evaluation on cartesian products).** *Let $M$ and $M'$ be closed oriented p-stratifolds. Suppose $\dim(M \times M') = k + k' - 1$. Let $\times$ be an external product in the sense of Definition 29. Then for differential characters $h \in \widehat{H}^k(M;\mathbb{Z})$ and $h' \in \widehat{H}^{k'}(M';\mathbb{Z})$, we have:*

$$(h \times h')([M \times M']) = \begin{cases} h([M])^{\langle c(h'),[M']\rangle} & \textit{if } \dim(M) = k-1 \\ h'([M'])^{(-1)^k \langle c(h),[M]\rangle} & \textit{if } \dim(M) = k \\ 1 & \textit{otherwise} \end{cases} \tag{56}$$

*Proof.* If $\dim(M)$ neither equals $k-1$ nor $k$, then either $\dim(M) < k-1$ or $\dim(M') < k'-1$. Thus we have $\widehat{H}^k(M;\mathbb{Z}) = \{0\}$ or $\widehat{H}^{k'}(M';\mathbb{Z}) = \{0\}$. Since $\times$ is bilinear, we have $h \times h' = 0$ in these cases.

Suppose $\dim(M) = k-1$, hence $\dim(M') = k'$. Then $h$ is topologically trivial for dimensional reasons. Thus we may choose $\varrho \in \Omega^{k-1}(M)$ such that $\iota(\varrho) = h$. By Definition 29, we then have:

$$\begin{aligned} (h \times h')([M \times M']) &= (\iota(\varrho) \times h')([M \times M']) \\ &\overset{(54)}{=} (\iota(\varrho \times \mathrm{curv}(h')))([M \times M']) \\ &= \exp\Big(2\pi i \int_{M \times M'} \varrho \times \mathrm{curv}(h')\Big) \\ &= \exp\Big(2\pi i\Big(\int_M \varrho \cdot \int_{M'} \mathrm{curv}(h')\Big)\Big) \\ &= \exp\Big(2\pi i \int_M \varrho\Big)^{\langle c(h'),[M']\rangle} \\ &= h([M])^{\langle c(h'),[M']\rangle}. \end{aligned}$$

Similarly, for $\dim(M) = k$ and $\dim(M') = k'-1$, we find $\varrho' \in \Omega^{k'-1}(M')$ such that $h' = \iota(\varrho')$. This yields

$$\begin{aligned} h \times h' &= h \times \iota(\varrho') \\ &= (-1)^{kk'} \iota(\varrho') \times h \\ &= (-1)^{kk'} \iota(\varrho' \times \mathrm{curv}(h)) \\ &= (-1)^{kk'} \iota((-1)^{k(k'-1)} \mathrm{curv}(h) \times \varrho) \\ &= (-1)^k \iota(\mathrm{curv}(h) \times \varrho') \end{aligned}$$

and hence

$$\begin{aligned} (h \times h')([M \times M']) &= \exp\Big(2\pi i \int_{M \times M'} (-1)^k \mathrm{curv}(h) \times \varrho'\Big) \\ &= h'([M'])^{(-1)^k \langle c(h),[M]\rangle}. \end{aligned}$$

□

Now we use this special case to show that the differential cohomology cross product is uniquely determined by the axioms in Definition 29. The main idea of the proof is to use a splitting of the Künneth sequence

$$0 \to \big[H_*(X;\mathbb{Z}) \otimes H_*(X';\mathbb{Z})\big]_n \xrightarrow{\times} H_n(X \times X';\mathbb{Z}) \to \mathrm{Tor}(H_*(X;\mathbb{Z}), H_*(X';\mathbb{Z}))_{n-1} \to 0$$

on the level of cycles. We use the well-known Alexander-Whitney and Eilenberg-Zilber maps $C_*(X \times X';\mathbb{Z}) \underset{EZ}{\overset{AW}{\rightleftarrows}} C_*(X;\mathbb{Z}) \otimes C_*(X';\mathbb{Z})$. These are chain homotopy inverses of each other with $EZ \circ AW$ chain homotopic to the identity on $C_*(X \times X';\mathbb{Z})$ and $AW \circ EZ = \mathrm{id}_{C_*(X;\mathbb{Z}) \otimes C_*(X';\mathbb{Z})}$, see [54, p. 167]. Let $i : Z_*(X;\mathbb{Z}) \to C_*(X;\mathbb{Z})$ be the inclusion and let $s : C_*(X;\mathbb{Z}) \to Z_*(X;\mathbb{Z})$ be a splitting as in Remark 6. Similarly, we have the inclusion $i'$ and a splitting $s'$ on $X'$. Set $S := (s \otimes s') \circ AW$ and $K := EZ \circ (i \otimes i')$. Denoting by $Z(C_*(X;\mathbb{Z}) \otimes C_*(X';\mathbb{Z}))$ the cycles of the tensor product complex, we obtain the following splitting of the Künneth sequence on the level of cycles:

$$
\begin{array}{ccccc}
0 \longrightarrow Z_*(X;\mathbb{Z}) \otimes Z_*(X';\mathbb{Z}) & \underset{s\otimes s'}{\overset{i\otimes i'}{\rightleftarrows}} & Z(C_*(X;\mathbb{Z}) \otimes C_*(X';\mathbb{Z})) & \longrightarrow & \dots \\
 & {\scriptstyle K \searrow \nwarrow S} & {\scriptstyle AW} \uparrow \downarrow {\scriptstyle EZ} & & \\
 & & Z_*(X \times X';\mathbb{Z}) & &
\end{array}
$$

In particular, we have $S \circ K = (s \otimes s') \circ AW \circ EZ \circ (i \otimes i') = \mathrm{id}_{Z_*(X;\mathbb{Z}) \otimes Z_*(X';\mathbb{Z})}$.

Now let $h \in \widehat{H}^k(X;\mathbb{Z})$ and $h' \in \widehat{H}^{k'}(X';\mathbb{Z})$ and $z \in Z_{k+k'-1}(X \times X';\mathbb{Z})$. We write $z = K \circ S(z) + (z - K \circ S(z))$. The Künneth sequence implies that $(z - K \circ S(z))$ represents a torsion class. Hence $(h \times h')(z - K \circ S(z))$ may be computed as in Remark 16. We compute $(h \times h')(K \circ S(z))$ as described in Remark 17 using geometric chains:

The splitting $S$ decomposes a cycle $z \in Z_{k+k'-1}(X \times X';\mathbb{Z})$ into a sum of tensor products of cycles with degrees adding up to $k + k' - 1$. We write

$$K \circ S(z) = \sum_{i+j=k+k'-1} \sum_m y_i^m \times y'^m_j , \tag{57}$$

where $y_i^m \in Z_i(X;\mathbb{Z})$ and $y'^m_j \in Z_j(X';\mathbb{Z})$. Now we are able to compute $(h \times h')(K \circ S(z))$.

**Theorem 31 (Uniqueness of Cross Product).** *The differential cohomology cross product is uniquely determined by the axioms in Definition 29.*

*Explicitly, for $h \in \widehat{H}^k(X;\mathbb{Z})$ and $h' \in \widehat{H}^{k'}(X';\mathbb{Z})$, the evaluation of $h \times h'$ on a cycle $z \in Z_{k+k'-1}(X \times X';\mathbb{Z})$ can be computed as follows: Decompose $K \circ S(z)$ as in (57). Choose $N \in \mathbb{N}$ and $x \in C_{k+k'}(X;\mathbb{Z})$ as in Remark 16 such that $N \cdot (z - K \circ S(z)) = \partial x$. Then we have:*

$$
\begin{aligned}
(h \times h')(z) = \prod_m & \left[ h(y^m_{k-1})^{\langle c(h'), y'^m_{k'} \rangle} \cdot h'(y'^m_{k'-1})^{(-1)^k \langle c(h), y^m_k \rangle} \right] \\
& \cdot \exp\Big( \frac{2\pi i}{N} \Big( \int_x \mathrm{curv}(h) \times \mathrm{curv}(h') - \langle c(h) \times c(h'), x \rangle \Big) \Big).
\end{aligned} \tag{58}
$$

*Proof.* As above, we write $z = K\circ S(z) + (z - K\circ S(z))$. We evaluate $(h\times h')$ on the two summands separately.

a) By Remark 16, we have:

$$
\begin{aligned}
(h\times h')(1-K{\circ}S(z)) \\
&\overset{(32)}{=} \exp\frac{2\pi i}{N}\Big(\int_x \operatorname{curv}(h\times h') - \langle c(h\times h'), x\rangle\Big) \\
&\overset{(52),(53)}{=} \exp\frac{2\pi i}{N}\Big(\int_x \operatorname{curv}(h)\times\operatorname{curv}(h') - \langle c(h)\times c(h'), x\rangle\Big)
\end{aligned}
$$

which yields the second contribution to (58). This shows in particular, that the value of $h\times h'$ on torsion cycles is uniquely determined by compatibility with curvature and characteristic class in Definition 29.

b) As in (57) write $K\circ S(z) = \sum_{i+j=k+k'-1}\sum_m y_i^m \times {y'}_j^m$, where $y_i^m \in Z_i(X;\mathbb{Z})$ and ${y'}_j^m \in Z_j(X';\mathbb{Z})$. Let $\zeta^X(y_i^m) = [M_i^m \xrightarrow{g_i^m} X]$ and $\zeta^{X'}({y'}_j^m) = [{M'}_j^m \xrightarrow{{g'}_j^m} X']$. Choose fundamental cycles $x_i^m$ of $M_i^m$ and ${x'}_j^m$ of ${M'}_j^m$. Then we have $y_i^m = {g_i^m}_* x_i^m + \partial a^X(y_i^m)$ and ${y'}_j^m = {{g'}_j^m}_* {x'}_j^m + \partial a^X({y'}_j^m)$ up to boundaries of thin chains. Thus we have

$$
\begin{aligned}
y_i^m \times {y'}_j^m &= {g_i^m}_* x_i^m \times {{g'}_j^m}_* {x'}_j^m + \partial a^X(y_m^i)\times {g'}_{j*}^m {x'}_j^m \\
&\quad + {g_i^m}_* x_i^m \times \partial a^{X'}({y'}_j^m) + \partial a^X(y_i^m)\times \partial a^{X'}({y'}_j^m) \\
&= \big(g_i^m \times {g'}_j^m\big)_*(x_i^m \times {x'}_j^m) \\
&\quad + \partial\big(a^X(y_m^i)\times {g'}_{j*}^m {x'}_j^m + (-1)^i {g_i^m}_* x_i^m \times a^{X'}({y'}_j^m)) \\
&\quad + \partial a^X(y_i^m)\times \partial a^{X'}({y'}_j^m))
\end{aligned} \tag{59}
$$

up to boundaries of thin chains.

The character $h\times h'$ vanishes on the third summand of (59) by (24) and (52) and the fact that $\operatorname{curv}(h)$ and $\operatorname{curv}(h')$ are closed forms. For degrees $(i,j)$ different from $(k-1,k')$ and $(k,k'-1)$, the character $h\times h'$ vanishes on the first summand of (59) by Lemma 30 and on the second summand by (24) and (52). Now for $(i,j)=(k-1,k')$, we have:

$$
\begin{aligned}
&(h\times h)(y_{k-1}^m \times {y'}_{k'}^m) \\
&= h([\zeta^X(y_{k-1}^m)]_{\partial S_k})^{\langle c(h'),{y'}_{k'}^m\rangle}\cdot \exp\Big(2\pi i \int_{-a^X(y_k^m)\times {y'}_{k'}^m} \operatorname{curv}(h\times h')\Big) \\
&= h([\zeta^X(y_{k-1}^m)]_{\partial S_k})^{\langle c(h'),{y'}_{k'}^m\rangle}\cdot \exp\Big(2\pi i \int_{-a^X(y_k^m)} \operatorname{curv}(h)\cdot \int_{{y'}_{k'}^m} \operatorname{curv}(h')\Big) \\
&= h([\zeta^X(y_{k-1}^m)]_{\partial S_k})^{\langle c(h'),{y'}_{k'}^m\rangle}\cdot h(-\partial a^X(y_k^m))^{\langle c(h'),{y'}_{k'}^m\rangle} \\
&= h(y_{k-1}^m)^{\langle c(h'),{y'}_{k'}^m\rangle}.
\end{aligned}
$$

Analogously, for $(i,j) = (k, k'-1)$, we have:

$$(h \times h)(y_k^m \times {y'}_{k'-1}^m) = h'({y'}_{k'-1}^m)^{(-1)^k \cdot \langle c(h), y_k^m \rangle} .$$

In particular, the evaluation of $h \times h'$ on $K \circ S(z)$ is uniquely determined by the axioms in Definition 29 (through Lemma 30). □

**Corollary 32 (Uniqueness of Ring Structure).** *The ring structure on differential cohomology is uniquely determined by the axioms in Definition 28.*

*Remark 33.* We have shown *uniqueness* of the ring structure. We could take (58) as definition of a differential cohomology cross product to prove *existence* of the cross product and ring structure on differential cohomology. This would require to verify the axioms in Definition 29.

Here we take existence of the ring structure and cross product for granted. In [3] we start from (58) to construct a cross product between relative and absolute differential characters. There we show that the product defined by (58) satsfies the axioms in Definition 29.

*Example 34.* Let $h_1, h_2 \in \widehat{H}^1(X;\mathbb{Z}) \cong C^\infty(X; \mathrm{U}(1))$. As in Example 18, we denote the corresponding smooth functions by $\bar{h}_1$, $\bar{h}_2$. Now $h_1 * h_2 \in \widehat{H}^2(X;\mathbb{Z})$. Hence, given two smooth functions $\bar{h}_j : X \to \mathrm{U}(1)$, we obtain a $\mathrm{U}(1)$-bundle with connection over $X$ (up to isomorphism). We now describe this bundle in classical geometric terms.

Let $i \in \widehat{H}^1(\mathrm{U}(1);\mathbb{Z})$ be the differential character that corresponds to the smooth function $\bar{i} = \mathrm{id}_{\mathrm{U}(1)} : \mathrm{U}(1) \to \mathrm{U}(1)$. Then we have $\bar{h}_j = \mathrm{id}_{\mathrm{U}(1)} \circ \bar{h}_j$ and thus $h_j = \bar{h}_j^* i$. We put $\bar{h} = (\bar{h}_1, \bar{h}_2) : X \to \mathrm{U}(1) \times \mathrm{U}(1) =: T^2$. Let $\Delta : \mathrm{U}(1) \to T^2$, $t \mapsto (t,t)$, the diagonal map. This yields

$$\begin{aligned} h_1 * h_2 &= \Delta^*(h_1 \times h_2) \\ &= \Delta^*(\bar{h}_1^* i \times \bar{h}_2^* i) \\ &\overset{(51)}{=} \Delta^*(\bar{h}_1 \times \bar{h}_2)^*(i \times i) \\ &= ((\bar{h}_1 \times \bar{h}_2) \circ \Delta)^*(i \times i) \\ &= \bar{h}^*(i \times i). \end{aligned}$$

The bundle corresponding to $h_1 * h_2$ is thus given by pull-back along $\bar{h}$ of a universal bundle with connection $(P, \nabla)$ on $T^2$ which represents $i \times i \in \widehat{H}^2(T^2;\mathbb{Z})$.

The bundle $(P, \nabla)$ was described in algebraic geometric terms in [5, Sec. 1] where it leads to the regulator map in algebraic $K$-theory. The total space is identified with the Heisenberg manifold $H(\mathbb{R})/H(\mathbb{Z})$. In [10, p. 60] it is called the *Poincaré bundle*. We now determine this bundle.

The curvature $\mathrm{curv}(i)$ is a volume form on $\mathrm{U}(1)$ with total volume 1. Thus by (52), the curvature $\mathrm{curv}(i \times i)$ is a volume form on $T^2$ with total volume 1. Since $H^2(T^2;\mathbb{Z})$ has no torsion, the characteristic class $c(i \times i)$ can be identified with the de Rham class of $\mathrm{curv}(i \times i)$. This class determines the $\mathrm{U}(1)$-bundle $P \to T^2$ topologically. It remains to determine the connection $\nabla$.

Let $\Theta_1, \Theta_2 : \mathbb{R}^2 \to \mathbb{R}$ denote the projection on the first and second factor, respectively. Let $p : \mathbb{R}^2 \to \mathbb{R}^2/\mathbb{Z}^2 \cong T^2$, $v = (v_1, v_2) \mapsto (\exp(2\pi i v_1), \exp(2\pi i v_2))$, denote the projection. Let $\nabla$ be any connection on $P$ with curvature $\frac{i}{2\pi}\mathrm{curv}(\nabla) = \mathrm{curv}(i \times i)$. Fix a trivialization $T : p^*P \to \mathbb{R}^2 \times \mathrm{U}(1)$. As in Example 19, we denote by $\vartheta(p^*\nabla, T) \in \Omega^1(\mathbb{R}^2)$ the 1-form that corresponds to the connection $p^*\nabla$. The trivialization can be chosen such that

$$\vartheta(p^*\nabla, T) = (\Theta_1/2 - w_1)d\Theta_2 - (\Theta_2/2 - w_2)d\Theta_1 =: A_w$$

for some $w = (w_1, w_2) \in \mathbb{R}^2$. Two forms $A_w$ and $A_{w'}$ describe the same connection $\nabla$ on $P$ if and only if $w - w' \in \mathbb{Z}^2$.

The parameter $w$, and hence the connection $\nabla$, can be determined by the holonomy along two particular curves in $T^2$. We consider the curves $\gamma_1 : [0,1] \to T^2$, $t \mapsto (\exp(2\pi i t), 1)$, and $\gamma_2 : [0,1] \to T^2$, $t \mapsto (1, \exp(2\pi i t))$. We set $\Gamma_1 : [0,1] \to \mathbb{R}^2$, $t \mapsto (t, 0)$, and $\Gamma_2 : [0,1] \mapsto \mathbb{R}^2$, $t \mapsto (0, t)$, so that $\Gamma_j$ lifts $\gamma_j$. Then we have:

$$\begin{aligned}\mathrm{Hol}^\nabla(\gamma_1) &= \exp\Big(2\pi i \int_{\Gamma_1} A_w\Big)\\ &= \exp\Big(2\pi i \int_{\Gamma} (-0/2 + w_2) d\Theta_1\Big)\\ &= \exp\Big(2\pi i w_2\Big)\end{aligned}$$

and similarly

$$\mathrm{Hol}^\nabla(\gamma_2) = \exp\Big(-2\pi i w_1\Big).$$

To determine the connection, we evaluate $i \times i$ on the cycles $\gamma_1$ and $\gamma_2$. Denote the fundamental cycle $[0,1] \to \mathrm{U}(1)$, $t \mapsto \exp(2\pi i t)$, of $\mathrm{U}(1)$ by $y$. Then the decomposition (57) of $\gamma_1$ is given by

$$\gamma_1 = y \times 1 = K(y \otimes 1).$$

We apply Theorem 31 with $z = \gamma_1$ and observe that we can choose $x = 0$ because $\gamma_1 = K(S(\gamma_1))$. Since $[\zeta^{\mathrm{U}(1)}(y)]_{\partial S_2} = [\gamma_1]_{\partial S_2}$ we may choose $a(\gamma_1) = 0$. Now (58) says

$$i \times i(\gamma_1) = i(1)^{-1} = 1.$$

Similarly, we get $i \times i(\gamma_2) = 1$. Hence our connection $\nabla$ is given by $A_0 = \Theta_1/2d\Theta_2 - \Theta_2/2d\Theta_1$.

*Remark 35. Künneth sequence.* The exactness of the Künneth sequence for singular cohomology

$$0 \to \left[H^*(X;\mathbb{Z}) \otimes H^*(X';\mathbb{Z})\right]_n \xrightarrow{\times} H^n(X \times X';\mathbb{Z}) \to \mathrm{Tor}(H^*(X;\mathbb{Z}),H^*(X';\mathbb{Z}))_{n+1} \to 0$$

implies that the cohomology cross product is injective. The Künneth sequence is usually constructed in two steps: the first one is purely algebraic and relates the homology of tensor products of chain complexes with the tensor product of the homologies; the second one identifies the singular homology of the cartesian product of spaces with the homology of the tensor product of the singular chain complexes.

The question arises whether there is a Künneth sequence for differential cohomology. As to the above mentioned first step, the differential cohomology groups of a space $X$ can be constructed as the homology groups of a chain complex using a modification of the Hopkins-Singer complex, as described in [8, p. 271]. This way one obtains the homological algebraic Künneth sequence for that complex. The middle term of that sequence is the homology of the tensor product complex. The relation of this tensor product homology to the differential cohomology of the cartesian product seems to be unknown.

The following example illustrates that the differential cohomology cross product is in general not injective:

*Example 36.* Let $X, X'$ be closed manifolds of dimensions $k-1$ and $k'$, respectively. Let $\varrho \in \Omega^{k-1}(X)$ and $\varrho \in \Omega^{k'}(X')$ be volume forms for some Riemannian metrics on $X$ and $X'$ with total volume 1. In particular, $\varrho$ and $\varrho'$ are closed with integral periods, and $\frac{1}{2}\varrho$ does not have integral periods. Choose a differential character $h' \in \widehat{H}^{k'}(X';\mathbb{Z})$ with $\mathrm{curv}(h') = 2\varrho'$. Set $h := \iota(\frac{1}{2} \cdot \varrho) \neq 0 \in \widehat{H}^k(X)$. Then we have $h \times h' \overset{(54)}{=} \iota(\frac{1}{2} \cdot \varrho \times 2\varrho') = \iota(\varrho \times \varrho')$. This vanishes since $\int_{X \times X'} \varrho \times \varrho' = 1$ and $\varrho \times \varrho'$ thus has integral periods.

## 7 Fiber Integration

In this section we construct the fiber integration map for differential characters. Fiber integration has been described in some of the various models for differential cohomology. The construction of Hopkins and Singer in [45] is based on their own model and uses embeddings into high-dimensional Euclidean spaces. Fiber integration for smooth Deligne cohomology is constructed in [37]. In [27] and [49] Dupont and Ljungmann give a geometric construction of fiber integration for smooth Deligne cohomology where the combinatorial complications are taken care of by the calculus of simplicial forms. Uniqueness of fiber integration is discussed in [49, Ch. 6]. A model for differential characters involving stratifolds is described in [12] where fiber integration is also discussed. The fiber integration or Gysin map for de Rham-Federer currents is described in [43, Sec. 10].

We use the original definition of differential characters due to Cheeger and Simons. Our construction of the fiber integration map works for fiber bundles (with compact oriented fibers) on all smooth spaces in the sense of Sect. 2. The approaches in [12] and [27, 49] seem to be limited to fiber bundles over finite dimensional bases. However, allowing infinite-dimensional manifolds is important. For example, the transgression map from equivalence classes of gerbes with connection on $X$ to equivalence classes of line bundles with connection on the free loop space $\mathcal{L}(X)$ is constructed using fiber integration in the trivial bundle $S^1 \times \mathcal{L}(X) \to \mathcal{L}(X)$, compare Sect. 9.

We show that fiber integration (for fiber bundles whose fibers are closed oriented manifolds) is uniquely determined by certain naturality conditions. This yields an explicit formula for the fiber integration map which we then use for its definition. We show that this yields a well-defined fiber integration map that has the required properties. Finally, we discuss fiber integration in the case where the fiber has a boundary.

Similar approaches to our construction of the fiber integration map have been sketched briefly in [30, Prop. 2.1], in [19, Sec. 3.6] and in [10, Thm. 3.135].

## *7.1 Fiber Integration for Closed Fibers*

In this section we discuss fiber integration of differential characters for fiber bundles $\pi : E \to X$ with closed oriented fibers. Since the class of smooth spaces as defined in 2 may contain diffeological spaces for which there are different inequivalent notions of fiber bundles, we clarify the notion to be used in the sequel.

**Definition 37.** A fiber bundle is a surjective smooth map $\pi : E \to X$ between smooth spaces such that for any smooth map $M \xrightarrow{f} X$ from a (finite dimensional) stratifold $M$ we have a pull-back square of smooth maps

$$\begin{array}{ccc} f^*E & \xrightarrow{F} & E \\ {\scriptstyle\pi}\downarrow & & \downarrow{\scriptstyle\pi} \\ M & \xrightarrow[f]{} & X \end{array}$$

and the pull-back $\pi : f^*E \to M$ is a locally trivial fiber bundle in the usual sense.

Next we fix our notion of fiber integration for differential characters.

**Definition 38.** Let $F \hookrightarrow E \overset{\pi}{\twoheadrightarrow} X$ be a fiber bundle over a smooth space $X$ whose fibers are closed (i.e., finite-dimensional, compact and boundaryless) oriented manifolds. Fiber integration for differential characters associates to each such bundle a group homomorphism $\widehat{\pi}_! : \widehat{H}^*(E;\mathbb{Z}) \to \widehat{H}^{*-\dim F}(X;\mathbb{Z})$ such that the following holds:

1. *Naturality.* For any smooth map $g : Y \to X$ the fiber integration map commutes with the maps in the pull-back diagram

$$\begin{array}{ccc} g^*E & \xrightarrow{G} & E \\ \downarrow{\scriptstyle \pi} & & \downarrow{\scriptstyle \pi} \\ Y & \xrightarrow{g} & X\,. \end{array}$$

This means that for any $h \in \widehat{H}^k(E;\mathbb{Z})$, we have

$$\widehat{\pi}_!(G^*h) = g^*\widehat{\pi}_!(h)\,. \tag{60}$$

In other words, the following diagram is commutative for all $k$:

$$\begin{array}{ccc} \widehat{H}^k(E;\mathbb{Z}) & \xrightarrow{G^*} & \widehat{H}^k(g^*E;\mathbb{Z}) \\ \downarrow{\scriptstyle \widehat{\pi}_!} & & \downarrow{\scriptstyle \widehat{\pi}_!} \\ \widehat{H}^{k-\dim F}(X;\mathbb{Z}) & \xrightarrow{g^*} & \widehat{H}^{k-\dim F}(Y;\mathbb{Z}) \end{array} \tag{61}$$

2. *Compatibility with curvature.* Let $\fint_F : \Omega^*(E) \to \Omega^{*-\dim F}(X)$ be the usual fiber integration of differential forms, see [39, Ch. VII]. We require that the fiber integration of differential characters is compatible with the fiber integration of the curvature form, i.e.,

$$\begin{array}{ccc} \widehat{H}^k(E;\mathbb{Z}) & \xrightarrow{\mathrm{curv}} & \Omega^k_0(E) \\ \downarrow{\scriptstyle \widehat{\pi}_!} & & \downarrow{\scriptstyle \fint} \\ \widehat{H}^{k-\dim F}(X;\mathbb{Z}) & \xrightarrow{\mathrm{curv}} & \Omega^{k-\dim F}_0(X) \end{array} \tag{62}$$

commutes.

3. *Compatibility with topological trivializations of flat characters.* We demand that the following diagram commutes:

$$\begin{array}{ccc} \Omega^{k-1}_{cl}(E) & \xrightarrow{\iota} & \widehat{H}^k(E;\mathbb{Z}) \\ \downarrow{\scriptstyle \fint} & & \downarrow{\scriptstyle \widehat{\pi}_!} \\ \Omega^{k-1-\dim F}_{cl}(X) & \xrightarrow{\iota} & \widehat{H}^{k-\dim F}(X;\mathbb{Z})\,. \end{array} \tag{63}$$

Before we construct fiber integration for differential characters using geometric chains, we first show that it is uniquely determined by the above conditions:

**Theorem 39 (Uniqueness of Fiber Integration).** *If fiber integration for differential characters exists, then it is uniquely determined by the conditions of naturality and compatibility in Definition 38.*

*Proof.* Let $F \hookrightarrow E \stackrel{\pi}{\twoheadrightarrow} X$ be a fiber bundle with closed oriented fibers over a smooth space $X$. Let $\widehat{\pi}_! : \widehat{H}^k(E;\mathbb{Z}) \to \widehat{H}^{k-\dim F}(X;\mathbb{Z})$ be a fiber integration map as in Definition 38.

For $k < \dim F$ the map $\widehat{\pi}_!$ is uniquely determined, since in this case $\widehat{H}^{k-\dim F}(X;\mathbb{Z}) = \{0\}$ by (31). For $k = \dim F$, the compatibility with curvature implies that $\mathrm{curv}(\widehat{\pi}_! h) = \fint_F \mathrm{curv}(h) \in \Omega^0_0(X)$. For degree 0, the diagram (30) yields the isomorphisms

$$\begin{array}{ccc} \widehat{H}^0(X;\mathbb{Z}) & \xrightarrow[\cong]{\mathrm{curv}} & \Omega^0_0(X) \\ {\scriptstyle c=\mathrm{id}}\downarrow\cong & & \cong\downarrow \\ H^0(X;\mathbb{Z}) & \xrightarrow[\cong]{} & \mathrm{Hom}(H_0(X;\mathbb{Z}),\mathbb{Z}) \end{array} \tag{64}$$

Thus $\widehat{\pi}_! h$ is uniquely determined by its curvature.

Now let $k > \dim F$. Let $h \in \widehat{H}^k(E;\mathbb{Z})$ be a differential character on the total $E$ and $z \in Z_{k-1-\dim F}(X;\mathbb{Z})$ a smooth singular cycle in the base $X$. We show that the value of $\widehat{\pi}_! h$ on $z$ is uniquely determined by the conditions in Definition 38.

As in Lemma 7 we choose a geometric cycle $\zeta(z) = [M \xrightarrow{g} X] \in \mathcal{Z}_{k-1-\dim F}(X)$ and a smooth singular chain $a(z) \in C_{k-\dim F}(X;\mathbb{Z})$ such that $[z - \partial a(z)]_{\partial S_{k-\dim F}} = [\zeta(z)]_{\partial S_{k-\dim F}}$. We then have:

$$\begin{aligned} (\widehat{\pi}_! h)(z) &\stackrel{(34)}{=} (g^*\widehat{\pi}_! h)[M] \cdot \exp\Big(2\pi i \int_{a(z)} \mathrm{curv}(\widehat{\pi}_! h)\Big) \\ &\stackrel{(62)}{=} (g^*\widehat{\pi}_! h)[M] \cdot \exp\Big(2\pi i \int_{a(z)} \fint_F \mathrm{curv}(h)\Big) \\ &\stackrel{(60)}{=} (\widehat{\pi}_! G^* h)[M] \cdot \exp\Big(2\pi i \int_{a(z)} \fint_F \mathrm{curv}(h)\Big) . \end{aligned}$$

The differential character $G^*h \in \widehat{H}^k(g^*E;\mathbb{Z})$ is topologically trivial and flat for dimensional reasons (note that $\dim(g^*E) = k-1$). Hence $G^*h = \iota(\chi)$ for some closed differential form $\chi \in \Omega^{k-1}(g^*E)$. From the commutative diagram (63) we then have

$$
\begin{aligned}
(\widehat{\pi}_! h)(z) &= (\widehat{\pi}_! G^* h)[M] \cdot \exp\Big(2\pi i \int_{a(z)} \fint_F \mathrm{curv}(h)\Big) \\
&= (\widehat{\pi}_! \iota(\chi))[M] \cdot \exp\Big(2\pi i \int_{a(z)} \fint_F \mathrm{curv}(h)\Big) \\
&\overset{(63)}{=} \iota(\fint_F \chi)[M] \cdot \exp\Big(2\pi i \int_{a(z)} \fint_F \mathrm{curv}(h)\Big) \\
&= \exp\Big(2\pi i \int_M \fint_F \chi\Big) \cdot \exp\Big(2\pi i \int_{a(z)} \fint_F \mathrm{curv}(h)\Big) . \qquad (65)
\end{aligned}
$$

We thus obtained an expression for the value of $\widehat{\pi}_! h$ on $z$, which is uniquely determined by the conditions of naturality and compatibility. □

We can rewrite formula (65) more elegantly in terms of the pull-back operation $\mathrm{PB}_\bullet$ from Sect. 4: As above, let $h \in \widehat{H}^k(E;\mathbb{Z})$ be a differential character on the total space and $z \in Z_{k-1-\dim F}(X;\mathbb{Z})$ a smooth singular cycle in the base. As above we get the geometric cycle $\zeta(z) = [M \xrightarrow{g} X] \in \mathcal{Z}_{k-1-\dim F}(X)$ and the smooth singular chain $a(z) \in C_{k-\dim F}(X;\mathbb{Z})$ such that $[z - \partial a(z)]_{\partial S_{k-\dim F}} = [\zeta(z)]_{\partial S_{k-\dim F}}$. We then have:

$$
\begin{aligned}
(\widehat{\pi}_! h)(z) &\overset{(65)}{=} \exp\Big(2\pi i \int_M \fint_F \chi\Big) \cdot \exp\Big(2\pi i \int_{a(z)} \fint_F \mathrm{curv}(h)\Big) \\
&= \exp\Big(2\pi i \int_{g^*E} \chi\Big) \cdot \exp\Big(2\pi i \int_{a(z)} \fint_F \mathrm{curv}(h)\Big) \\
&= \iota(\chi)([g^*E]) \cdot \exp\Big(2\pi i \int_{a(z)} \fint_F \mathrm{curv}(h)\Big) \\
&= G^* h([g^*E]) \cdot \exp\Big(2\pi i \int_{a(z)} \fint_F \mathrm{curv}(h)\Big) \\
&= h([g^*E \xrightarrow{G} E]_{\partial S_k}) \cdot \exp\Big(2\pi i \int_{a(z)} \fint_F \mathrm{curv}(h)\Big) \\
&= h([\mathrm{PB}_E([M \xrightarrow{g} X])]_{\partial S_k}) \cdot \exp\Big(2\pi i \int_{a(z)} \fint_F \mathrm{curv}(h)\Big) \\
&= h([\mathrm{PB}_E \zeta(z)]_{\partial S_k}) \cdot \exp\Big(2\pi i \int_{a(z)} \fint_F \mathrm{curv}(h)\Big) .
\end{aligned}
$$

Hence we obtain the following constructive definition for the fiber integration map on differential characters:

**Definition 40.** Let $F \hookrightarrow E \twoheadrightarrow X$ be a fiber bundle with closed oriented fibers over a smooth space $X$. For $k < \dim F$, the fiber integration map $\widehat{\pi}_! : \widehat{H}^k(E;\mathbb{Z}) \to \widehat{H}^{k-\dim F}(X;\mathbb{Z}) = \{0\}$ is trivial. For $k = \dim F$, the fiber integration map $\widehat{\pi}_! : \widehat{H}^{\dim F}(E;\mathbb{Z}) \to \widehat{H}^0(X;\mathbb{Z}) = H^0(X;\mathbb{Z})$ is defined as:

$$\widehat{\pi}_! h := \pi_! c(h) .$$

For $k > \dim F$, the fiber integration map $\widehat{\pi}_! : \widehat{H}^k(E;\mathbb{Z}) \to \widehat{H}^{k-\dim F}(X;\mathbb{Z})$ is defined as:

$$(\widehat{\pi}_! h)(z) := h([\mathrm{PB}_E \zeta(z)]_{\partial S_k}) \cdot \exp\Big(2\pi i \int_{a(z)} \fint_F \mathrm{curv}(h)\Big) \tag{66}$$

$$= (G^* h)([g^* E]) \cdot \exp\Big(2\pi i \int_{a(z)} \fint_F \mathrm{curv}(h)\Big). \tag{67}$$

Using the transfer map $\lambda$ constructed in Remark 8, we obtain the following expression for fiber integration:

**Lemma 41 (Fiber Integration via Transfer Map).** *Let* $k > \dim F$. *Let* $h \in \widehat{H}^k(E;\mathbb{Z})$ *and let* $\lambda : C_{k-1-\dim F}(X;\mathbb{Z}) \to C_{k-1}(E;\mathbb{Z})$ *as defined in Remark 8. Then we have for any* $z \in Z_{k-1-\dim F}(X;\mathbb{Z})$*:*

$$(\widehat{\pi}_! h)(z) = h(\lambda(z)) \cdot \exp\Big(2\pi i \int_{a(z)} \fint_F \mathrm{curv}(h)\Big). \tag{68}$$

*Proof.* By (66) and the construction of $\lambda$, we find:

$$\begin{aligned}(\widehat{\pi}_! h)(z) &:= h([\mathrm{PB}_E \zeta(z)]_{\partial S_k}) \cdot \exp\Big(2\pi i \int_{a(z)} \fint_F \mathrm{curv}(h)\Big) \\ &\overset{(15)}{=} h([\lambda(z)]_{\partial S_k}) \cdot \exp\Big(2\pi i \int_{a(z)} \fint_F \mathrm{curv}(h)\Big) \\ &= h(\lambda(z)) \cdot \exp\Big(2\pi i \int_{a(z)} \fint_F \mathrm{curv}(h)\Big).\end{aligned}$$

In the last equation we used thin invariance of differential characters. □

**Lemma 42.** *The fiber integration* $\widehat{\pi}_!$ *as defined in* (66) *is a group homomorphism* $\widehat{H}^k(E;\mathbb{Z}) \to \widehat{H}^{k-\dim F}(X;\mathbb{Z})$. *In particular, for* $k > \dim F$ *the map* $z \mapsto \widehat{\pi}_! h(z)$ *is indeed a differential character.*

*Proof.* For $k < \dim F$, we obtain the trivial map $\widehat{H}^k(E;\mathbb{Z}) \to \widehat{H}^{k-\dim F}(X;\mathbb{Z}) = \{0\}$, which is a group homomorphism. For $k = \dim F$, the fiber integration map $\widehat{\pi}_! = \pi_! \circ c : \widehat{H}^{\dim F}(E;\mathbb{Z}) \to \widehat{H}^0(X;\mathbb{Z})$ is the composition of the group homomorphisms $c : \widehat{H}^{\dim F}(E;\mathbb{Z}) \to H^{\dim F}(E;\mathbb{Z})$ and $\pi_! : H^{\dim F}(E;\mathbb{Z}) \to H^0(X;\mathbb{Z})$.

Now let $k > \dim F$. We show that $\widehat{\pi}_! h$ is indeed a differential character. The map $z \mapsto \widehat{\pi}_! h(z)$ is a group homomorphism $Z_{k-1-\dim F}(X;\mathbb{Z}) \to \mathrm{U}(1)$ because all ingredients of the right hand side of (66) are homomorphisms.

We check that the evaluation of $\widehat{\pi}_! h$ on a boundary is given by the integral of a differential form. Let $z = \partial c \in B_{k-1-\dim F}(X;\mathbb{Z})$ be a smooth singular boundary on $X$. As in Lemma 7, we choose geometric chains $\zeta(\partial c) \in \mathcal{B}_{k-\dim F-1}(X)$ and $\zeta(c) \in \mathcal{C}_{k-\dim F}(X)$, and smooth singular chains $a(z) \in C_{k-\dim F}(X;\mathbb{Z})$ and $y(c) \in Z_{k-\dim F}(X;\mathbb{Z})$ such that $\partial\zeta(c) = \zeta(\partial c)$

and $[c - a(\partial c) - \partial a(c + y(c))]_{S_{k-\dim F}} = [\zeta(c)]_{S_{k-\dim F}}$. Using (20) for the transfer map $\lambda$, we obtain:

$$\begin{aligned}
(\widehat{\pi}_! h)(\partial c) &= h(\lambda(\partial c)) \cdot \exp\Big(2\pi i \int_{a(\partial c)} \fint_F \operatorname{curv}(h)\Big) \\
&\overset{(17)}{=} h(\partial \lambda(c)) \cdot \exp\Big(2\pi i \int_{a(\partial c)} \fint_F \operatorname{curv}(h)\Big) \\
&= \exp\Big(2\pi i \cdot \Big(\int_{\lambda(c)} \operatorname{curv}(h) + \int_{a(\partial c)} \fint_F \operatorname{curv}(h)\Big)\Big) \\
&\overset{(20)}{=} \exp\Big(2\pi i \cdot \Big(\int_{c - a(\partial c)} \fint_F \operatorname{curv}(h) + \int_{a(\partial c)} \fint_F \operatorname{curv}(h)\Big)\Big) \\
&= \exp\Big(2\pi i \int_c \fint_F \operatorname{curv}(h)\Big). \qquad (69)
\end{aligned}$$

Thus $\widehat{\pi}_! h$ is indeed a differential character. From (66) it is now clear that $h \mapsto \widehat{\pi}_! h$ is a homomorphism $\widehat{H}^k(E;\mathbb{Z}) \to \widehat{H}^{k-\dim F}(X;\mathbb{Z})$. □

We show that the definition of $\widehat{\pi}_! h$ in (66) is independent of the choices.

**Lemma 43.** *Let $k > \dim F$. Let $\zeta' : Z_{k-1-\dim F}(X;\mathbb{Z}) \to \mathcal{Z}_{k-1-\dim F}(X)$ and $a' : Z_{k-1-\dim F}(X;\mathbb{Z}) \to C_{k-\dim F}(X;\mathbb{Z})$ be any maps (not necessarily homomorphisms) such that* (5) *in Lemma 7 holds, i.e.,*

$$[z - \partial a'(z)]_{\partial S_{k-\dim F}} = [\zeta'(z)]_{k-1-\dim F}$$

*is true for all $z \in Z_{k-1-\dim F}(X;\mathbb{Z})$. Then* (66) *remains valid, i.e.,*

$$(\widehat{\pi}_! h)(z) := h([\mathrm{PB}_E \zeta'(z)]_{\partial S_k}) \cdot \exp\Big(2\pi i \int_{a'(z)} \fint_F \operatorname{curv}(h)\Big)$$

*holds for all $z \in Z_{k-1-\dim F}(X;\mathbb{Z})$ and all $h \in \widehat{H}^k(E;\mathbb{Z})$.*

*Proof.* Let $z \in Z_{k-1-\dim F}(X;\mathbb{Z})$ be a cycle. Then we find a geometric boundary $\partial\beta(z) \in \mathcal{B}_{k-1-\dim F}(X)$ such that $\zeta'(z) - \zeta(z) = \partial\beta(z)$. Since

$$[\partial a(z) - \partial a'(z)]_{\partial S_{k-\dim F}} = [\partial\beta(z)]_{\partial S_{k-\dim F}} = \partial[\beta(z)]_{S_{k-\dim F}},$$

we find a smooth singular cycle $w(z) \in Z_{k-\dim F}(X;\mathbb{Z})$ such that

$$[a(z) - a'(z) - w(z)]_{S_{k-\dim F}} = [\beta(z)]_{S_{k-\dim F}}. \qquad (70)$$

We then have:

$$\begin{aligned}
h([\mathrm{PB}_E \zeta'(z)]_{\partial S_k}) &\cdot h([\mathrm{PB}_E \zeta(z)]_{\partial S_k})^{-1} \\
&= h([\mathrm{PB}_E \partial\beta(z)]_{\partial S_k}) \\
&\overset{(10)}{=} h([\partial \mathrm{PB}_E \beta(z)]_{\partial S_k})
\end{aligned}$$

$$
\begin{aligned}
&\overset{(2)}{=} h(\partial[\mathrm{PB}_E(\beta(z)]_{S_k}) \\
&\overset{(25)}{=} \exp\Big(2\pi i \int_{[\mathrm{PB}_E\beta(z)]_k} \mathrm{curv}(h)\Big) \\
&\overset{(12)}{=} \exp\Big(2\pi i \int_{[\beta(z)]_{S_{k-\dim F}}} \fint_F \mathrm{curv}(h)\Big) \\
&\overset{(70)}{=} \exp\Big(2\pi i \int_{a(z)-a'(z)-w(z)} \fint_F \mathrm{curv}(h)\Big) \\
&= \exp\Big(2\pi i \int_{a(z)-a'(z)} \fint_F \mathrm{curv}(h)\Big) \cdot \exp\Big(2\pi i \underbrace{\int_{w(z)} \fint_F \mathrm{curv}(h)}_{\in \mathbb{Z}}\Big)^{-1} \\
&= \exp\Big(2\pi i \int_{a(z)} \fint_F \mathrm{curv}(h)\Big) \cdot \exp\Big(2\pi i \int_{a'(z)} \fint_F \mathrm{curv}(h)\Big)^{-1}.
\end{aligned}
$$

This proves the lemma. □

**Theorem 44 (Existence of Fiber Integration).** *Fiber integration* $\widehat{\pi}_!$ *as defined in* (66) *satisfies the axioms in Definition 38.*

*Proof.* For $k < \dim F$, the trivial map $\widehat{\pi}_! : \widehat{H}^k(X;\mathbb{Z}) \to \widehat{H}^{k-\dim F}(X;\mathbb{Z}) = \{0\}$ obviously satisfies the axioms in Definition 38. For $k = \dim F$, the fiber integration $\widehat{\pi}_! = \pi_! \circ c$ is natural since $\pi_! : H^k(E;\mathbb{Z}) \to H^{k-\dim F}(X;\mathbb{Z})$ is natural with respect to bundle maps and $c : \widehat{H}^k(E;\mathbb{Z}) \to H^k(E;\mathbb{Z})$ is natural with respect to any smooth maps. Compatibility with curvature follows from the commutative diagram (64). To show compatibility with topological trivializations, let $h = \iota(\varrho)$ for some $\varrho \in \Omega^{\dim F-1}(E)$. Then we have $c(\iota(\varrho)) = 0$. For dimensional reasons, we have $\fint_F \varrho = 0$. Thus $\widehat{\pi}_!\iota(\varrho) = \pi_! c(\iota(\varrho)) = 0 = \iota(\fint_F \varrho)$.

Now let $k > \dim F$. Equation (69) yields for the curvature of $\widehat{\pi}_! h$:

$$\mathrm{curv}(\widehat{\pi}_! h) = \fint_F \mathrm{curv}(h)\,. \tag{71}$$

This is compatibility with curvature (62).

Now let $h = \iota(\eta)$ for some $\eta \in \Omega^{k-1}(E)$. Let $z \in Z_{k-1-\dim F}(X;\mathbb{Z})$. Using the Stokes theorem we find:

$$
\begin{aligned}
(\widehat{\pi}_! h)(z) &= h(\lambda(z)) \cdot \exp\Big(2\pi i \int_{a(z)} \fint_F \mathrm{curv}(h)\Big) \\
&= \exp\Big(2\pi i \int_{\lambda(z)} \eta\Big) \cdot \exp\Big(2\pi i \int_{a(z)} \fint_F \mathrm{curv}(h)\Big) \\
&\overset{(19)}{=} \exp\Big(2\pi i \int_{[\zeta(z)]_{\partial S_{k-\dim F}}} \fint_F \eta\Big) \cdot \exp\Big(2\pi i \int_{a(z)} \fint_F d\eta\Big) \\
&\overset{(5)}{=} \exp\Big(2\pi i \cdot \Big(\int_{z-\partial a(z)} \fint_F \eta + \int_{a(z)} \fint_F d\eta\Big)\Big)
\end{aligned}
$$

$$= \exp\left(2\pi i \int_z \fint_F \eta\right).$$

Hence $\widehat{\pi}_! h = \iota(\fint_F \eta)$, as claimed in (63).

It remains to prove naturality. Let $g : Y \to X$ be a smooth map. We have the pull-back diagram

$$\begin{array}{ccc} g^*E & \xrightarrow{G} & E \\ \downarrow \pi & & \downarrow \pi \\ Y & \xrightarrow{g} & X. \end{array}$$

Let $z \in Z_{k-1-\dim F}(Y;\mathbb{Z})$. As in Lemma 7 we choose $\zeta(z) \in \mathcal{Z}_{k-1-\dim F}(Y)$ and $a(z) \in C_{k-\dim F}(Y;\mathbb{Z})$ such that $[z - \partial a(z)]_{\partial S_{k-\dim F}} = [\zeta(z)]_{\partial S_{k-\dim F}}$. Hence

$$\begin{aligned} [g_*\zeta(z)]_{\partial S_{k-\dim F}} &= g_*[\zeta(z)]_{\partial S_{k-\dim F}} \\ &= g_*[z - \partial a(z)]_{\partial S_{k-\dim F}} \\ &= [g_* z - \partial g_* a(z)]_{\partial S_{k-\dim F}}. \end{aligned}$$

Now let $h \in \widehat{H}^k(E;\mathbb{Z})$. By Lemma 43, we may choose $\zeta(g_*z) = g_*\zeta(z)$ and $a(g_*z) = g_*a(z)$ to compute $g^*(\widehat{\pi}_! h)(z)$. This yields:

$$\begin{aligned} \widehat{\pi}_! G^*h(z) &= G^*h([\mathrm{PB}_{g^*E}\zeta(z)]_{\partial S_k}) \cdot \exp\left(2\pi i \int_{a(z)} \fint_F \mathrm{curv}(G^*h)\right) \\ &= h(G_*[\mathrm{PB}_{g^*E}\zeta(z)]_{\partial S_k}) \cdot \exp\left(2\pi i \int_{a(z)} \fint_F G^*\mathrm{curv}(h)\right) \\ &= h([G_*\mathrm{PB}_{g^*E}\zeta(z)]_{\partial S_k}) \cdot \exp\left(2\pi i \int_{a(z)} g^* \fint_F \mathrm{curv}(h)\right) \\ &\overset{(11)}{=} h([\mathrm{PB}_E(g_*\zeta(z))]_{\partial S_k}) \cdot \exp\left(2\pi i \int_{g_*a(z)} \fint_F \mathrm{curv}(h)\right) \\ &= h([\mathrm{PB}_E\zeta(g_*z)]_{\partial S_k}) \cdot \exp\left(2\pi i \int_{a(g_*z)} \fint_F \mathrm{curv}(h)\right) \\ &= \widehat{\pi}_! h(g_*z) \\ &= g^*(\widehat{\pi}_! h)(z). \end{aligned}$$

For the third equality we use compatibility of fiber integration and pull-back of differential forms, see [39, Ch. VII, Prop. VIII]. This proves (60). □

**Corollary 45 (Existence and Uniqueness of Fiber Integration).** *There is a unique fiber integration of differential characters satisfying the axioms in Definition 38.*

*Remark 46.* The proof of Theorem 44 shows more than compatibility with topological trivializations of flat characters. Namely, (63) commutes for all

$\eta \in \Omega^{k-1}(E)$, not necessarily closed. In other words, we have shown compatibility with topological trivializations of characters, not necessarily flat.

**Proposition 47 (Compatibility of Fiber Integration with Characteristic Class).** *Fiber integration of differential characters is compatible with the characteristic class, i.e., the diagram*

$$\begin{array}{ccc} \widehat{H}^k(E;\mathbb{Z}) & \xrightarrow{c} & H^k(E;\mathbb{Z}) \\ \downarrow{\scriptstyle \widehat{\pi}_!} & & \downarrow{\scriptstyle \pi_!} \\ \widehat{H}^{k-\dim F}(X;\mathbb{Z}) & \xrightarrow{c} & H^{k-\dim F}(X;\mathbb{Z}) \end{array} \tag{72}$$

*commutes.*

*Proof.* For $k < \dim F$, there is nothing to show. For $k = \dim F$, this follows from the commutative diagram (64).

Thus let $k > \dim F$. We compute the characteristic class $c(\widehat{\pi}_! h)$. Let $\tilde{h} \in \mathrm{Hom}(Z_{k-1}(E;\mathbb{Z}),\mathbb{R})$ be a real lift of the differential character $h \in \widehat{H}^k(E;\mathbb{Z})$ and denote by

$$\mu^{\tilde{h}} : c \mapsto \int_c \mathrm{curv}(h) - \tilde{h}(\partial c)$$

the corresponding cocycle representing the characteristic class $c(h)$.

Let $z \in Z_{k-1-\dim F}(X;\mathbb{Z})$ be a smooth singular cycle in the base $X$. As in Lemma 7, we get the geometric cycle $\zeta(z) \in \mathcal{Z}_{k-1-\dim F}(X)$ and the smooth singular chain $a(z) \in C_{k-\dim F}(X;\mathbb{Z})$ such that $[z - \partial a(z)]_{\partial S_{k-\dim F}} = [\zeta(z)]_{\partial S_{k-\dim F}}$. By definition, the $(k-1)$-chain $\lambda(z)$ represents the fundamental class of the pull-back $\mathrm{PB}_E\zeta(z)$, i.e., $[\lambda(z)]_{\partial S_k} = [\mathrm{PB}_E\zeta(z)]_{\partial S_k}$, where $\lambda : Z_{k-1-\dim F}(X;\mathbb{Z}) \to Z_{k-1}(E;\mathbb{Z})$ is the transfer map constructed in Remark 8. We obtain a real lift $\widetilde{\widehat{\pi}_! h}$ of the differential character $\widehat{\pi}_! h$ by setting

$$\widetilde{\widehat{\pi}_! h}(z) := \tilde{h}(\lambda(z)) + \int_{a(z)} \fint_F \mathrm{curv}(h)\,. \tag{73}$$

Hence the characteristic class of the differential character $\widehat{\pi}_! h$ is represented by the cocycle

$$\begin{aligned} c \mapsto & \int_c \mathrm{curv}(\widehat{\pi}_! h) - \widetilde{\widehat{\pi}_! h}(\partial c) \\ &= \int_c \fint_F \mathrm{curv}(h) - \tilde{h}(\lambda(\partial c)) - \int_{a(\partial c)} \fint_F \mathrm{curv}(h) \\ &\overset{(17)}{=} \int_{c - a(\partial c)} \fint_F \mathrm{curv}(h) - \tilde{h}(\partial \lambda(c)) \\ &\overset{(20)}{=} \int_{\lambda(c)} \mathrm{curv}(h) - \tilde{h}(\partial \lambda(c)) \end{aligned}$$

$$= (\mu^{\tilde{h}} \circ \lambda)(c)\,. \tag{74}$$

By Remark 10 this cocycle represents the cohomology class $\pi_!(c(h))$, hence

$$c(\widehat{\pi}_!(h)) = \pi_!(c(h)). \qquad \square$$

**Proposition 48.** *Fiber integration of differential characters is compatible with the inclusion of cohomology classes with coefficients in* $\mathrm{U}(1)$, *i.e., the diagram*

$$\begin{array}{ccc} H^{k-1}(E;\mathrm{U}(1)) & \xrightarrow{\quad j \quad} & \widehat{H}^k(E;\mathbb{Z}) \\ \Big\downarrow{\scriptstyle \pi_!} & & \Big\downarrow{\scriptstyle \widehat{\pi}_!} \\ H^{k-1-\dim F}(X;\mathrm{U}(1)) & \xrightarrow{\quad j \quad} & \widehat{H}^{k-\dim F}(X;\mathbb{Z}) \end{array}$$

*commutes.*

*Proof.* For $k < \dim F$ there is nothing to show since both fiber integration maps are trivial for dimensional reasons. Let $k = \dim F$ and $u \in H^{\dim F-1}(E;\mathrm{U}(1))$. Diagram (30) shows that $c(j(u))$ is a torsion class. Thus $\widehat{\pi}_! j(u) = \pi_! c(j(u)) = 0$ since $\widehat{H}^0(X;\mathbb{Z}) = H^0(X;\mathbb{Z})$ is torsion free. On the other hand, we have $\pi_! u \in H^{-1}(X,\mathrm{U}(1)) = \{0\}$ and hence $j(\pi_! u) = 0$.

Now let $k > \dim F$ and $u \in H^{k-1}(E;\mathrm{U}(1))$. Diagram (30) shows $\mathrm{curv}(j(u)) = 0$. As explained in Remark 10, fiber integration $\pi_! : H^{k-1}(E;\mathrm{U}(1)) \to H^{k-1-\dim F}(X;\mathrm{U}(1))$ for singular cohomology is induced by pre-composition of cocycles with the transfer map $\lambda : C_{k-1-\dim F}(X;\mathbb{Z}) \to C_{k-1}(E;\mathbb{Z})$ constructed in Remark 8. Thus for any $z \in Z_{k-1-\dim F}(X;\mathbb{Z})$ we have:

$$\begin{aligned} \widehat{\pi}_! j(u)(z) &\overset{(68)}{=} j(u)(\lambda(z)) \cdot \exp\Big(2\pi i \int_{a(z)} \fint_F \underbrace{\mathrm{curv}(j(u))}_{=0}\Big) \\ &\overset{(29)}{=} \langle u, [\lambda(z)]\rangle \\ &\overset{(23)}{=} \langle \pi_! u, [z]\rangle \\ &\overset{(29)}{=} (j(\pi_! u))(z). \end{aligned} \qquad \square$$

**Proposition 49 (Orientation Reversal).** *Let* $E \xrightarrow{\pi} X$ *be a fiber bundle with closed oriented fibers over a smooth space* $X$. *Let* $\overline{\pi} : \overline{E} \to X$ *denote the bundle with fiber orientation reversed and* $\widehat{\overline{\pi}}_!$ *the corresponding fiber integration. For every* $h \in \widehat{H}^k(E;\mathbb{Z})$ *we have*

$$\widehat{\overline{\pi}}_! h = -\widehat{\pi}_! h.$$

*Proof.* There is nothing to show in case $k < \dim F$. For $k = \dim F$, we have $\widehat{\overline{\pi}}_! = \overline{\pi}_! \circ c = -\pi_! \circ c = -\widehat{\pi}_!$.

Now let $k > \dim F$. Let $h \in \widehat{H}^k(E;\mathbb{Z})$ and $z \in Z_{k-1-\dim F}(X;\mathbb{Z})$. Choose $\zeta(z) \in \mathcal{Z}_{k-1-\dim F}(X)$ and $a(z) \in C_{k-\dim F}(X;\mathbb{Z})$ as in Definition 40. We have $\mathrm{PB}_{\overline{E}}(\zeta(z)) = \overline{\mathrm{PB}_E(\zeta(z))}$ and $⨍_{\overline{F}} = -⨍_F$. This yields

$$
\begin{aligned}
(\widehat{\pi}_! h)(z) &= h([\mathrm{PB}_{\overline{E}}\zeta(z)]_{\partial S_k}) \cdot \exp\Big(2\pi i \int_{a(z)} ⨍_{\overline{F}} \mathrm{curv}(h)\Big) \\
&= h([\overline{\mathrm{PB}_E\zeta(z)}]_{\partial S_k}) \cdot \exp\Big(-2\pi i \int_{a(z)} ⨍_{F} \mathrm{curv}(h)\Big) \\
&= h(-[\mathrm{PB}_E\zeta(z)]_{\partial S_k}) \cdot \exp\Big(-2\pi i \int_{a(z)} ⨍_{F} \mathrm{curv}(h)\Big) \\
&= (\widehat{\pi}_! h(z))^{-1} \\
&= (-\widehat{\pi}_! h)(z)\,.
\end{aligned}
$$

□

*Example 50.* We consider the case $k = 1$ and $\dim F = 0$. Then $h \in \widehat{H}^1(E;\mathbb{Z}) = C^\infty(E, \mathrm{U}(1))$ is a smooth $\mathrm{U}(1)$-valued function on $E$ and $\pi : E \to X$ is a finite covering. We orient the fibers such that each point is positively oriented. It is easy to see that the function $\widehat{\pi}_! h \in \widehat{H}^1(X;\mathbb{Z}) = C^\infty(X, \mathrm{U}(1))$ is given by

$$\widehat{\pi}_! h(x) = \prod_{e \in \pi^{-1}(x)} h(e).$$

*Example 51.* Again consider a finite covering $\pi : E \to X$, i.e., $\dim F = 0$, but now $k = 2$. Let $P \to E$ be a $\mathrm{U}(1)$-bundle with connection whose isomorphism class corresponds to a differential character $h \in \widehat{H}^2(X;\mathbb{Z})$. Here it is convenient to take for $P$ the Hermitian line bundle rather than the $\mathrm{U}(1)$-principal bundle. Then $\widehat{\pi}_! h$ is given by the bundle whose fibers over $x \in X$ is

$$(\widehat{\pi}_! P)_x = \bigotimes_{e \in \pi^{-1}(x)} P_e.$$

This bundle inherits a natural tensor product connection from $P$.

*Example 52.* Now let $\pi : E \to X$ be a circle bundle with oriented fibers, hence $\dim F = 1$. The fiber integration map $\widehat{\pi}_! : \widehat{H}^2(E;\mathbb{Z}) \to \widehat{H}^1(X;\mathbb{Z})$ can be described as follows: Let $P \to E$ be a $\mathrm{U}(1)$-bundle with connection. For any $x \in X$ the holonomy of $P$ along the oriented fiber $E_x$ yields an element in $\mathrm{U}(1)$. In this way, we obtain a smooth function $X \to \mathrm{U}(1)$.

We show that fiber integration is functorial with respect to composition of fiber bundle projections, compare [12, p. 12].

**Proposition 53 (Functoriality of Fiber Integration).** *Let $\kappa : N \to E$ and $\pi : E \to X$ be fiber bundles with compact oriented fibers $L$ and $F$, respectively. Let $\pi \circ \kappa : N \to X$ be the composite fiber bundle with the composite orientation. Then we have*

$$\widehat{(\pi \circ \kappa)}_! = \widehat{\pi}_! \circ \widehat{\kappa}_! . \tag{75}$$

*Proof.* Denote the fibers of $\pi \circ \kappa$ by $Q$. The bundle projection $\kappa$ restricts to a fiber bundle $\kappa|_Q : Q \to F$ with fibers $L$.

For $k < \dim F + \dim L$, we have $\widehat{(\pi \circ \kappa)}_! \equiv 0$. Now let $h \in \widehat{H}^k(N;\mathbb{Z})$. Then we have $\widehat{\kappa}_! h \in H^{k-\dim L}(E;\mathbb{Z})$ and $k - \dim L < \dim F$. Thus $\widehat{\pi}_!(\widehat{\kappa}_! h) = 0$.

For $k = \dim F + \dim L$, we have:

$$\widehat{(\pi \circ \kappa)}_! = (\pi \circ \kappa)_! \circ c = \pi_! \circ \kappa_! \circ c \overset{(72)}{=} \pi \circ c \circ \widehat{\kappa}_! = \widehat{\pi}_! \circ \widehat{\kappa}_! .$$

In the second equality we have used the functoriality of fiber integration for singular cohomology, see [6, p. 484].

Now let $k > \dim F + \dim L$. Let $h \in \widehat{H}^k(N;\mathbb{Z})$ be a character. For any cycle $z \in Z_{k-\dim(F)-\dim(L)-1}(X;\mathbb{Z})$ choose the geometric cycle $\zeta(z) = [M \xrightarrow{g} X] \in \mathcal{Z}_{k-\dim(F)-\dim(L)-1}(X)$ and $a(z) \in C_{k-\dim(F)-\dim(L)}(X;\mathbb{Z})$ as in Lemma 7. Then we have the pull-back bundles

$$\begin{array}{ccc} G^*N & \xrightarrow{\mathbf{G}} & N \\ \downarrow & & \downarrow \kappa \\ g^*E & \xrightarrow{G} & E \\ \downarrow & & \downarrow \pi \\ M & \xrightarrow{g} & X \end{array}$$

which define the geometric cycles $\mathrm{PB}_\pi(\zeta(z)) = [g^*E \xrightarrow{G} E]$ as well as $\mathrm{PB}_\kappa(\mathrm{PB}_\pi(\zeta(z))) = \mathrm{PB}_{\pi\circ\kappa}(\zeta(z)) = [G^*N \xrightarrow{\mathbf{G}} N]$. We pull back $h$ to the stratifold $G^*N$, where it is topologically trivial for dimensional reasons. Thus we find a differential form $\chi \in \Omega^{k-1}(G^*N)$ such that $\mathbf{G}^*h = \iota(\chi)$. By the compatibility conditions (60) and (63), we have $\widehat{\kappa}_!(\mathbf{G}^*h) = G^*(\widehat{\kappa}_! h) = \iota(\fint_L \chi)$. In particular, $(\mathbf{G}^*h)([G^*N]) = \iota(\chi)([G^*N]) = \iota(\fint_L \chi)([g^*E]) = (\widehat{\kappa}_!(\mathbf{G}^*h))([g^*E])$.

This yields:

$$\begin{aligned} (\widehat{\pi \circ \kappa}_! h)(z) &\overset{(67)}{=} (\mathbf{G}^*h)([G^*N]) \cdot \exp\Big(2\pi i \int_{a(z)} \fint_Q \mathrm{curv}(h)\Big) \\ &= (G^*(\widehat{\kappa}_! h))([g^*E]) \cdot \exp\Big(2\pi i \int_{a(z)} \fint_F \mathrm{curv}(\widehat{\kappa}_! h)\Big) \\ &\overset{(67)}{=} \big(\widehat{\pi}_!(\widehat{\kappa}_! h)\big)(z). \end{aligned}$$

□

## 7.2 Fiber Integration for Fibers with Boundary

Let $(F, \partial F) \hookrightarrow (E, \partial E) \xrightarrow{(\pi^E, \pi^{\partial E})} X$ be a fiber bundle bundle whose fibers are compact oriented manifolds with boundary. For any differential form $\omega \in \Omega^*(E)$ on the total space $E$ we have the fiberwise Stokes theorem [39, p. 311]:

$$\fint_F d\omega = d \fint_F \omega + (-1)^{\deg \omega + \dim \partial F} \fint_{\partial F} \omega \,. \tag{76}$$

In particular, if $\omega \in \Omega^k(E)$ is a closed form, then $\fint_{\partial F} \omega \in \Omega^{k-\mathrm{im}\partial F}(X)$ is exact. Thus fiber integration of differential forms in the bundle $\pi^{\partial E} : \partial E \to X$ induces the trivial map on de Rham cohomology. The same holds true for fiber integration on singular cohomology.

Denote by $\widehat{\pi}_!^{\partial E} : \widehat{H}^k(\partial E; \mathbb{Z}) \to \widehat{H}^{k-\dim \partial F}(X; \mathbb{Z})$ the fiber integration map for the bundle $\partial F \hookrightarrow \partial E \xrightarrow{\pi^{\partial E}} X$ as constructed in the previous section. In the following, we do not distinguish in notation between a differential character $h \in \widehat{H}^k(E; \mathbb{Z})$ and its pull-back to $\partial E$. Applying the fiber integration map $\widehat{\pi}_!^{\partial E}$ to $h \in \widehat{H}^k(E; \mathbb{Z})$ yields the following (compare also [45, p. 363] and [30, p. 305]):

**Proposition 54 (Fiber Integration for Fibers that Bound).** *Let $(F, \partial F) \hookrightarrow (E, \partial E) \xrightarrow{(\pi^E, \pi^{\partial E})} X$ be a fiber bundle with compact oriented fibers with boundary over $X$ and let $h \in \widehat{H}^k(E; \mathbb{Z})$ be a differential character. Then $\widehat{\pi}_!^{\partial E} h \in \widehat{H}^{k-\dim \partial F}(X; \mathbb{Z})$ is topologically trivial. A topological trivialization is given by:*

$$\widehat{\pi}_!^{\partial E} h = \iota\Big((-1)^{k-\dim F} \fint_F \mathrm{curv}(h)\Big) \,. \tag{77}$$

*In particular, for $k = \dim \partial F$, we have $\widehat{\pi}_!^{\partial E} h = 0 \in \widehat{H}^0(X; \mathbb{Z})$.*

*Proof.* As explained in Remark 8, we construct transfer maps $\lambda^E$ and $\lambda^{\partial E}$ for the bundles $\pi^E : E \to X$ and $\pi^{\partial E} : \partial E \to X$, respectively. By (9), we have

$$\mathrm{PB}_{\partial E}(\zeta(z)) = \begin{cases} \partial(\mathrm{PB}_E \zeta(z)) & \text{for } z \in Z_n(X; \mathbb{Z}),\ n \text{ even}, \\ \overline{\partial(\mathrm{PB}_E \zeta(z))} & \text{for } z \in Z_n(X; \mathbb{Z}),\ n \text{ odd}. \end{cases}$$

Thus we can arrange the choices in the construction of the transfer maps $\lambda^E$ and $\lambda^{\partial E}$ in such a way that we have:

$$\lambda^{\partial E} = (-1)^n \cdot \partial \circ \lambda^E : Z_n(X; \mathbb{Z}) \to B_{n+\dim F}(E; \mathbb{Z}) \,. \tag{78}$$

Now we prove the claim:

For $k < \dim \partial F$, there is nothing to show. Let $k = \dim \partial F$ and $h \in H^{\dim \partial F}(E; \mathbb{Z})$. Let $\tilde{h}$ be a real lift of $h$ and $\mu^{\tilde{h}} \in C^{\dim F}(E; \mathbb{Z})$ the corresponding cocycle representing $c(h)$. Since $Z_0(X; \mathbb{Z}) = C_0(X; \mathbb{Z})$, we may use (78) and (23) to conclude:

$$\widehat{\pi}^{\partial E} h = \pi^{\partial E} c(h) \overset{(23)}{=} [\mu^{\tilde{h}} \circ \lambda^{\partial E}] \overset{(78)}{=} [\mu^{\tilde{h}} \circ \partial \circ \lambda^{E}] = [0].$$

Now let $k > \dim \partial F$. Let $h \in \widehat{H}^k(E;\mathbb{Z})$ and $z \in Z_{k-1-\dim \partial F}(X;\mathbb{Z})$. Choose $\zeta(z) \in \mathcal{C}_{k-1-\dim \partial F}(X)$ and $a(z) \in C_{k-\dim \partial F}(X;\mathbb{Z})$ such that $[z - \partial a(z)]_{\partial S_{k-\dim \partial F}} = [\zeta(z)]_{\partial S_{k-\dim \partial F}}$. Then we compute:

$$
\begin{aligned}
&\widehat{\pi}_!^{\partial E} h(z) \\
&\overset{(68)}{=} h(\lambda^{\partial E}(z)) \cdot \exp\Big(2\pi i \int_{a(z)} \fint_{\partial F} \mathrm{curv}(h)\Big) \\
&\overset{(78),(76)}{=} h((-1)^{\deg(z)} \partial \lambda^E(z)) \cdot \exp\Big(2\pi i \int_{a(z)} d \fint_F \mathrm{curv}(h)\Big) \\
&\overset{(76)}{=} \exp\Big[2\pi i (-1)^{k-\dim F}\Big(\int_{\lambda^E(z)} \mathrm{curv}(h) + \int_{\partial a(z)} \fint_F \mathrm{curv}(h)\Big)\Big] \\
&\overset{(21)}{=} \exp\Big[2\pi i (-1)^{k-\dim F}\Big(\int_{[\zeta(z)]_{\partial S_{k-\dim \partial F}}} \fint_F \mathrm{curv}(h) + \int_{\partial a(z)} \fint_F \mathrm{curv}(h)\Big)\Big] \\
&\overset{(5)}{=} \exp\Big(2\pi i \int_z (-1)^{k-\dim F} \fint_F \mathrm{curv}(h)\Big) \\
&= \iota\Big((-1)^{k-\dim F} \fint_F \mathrm{curv}(h)\Big)(z)\,.
\end{aligned}
$$

□

*Remark 55.* Proposition 54 says that $\widehat{\pi}_!^{\partial E}(h)$ is topologically trivial. However, $\widehat{\pi}_!^{\partial E}(h)$ is in general not flat, since

$$\mathrm{curv}(\widehat{\pi}_!^{\partial E} h) = \fint_{\partial F} \mathrm{curv}(h) \overset{(76)}{=} (-1)^{k-\dim F} d \fint_F \mathrm{curv}(h)\,,$$

is an exact form, but need not be 0.

As a special case of fiber integration for fibers with boundary, we obtain the well-known homotopy formula:

*Example 56.* Differential cohomology is not a generalized cohomology theory, in particular, it is not homotopy invariant. Let $f : [0,1] \times X \to Y$ be a homotopy between smooth maps $f_0, f_1 : X \to Y$ and $h \in \widehat{H}^k(Y;\mathbb{Z})$ a differential character. Then we have the well-known homotopy formula [10, Prop. 3.28]:

$$f_1^* h - f_0^* h = \iota\Big(\int_0^1 f_s^* \mathrm{curv}(h) ds\Big)\,.$$

This is a special case of (77) for the trivial bundle $X \times [0,1] \to X$: for the left hand side we have $f_1^* h - f_0^* h = \widehat{\pi}_!^{\partial E} f^* h$. By the orientation conventions, we obtain for the right hand side $\fint_[ 0,1] f^* \mathrm{curv}(h) = (-1)^{k-1} \int_0^1 f_s^* \mathrm{curv}(h) ds$.

*Example 57.* Let the fibers of $\pi : E \to X$ be diffeomorphic to compact intervals and carry an orientation. Hence $\dim F = 1$. The boundary of $E$

decomposes as $\partial E = \partial^+ E \sqcup \partial^- E$ where $\partial^+ E$ consists of the endpoints of the oriented fibers and $\partial^- E$ of the initial points. The restriction of $\pi$ to $\partial^+ E$ is a diffeomorphism whose inverse we denote by $j^+ := (\pi|_{\partial^+ E})^{-1} : X \to \partial^+ E$, and similarly for $j^-$.

We consider the case $k = 1$. Then for any $h \in \widehat{H}^1(E;\mathbb{Z}) = C^\infty(E, \mathrm{U}(1))$ we have $\widehat{\pi}_!^{\partial E} h \in C^\infty(X, \mathrm{U}(1))$ where

$$\widehat{\pi}_!^{\partial E} h = (h \circ j^+) \cdot (h \circ j^-)^{-1}.$$

The exponent $-1$ in this formula is due to the fact that the points in $\partial^- E$ inherit a negative orientation.

Recall from Example 18 that $\mathrm{curv}(h) = d\tilde{h}$ where $\tilde{h}$ is a local lift of $h$. Integration along the fiber $E_x$ over $x \in X$ yields $\rho(x) = \tilde{h}(j^+(x)) - \tilde{h}(j^-(x))$. The ambiguity in the choice of $\tilde{h}$ cancels and we obtain a global smooth function $\rho : X \to \mathbb{R}$. Obviously, $\rho$ is a lift of $\widehat{\pi}_!^{\partial E} h$.

*Example 58.* Let $\pi : E \to X$ be as in Example 57. Now we consider the case $k = 2$. Let $P \to E$ be a U(1)-bundle with connection $\nabla$ corresponding to $h \in \widehat{H}^2(X;\mathbb{Z})$. Fiber integration along $\partial F$ yields the U(1)-bundle with connection over $X$ whose fiber over $x$ is

$$(\widehat{\pi}_!^{\partial E} P)_x = P_{j^+(x)} \otimes P^*_{j^-(x)}\,.$$

Fiber integration of $\mathrm{curv}(h)$ yields the 1-form $\rho$ on $X$. Integrating $\rho$ along a closed curve $c$ in $X$ yields

$$\exp\Big(2\pi i \int_c \rho\Big) = \exp\Big(2\pi i \int_{\pi^{-1}(c)} \mathrm{curv}(h)\Big) = h(j_*^- c - j_*^+ c) = \widehat{\pi}_! h(c)^{-1}\,.$$

As explained in Example 19, the 1-form $\varrho$ corresponds to the parallel transport in $(P, \nabla) \to E$ along $F$.

## *7.3 Fiber Products and the Up-Down Formula*

In this section we prove that the fiber integration in a fiber product is the external product of the fiber integrations. The up-down formula is an immediate consequence.

Let $E \to X$ and $E' \to X'$ be fiber bundles over smooth spaces $X$ and $X'$ with compact oriented fibers $F$ and $F'$, respectively. We consider the fiber product $E \times E' \xrightarrow{\pi^E \times \pi^{E'}} X \times X'$ as the composition of fiber bundles $E \times E' \xrightarrow{\mathrm{id} \times \pi^{E'}} E \times X' \xrightarrow{\pi^E \times \mathrm{id}} X \times X'$. Fiber integration on singular cohomology commutes up to sign with the external product. Explicitly, for singular cohomology classes $u \in H^k(E;\mathbb{Z})$ and $u' \in H^{k'}(E';\mathbb{Z})$, we have:

$$\begin{aligned}\pi_!^{E\times E'}(u\times u') &= (\pi^E\times \mathrm{id})_!((\mathrm{id}\times\pi^{E'})_!(u\times u'))\\ &= (\pi^E\times \mathrm{id})_!(u\times \pi_!^{E'}u')\\ &= (-1)^{(k'-\dim F')\dim F}\pi_!^E u\times \pi_!^{E'}u'\,. \end{aligned}\tag{79}$$

This follows from [25, p. 585] and the functoriality of fiber integration for singular cohomology, proved in [6, p. 484].

Similarly, for differential forms $\omega\in\Omega^k(E)$ and $\omega'\in\Omega^{k'}(E')$, we have:

$$\fint_{F\times F'}\omega\times\omega' = (-1)^{(k'-\dim F')\dim F}\Big(\fint_F\omega\Big)\times\Big(\fint_{F'}\omega'\Big)\,. \tag{80}$$

The analogous result for differential characters is the following:

**Theorem 59 (Fiber Integration on Fiber Products).** *Let $E\to X$ and $E'\to X'$ be fiber bundles over smooth spaces $X$ and $X'$ with closed oriented fibers $F$ and $F'$, respectively. Let $h\in\widehat H^k(E;\mathbb{Z})$ and $h'\in\widehat H^{k'}(E';\mathbb{Z})$. Then we have:*

$$\widehat\pi_!^{E\times E'}(h\times h') = (-1)^{(k'-\dim F')\dim F}\cdot\widehat\pi_!^E h\times\widehat\pi_!^{E'}h'\,. \tag{81}$$

*Proof.* Conceptually, the proof is just a computation using the explicit formulas we derived for fiber integration and external product. The crucial point is the construction transfer maps commuting with external products.

a) We compute the curvature of the characters in (81):

$$\begin{aligned}&\mathrm{curv}(\widehat\pi_!^{E\times E'}(h\times h'))\\ &\quad\overset{(62)}{=}\fint_{F\times F'}\mathrm{curv}(h\times h')\\ &\quad\overset{(52)}{=}\fint_{F\times F'}\mathrm{curv}(h)\times\mathrm{curv}(h')\\ &\quad\overset{(80)}{=}(-1)^{(k'-\dim F')\dim F}\cdot\Big(\fint_F\mathrm{curv}(h)\Big)\times\Big(\fint_{F'}\mathrm{curv}(h')\Big)\\ &\quad\overset{(62)}{=}(-1)^{(k'-\dim F')\dim F}\cdot\mathrm{curv}(\widehat\pi_!^E h)\times\mathrm{curv}(\widehat\pi_!^{E'}h')\\ &\quad\overset{(52)}{=}(-1)^{(k'-\dim F')\dim F}\cdot\mathrm{curv}(\widehat\pi_!^E h\times\widehat\pi_!^{E'}h')\,. \end{aligned}\tag{82}$$

Similarly, we find for the characteristic class:

$$\begin{aligned}c(\widehat\pi_!^{E\times E'}(h\times h')) &\overset{(72)}{=}\pi_!^{E\times E'}(c(h\times h'))\\ &\overset{(53)}{=}\pi_!^{E\times E'}(c(h)\times c(h'))\\ &\overset{(79)}{=}(-1)^{(k'-\dim F')\dim F}\cdot\pi_!^E(c(h))\times\pi_!^{E'}(c(h'))\\ &\overset{(72)}{=}(-1)^{(k'-\dim F')\dim F}\cdot c(\widehat\pi_!^E h)\times c(\widehat\pi_!^{E'}h')\end{aligned}$$

$$\overset{(53)}{=} (-1)^{(k'-\dim F')\dim F} \cdot c(\widehat{\pi}_!^E h \times \widehat{\pi}_!^{E'} h'). \tag{83}$$

Thus the differential characters $\widehat{\pi}_!^{E\times E'}(h \times h')$ and $(\widehat{\pi}_!^E h \times \widehat{\pi}_!^{E'} h')$ have the same curvature and characteristic class. By Remark 16 this implies that they coincide on cycles $z \in Z_{k+k'-\dim F\times F'-1}(X \times X';\mathbb{Z})$ that represent torsion classes.

b) Let $z \in Z_{k+k'-\dim F\times F'-1}(X \times X';\mathbb{Z})$ be a cycle. As in Sect. 6, we choose a splitting $S$ of the cycles in the Künneth sequence. Composing the homomorphism $\zeta^X$ with the pull-back operation $\mathrm{PB}_E$, we construct a transfer map $\lambda^X : Z_{*-\dim F}(X;\mathbb{Z}) \to Z_*(E;\mathbb{Z})$ as in Remark 8, and similarly for $\lambda^{X'}$. We use the splitting to extend $\lambda^X \otimes \lambda^{X'}$ to a transfer map $\lambda^{X\times X'}$ : $Z_{*-\dim F\times F'}(X \times X';\mathbb{Z}) \to Z_*(E \times E';\mathbb{Z})$ such that the following diagram is graded commutative:

$$\begin{array}{ccc}
Z_*(E;\mathbb{Z}) \otimes Z_*(E';\mathbb{Z}) & \xrightarrow{\times} & Z_*(E\times E';\mathbb{Z}) \\
\downarrow & & \downarrow \\
\frac{Z_*(E;\mathbb{Z})}{\partial S_{*+1}(E;\mathbb{Z})} \otimes \frac{Z_*(E';\mathbb{Z})}{\partial S_{*+1}(E';\mathbb{Z})} & \xrightarrow{\times} & \frac{Z_*(E\times E';\mathbb{Z})}{\partial S_{*+1}(E\times E';\mathbb{Z})} \\
\uparrow{\scriptstyle \psi_*^E \otimes \psi_*^{E'}} & & \uparrow{\scriptstyle \psi_*^{E\times E'}} \\
\mathcal{Z}_*(E) \otimes \mathcal{Z}_*(E') & \xrightarrow{\times} & \mathcal{Z}_*(E\times E') \\
\uparrow{\scriptstyle \mathrm{PB}_E \otimes \mathrm{PB}_{E'}} & & \uparrow{\scriptstyle \mathrm{PB}_{E\times E'}} \\
\mathcal{Z}_{*-\dim F}(X) \otimes \mathcal{Z}_{*-\dim F'}(X') & \xrightarrow{\times} & \mathcal{Z}_{*-\dim F\times F'}(X\times X') \\
\uparrow{\scriptstyle \zeta^X \otimes \zeta^{X'}} & & \uparrow{\scriptstyle \zeta^{X\times X'}} \\
Z_{*-\dim F}(X;\mathbb{Z}) \otimes Z_{*-\dim F'}(X';\mathbb{Z}) & \underset{S}{\overset{K}{\rightleftarrows}} & Z_{*-\dim F\times F'}(X\times X';\mathbb{Z})
\end{array}$$

(with the outer arrows $\lambda^X \otimes \lambda^{X'}$ from the bottom left to the top left, and $\lambda^{X\times X'}$ from the bottom right to the top right)

(84)

The graded commutativity is caused by the orientation conventions. As in (14), we have $\mathrm{PB}_{E\times E'}(\zeta_i \times \zeta_j') = (-1)^{j\cdot\dim F}\mathrm{PB}_E(\zeta_i)\times\mathrm{PB}_{E'}(\zeta_j')$ for geometric cycles $\zeta_i \in \mathcal{Z}_i(X)$ and $\zeta_j' \in \mathcal{Z}_j(X')$. Consequently,

$$\lambda^{X\times X'}(y_i \times y_j') = (-1)^{j\cdot\dim F}\lambda^X(y_i) \times \lambda^{X'}(y_j'). \tag{85}$$

Now write $z = K \circ S(z) + (z - K \circ S(z))$. By the Künneth sequence, the cycle $z - K \circ S(z)$ represents a torsion class. Thus by part a) the differential characters $\widehat{\pi}_!^{E\times E'}(h \times h')$ and $(\widehat{\pi}_!^E h \times \widehat{\pi}_!^{E'} h')$ coincide on $z - K \circ S(z)$. Hence it suffices to evaluate them on $K \circ S(z)$. By (68), we have:

$$\begin{aligned}\widehat{\pi}_!^{E\times E'}(h \times h')(K \circ S(z)) &= (h \times h')(\lambda^{X\times X'}(K \circ S(z))) \\ &\quad \cdot \exp\Big(2\pi i \int_{a(K\circ S(z))} \fint_{F\times F'} \mathrm{curv}(h \times h')\Big). \end{aligned} \tag{86}$$

As in (57), we write $K \circ S(z) = \sum_{i+j=k+k'-\dim F \times F'-1} \sum_m y_i^m \times y'^m_j$. As in the proof of Theorem 31, we write $\zeta^X(y_i^m) = [M_i^m \xrightarrow{g_i^m} X]$ and $\zeta^{X'}(y'^m_j) = [M'^m_j \xrightarrow{g'^m_j} X]$. Choose fundamental cycles $x_i^m$ of $M_i^m$ and $x'^m_j$ of $M'^m_j$. As in (59) we have

$$\begin{aligned} y_i^m \times y'^m_j &= (g_i^m \times g'^m_j)_*(x_i^m \times x'^m_j) \\ &\quad + \partial(a^X(y^i_m) \times g'^m_{j*} x'^m_j + (-1)^i g_i^m{}_* x_i^m \times a^{X'}(y'^m_j)) \\ &\quad + \partial a^X(y_i^m) \times \partial a^{X'}(y'^m_j))\,. \end{aligned} \tag{87}$$

Now we evaluate the characters $\widehat{\pi}_!^E h \times \widehat{\pi}_!^{E'} h'$ and $\widehat{\pi}_!^{E\times E'}(h \times h')$ on the cycle $y_i^m \times y'^m_j$. By part a), they agree on the second and third summand, since these are boundaries. By Lemma 30 both characters in (81) vanish on the first summand of (87) for degrees $(i,j)$ different from $(k - \dim F - 1, k' - \dim F')$ and $(k - \dim F, k' - \dim F' - 1)$.

We compute the remaining cases. Let $(i,j) = (k - \dim F - 1, k' - \dim F')$. Then we have $\widehat{\pi}_!^E h(g_i^m{}_* x_i^m) = h(\lambda^E(g_i^m{}_* x_i^m))$. This yields:

$$\begin{aligned} &\widehat{\pi}^{E\times E'}(h \times h')(g_i^m{}_* x_i^m \times g'^m_j{}_* x'^m_j) \\ &\quad = (h \times h')(\lambda^{E\times E'}(g_i^m{}_* x_i^m \times g'^m_j{}_* x'^m_j)) \\ &\quad \overset{(85)}{=} ((-1)^{j\cdot\dim F} \cdot h \times h')(\lambda^E(g_i^m{}_* x_i^m) \times \lambda^{E'}(g'^m_j{}_* x'^m_j)) \\ &\quad \overset{(56)}{=} ((-1)^{j\cdot\dim F} \cdot h)(\lambda^E(g_i^m{}_* x_i^m))^{\langle c(h'), \lambda^{E'}(g'^m_j{}_* x'^m_j)\rangle} \\ &\quad = ((-1)^{j\cdot\dim F} \cdot \widehat{\pi}_!^E h)(g_i^m{}_* x_i^m)^{\langle c(\widehat{\pi}_!^{E'} h'), g'^m_j{}_* x'^m_j\rangle} \\ &\quad \overset{(56)}{=} ((-1)^{(k'-\dim F')\cdot\dim F} \cdot (\widehat{\pi}_!^E h \times \widehat{\pi}_!^{E'} h'))(g_i^m{}_* x_i^m \times g'^m_j{}_* x'^m_j)\,. \end{aligned}$$

Similary, for $(i,j) = (k - \dim F, k' - \dim F' - 1)$ we have $\widehat{\pi}_!^{E'} h'(g'^m_j{}_* x'^m_j) = h'(\lambda^{E'}(g'^m_j{}_* x'^m_j))$.This yields:

$$\begin{aligned} &\widehat{\pi}^{E\times E'}(h \times h')(g_i^m{}_* x_i^m \times g'^m_j{}_* x'^m_j) \\ &\quad \overset{(85)}{=} (h \times h')((-1)^{j\cdot\dim F} \cdot \lambda^E(g_i^m{}_* x_i^m) \times \lambda^{E'}(g'^m_j{}_* x'^m_j)) \\ &\quad \overset{(56)}{=} ((-1)^{j\cdot\dim F} \cdot h')(\lambda^{E'}(g'^m_j{}_* x'^m_j))^{(-1)^k \langle c(h), \lambda^E(g_i^m{}_* x_i^m)\rangle} \\ &\quad = ((-1)^{(j+1)\cdot\dim F} \cdot \widehat{\pi}_!^{E'} h')(g'^m_j{}_* x'^m_j)^{(-1)^{k-\dim F} \langle c(\widehat{\pi}_!^E h), g_i^m{}_* x_i^m\rangle} \\ &\quad \overset{(56)}{=} ((-1)^{(k'-\dim F')\cdot\dim F} \cdot (\widehat{\pi}_!^E h \times \widehat{\pi}_!^{E'} h'))(g_i^m{}_* x_i^m \times g'^m_j{}_* x'^m_j)\,. \end{aligned}$$

This completes the proof. □

Fiber integration for differential forms satisfies the following up-down formula: for any $\eta \in \Omega^k(X)$ and $\omega \in \Omega^l(E)$, we have:

$$\fint_F \pi^*\eta \wedge \omega = \eta \wedge \fint_F \omega. \tag{88}$$

Likewise, fiber integration on singular cohomology satisfies the corresponding up-down formula: for any $u \in H^k(X;\mathbb{Z})$ and $w \in H^l(E;\mathbb{Z})$, we have:

$$\pi_!(\pi^* u \cup w) = u \cup \pi_! w.$$

For a proof, see [25, p. 585] or [6, p. 483].

Now we prove the corresponding up-down formula for fiber integration of differential characters. The idea of the proof is due to Chern who proved the up-down formula for singular cohomology in [25, p. 585]. The same idea has been used in [12] along the lines of a representation of differential cohomology by cohomology stratifolds.

**Theorem 60 (Up-Down Formula).** *Let $E \to X$ be a fiber bundle over a smooth space $X$ with closed oriented fibers $F$. Let $h \in \widehat{H}^k(X;\mathbb{Z})$ and $f \in \widehat{H}^l(E;\mathbb{Z})$. Then we have*

$$\widehat{\pi}_!(\pi^* h * f) = h * (\widehat{\pi}_! f) \in \widehat{H}^{k+l-\dim F}(X;\mathbb{Z}). \tag{89}$$

*Proof.* We decompose the fiber product $E \times E \to X \times X$ as the composite fiber bundle $E \times E \xrightarrow{\pi \times \mathrm{id}_E} X \times E \xrightarrow{\mathrm{id}_X \times \pi} X \times X$. Let $\Delta_E : E \to E \times E$ and $\Delta_X : X \to X \times X$ denote the diagonal maps. Then we have the bundle map

$$\begin{array}{ccc} E & \xrightarrow{(\pi \times \mathrm{id}_E)\circ \Delta_E} & X \times E \\ {\scriptstyle \pi}\downarrow & & \downarrow{\scriptstyle \mathrm{id}_X \times \pi} \\ X & \xrightarrow[\Delta_X]{} & X \times X \end{array}$$

The up-down formula now follows from the product formula (81) and the naturality (60) of fiber integration:

$$\begin{aligned} \widehat{\pi}_!^E(\pi^* h * f) &\overset{(55)}{=} \widehat{\pi}_!^E(\Delta_E^*(\pi^* h \times f)) \\ &\overset{(51)}{=} \widehat{\pi}_!^E(\Delta_E^*(\pi \times \mathrm{id}_E)^*(h \times f)) \\ &\overset{(60)}{=} \Delta_X^*(\widehat{\pi}_!^{X\times E}(h \times f)) \\ &\overset{(81)}{=} \Delta_X^*(h \times \widehat{\pi}_!^E f) \\ &\overset{(55)}{=} h * \widehat{\pi}_!^E f. \end{aligned}$$

There is no sign in the second last equation because the fiber over the first factor is zero-dimensional. □

# 8 Relative Differential Characters

In this chapter, we discuss several aspects of relative differential characters, defined in [8]. From a geometric point of view, relative differential characters are to be considered as topological trivializations or global sections of differential characters. We explain this point of view in Sect. 8.1.

From a topological point of view, the group of relative differential characters should be considered as a relative version of differential cohomology. However, differential cohomology is not a (generalized) cohomology theory in the sense of Eilenberg and Steenrod. In particular, one cannot expect to obtain the usual long exact sequence relating the groups of relative and absolute differential characters. In Sect. 8.2, we derive an exact sequence that relates the groups of relative and absolute differential characters. This sequence characterizes in particular the existence and uniqueness of global sections.

## *8.1 Definition and Examples*

Let $k \geq 1$ and $\varphi : A \to X$ a smooth map. Relative differential characters in $\widehat{H}^k(\varphi;\mathbb{Z})$ may be considered as differential characters on $X$ with sections along the map $\varphi$. We briefly recall the construction of $\widehat{H}^k(\varphi;\mathbb{Z})$ from [8]. Then we construct an exact sequence which characterizes those differential characters in $\widehat{H}^k(X;\mathbb{Z})$ which admit sections along the map $\varphi$, i.e., which are in the image of the natural map $\widehat{H}^k(\varphi;\mathbb{Z}) \to \widehat{H}^k(X;\mathbb{Z})$.

The *mapping cone complex* of a smooth map $\varphi : A \to X$ is the complex $C_k(\varphi;\mathbb{Z}) := C_k(X;\mathbb{Z}) \times C_{k-1}(A;\mathbb{Z})$ of pairs of smooth singular chains with the differential $\partial_\varphi(s,t) := (\partial s + \varphi_* t, -\partial t)$. The homology $H_k(\varphi;\mathbb{Z})$ of this complex coincides with the homology of the mapping cone of $\varphi$ in the topological sense. For the special case of an embedding $A \subset X$ it coincides with the relative homology $H_k(X,A;\mathbb{Z})$.

Similarly, we consider the mapping cone de Rham complex $\Omega^k(\varphi) := \Omega^k(X)\times\Omega^{k-1}(A)$ of pairs of differential forms with the differential $d_\varphi(\omega,\vartheta) := (d\omega, \varphi^*\omega - d\vartheta)$. The homology $H^k_{\mathrm{dR}}(\varphi)$ of this complex is the mapping cone de Rham cohomology for the map $\varphi$, as explained in [7, p. 78].

We denote by $Z_k(\varphi;\mathbb{Z})$ the group of cycles of the mapping cone complex and by $B_k(\varphi;\mathbb{Z})$ the space of boundaries. The group of relative differential characters is defined as:

$$\widehat{H}^k(\varphi;\mathbb{Z}) := \left\{ f \in \mathrm{Hom}(Z_{k-1}(\varphi;\mathbb{Z}), \mathrm{U}(1)) \,\middle|\, f \circ \partial_\varphi \in \Omega^k(\varphi) \right\}.$$

The notation $f \circ \partial_\varphi \in \Omega^k(\varphi)$ means that there exists a pair of differential forms $(\omega,\vartheta) \in \Omega^k(\varphi)$ such that for every pair of smooth singular chains $(x,y) \in C_k(\varphi;\mathbb{Z})$ we have

$$f(\partial_\varphi(x,y)) = \exp\left[2\pi i \cdot \Big(\int_x \omega + \int_y \vartheta\Big)\right]. \tag{90}$$

The form $\omega =: \mathrm{curv}(f)$ in the definition is called the *curvature* of the relative differential character $f$ and the form $\vartheta =: \mathrm{cov}(f)$ is called its *covariant derivative.* As in the absolute case, the curvature is uniquely determined by the differential character. For $k \geq 2$, this is also true for the covariant derivative. For $k = 1$, the function $\vartheta$ is unique only up to addition of a locally constant integer valued function, see Example 64.

It is shown in [8, p. 273f] that relative differential characters $f \in \widehat{H}^k(\varphi;\mathbb{Z})$ have characteristic classes $c(f) \in H^k(\varphi;\mathbb{Z})$, the $k$-th cohomology of the mapping cone complex. By [8, Thm. 2.4], the group $\widehat{H}^k(\varphi;\mathbb{Z})$ fits into short exact sequences similar to the ones in (30):

$$0 \longrightarrow \frac{\Omega^{k-1}(\varphi)}{\Omega_0^{k-1}(\varphi)} \xrightarrow{\ \iota\ } \widehat{H}^k(\varphi;\mathbb{Z}) \xrightarrow{\ c\ } H^k(\varphi;\mathbb{Z}) \longrightarrow 0,$$

$$0 \longrightarrow H^{k-1}(\varphi;\mathrm{U}(1)) \longrightarrow \widehat{H}^k(\varphi;\mathbb{Z}) \xrightarrow{(\mathrm{curv},\mathrm{cov})} \Omega_0^k(\varphi) \longrightarrow 0.$$

Here $\Omega_0^k(\varphi)$ denotes the space of all $d_\varphi$-closed pairs $(\omega,\vartheta) \in \Omega^k(\varphi)$ with integral periods, i.e., $\int_{(s,t)}(\omega,\vartheta) \in \mathbb{Z}$ for all relative cycles $(s,t) \in Z_k(\varphi;\mathbb{Z})$.

Furthermore, we have the obvious maps

$$\widehat{H}^{k-1}(A;\mathbb{Z}) \xrightarrow{\ \breve{\imath}\ } \widehat{H}^k(\varphi;\mathbb{Z}) \xrightarrow{\ \breve{p}\ } \widehat{H}^k(X;\mathbb{Z}) \tag{91}$$

which map a differential character $g \in \widehat{H}^{k-1}(A;\mathbb{Z})$ to $\breve{\imath}(g) : (s,t) \mapsto g(t)$ and a relative differential character $f \in \widehat{H}^k(\varphi;\mathbb{Z})$ to $\breve{p}(f) : z \mapsto f(z,0)$. One easily checks that $\mathrm{curv}(\breve{\imath}(g)) = 0$, $\mathrm{cov}(\breve{\imath}(g)) = -\mathrm{curv}(g)$, and $\mathrm{curv}(\breve{p}(f)) = \mathrm{curv}(f)$.

*Remark 61.* Note that $\breve{p}$ is defined also for $k = 1$. As in the absolute case we set $\widehat{H}^0(\varphi;\mathbb{Z}) := H^0(\varphi;\mathbb{Z})$ and $\widehat{H}^k(\varphi;\mathbb{Z}) := 0$ for $k < 0$. Moreover, $\breve{\imath} : \widehat{H}^0(\varphi;\mathbb{Z}) \to \widehat{H}^1(\varphi;\mathbb{Z})$ is defined to be zero while $\breve{p} : \widehat{H}^0(\varphi;\mathbb{Z}) \to \widehat{H}^0(X;\mathbb{Z})$ is defined to coincide with the usual map in the long exact sequence

$$0 \longrightarrow H^0(\varphi;\mathbb{Z}) \longrightarrow H^0(X;\mathbb{Z}) \longrightarrow \cdots$$

*Remark 62. Compatibility with characteristic class.* The above homomorphisms $\breve{\imath} : \widehat{H}^{k-1}(A;\mathbb{Z}) \to \widehat{H}^k(\varphi;\mathbb{Z})$ and $\breve{p} : \widehat{H}^k(\varphi;\mathbb{Z}) \to \widehat{H}^k(X;\mathbb{Z})$ are compatible with characteristic classes and the corresponding map in the long exact sequence for absolute and mapping cone cohomology. Thus we have the following commutative diagram:

$$\begin{array}{ccccc} \widehat{H}^{k-1}(A;\mathbb{Z}) & \xrightarrow{\breve{\imath}} & \widehat{H}^k(\varphi;\mathbb{Z}) & \xrightarrow{\breve{p}} & \widehat{H}^k(X;\mathbb{Z}) \\ \downarrow c & & \downarrow c & & \downarrow c \\ H^{k-1}(A;\mathbb{Z}) & \longrightarrow & H^k(\varphi;\mathbb{Z}) & \longrightarrow & H^k(X;\mathbb{Z}) \end{array}$$

**Definition 63.** Let $\varphi : A \to X$ be a smooth map. A differential character $h \in \widehat{H}^k(X;\mathbb{Z})$ is said to *admit sections* along $\varphi$ if it lies in the image of the map $\breve{p} : \widehat{H}^k(\varphi;\mathbb{Z}) \to \widehat{H}^k(X;\mathbb{Z})$.

Let $h \in \mathrm{im}(\breve{p}) \subset \widehat{H}^k(X;\mathbb{Z})$ be a differential character that admits sections along the map $\varphi$. Then any relative differential character $f \in \widehat{H}^k(\varphi;\mathbb{Z})$ with $\breve{p}(f) = h$ is called a *section* of $h$ along $\varphi$. A section $f \in \widehat{H}^k(\varphi;\mathbb{Z})$ of $\breve{p}(f) \in \widehat{H}^k(X;\mathbb{Z})$ along $\varphi$ is called *parallel* if $\mathrm{cov}(f) = 0$.

*Example 64.* Let $k = 1$. Since $Z_0(\varphi;\mathbb{Z}) = Z_0(X;\mathbb{Z})$, any relative differential character of degree 1 corresponds to a function $\bar{f} : X \to \mathrm{U}(1)$ as in the absolute case. Using (90) with $y = 0$ one sees that $\bar{f}$ is smooth and $\mathrm{curv}(f) = d\tilde{f}$ where $\tilde{f}$ is a local lift of $\bar{f}$ as in Example 18. Equation (90) with $x = 0$ shows that $\vartheta$ is a lift of $\bar{f} \circ \varphi$ on $A$. Such a lift is unique only up to addition of a locally constant integer valued function.

To summarize, $\widehat{H}^1(\varphi;\mathbb{Z})$ is the subgroup of $\widehat{H}^1(X;\mathbb{Z}) = C^\infty(X, \mathrm{U}(1))$ containing those functions $\bar{f}$ which are trivial along $\varphi$ in the sense that $\bar{f} \circ \varphi$ has a lift.

*Example 65.* Let $k = 2$. Given $f \in \widehat{H}^2(\varphi;\mathbb{Z})$ we have $\breve{p}(f) \in \widehat{H}^2(X;\mathbb{Z})$ which by Example 19 corresponds to a $\mathrm{U}(1)$-principal bundle $P \to X$ with connection $\nabla$. We pull back $P$ and $\nabla$ along $\varphi$ and we obtain a $\mathrm{U}(1)$-principal bundle $\varphi^*P \to A$ with connection $\varphi^*\nabla$.
*Relative characters determine sections.* Fix $x_0 \in A$. For any two curves $c$ and $c'$ emanating from $x_0$ and ending at the same point $x \in A$, we look at the cycle $c - c' \in Z_1(A;\mathbb{Z})$. Using (90) we compute:

$$\begin{aligned} \varphi^*(\breve{p}f)(c - c') &= (\breve{p}f)(\varphi_*(c - c')) \\ &= f(\varphi_*(c - c'), 0) \\ &= f(\partial_\varphi(0, c - c')) \\ &= \exp\left(2\pi i \int_{c-c'} \vartheta\right) \\ &= \exp\left(2\pi i \int_c \vartheta\right) \cdot \exp\left(2\pi i \int_{c'} \vartheta\right)^{-1}. \end{aligned}$$

We recall from Example 19 that for any $p_0 \in \varphi^*P$ over $x_0$ we have

$$(\mathcal{P}_{c'}^{\varphi^*\nabla})^{-1} \circ \mathcal{P}_c^{\varphi^*\nabla}(p_0) = \mathcal{P}_{c-c'}^{\varphi^*\nabla}(p_0) = p_0 \cdot \varphi^*(\breve{p}f)(c - c').$$

Therefore

$$\mathcal{P}_c^{\varphi^*\nabla}(p_0)\cdot \exp\left(2\pi i\int_c \vartheta\right)^{-1} = \mathcal{P}_{c'}^{\varphi^*\nabla}(p_0)\cdot \exp\left(2\pi i\int_{c'} \vartheta\right)^{-1}.$$

Hence the expression

$$\mathcal{P}_c^{\varphi^*\nabla}(p_0)\cdot \exp\left(2\pi i\int_c \vartheta\right)^{-1}$$

depends on $x$ but not on the choice of curve connecting $x_0$ and $x$. Fixing $x_0$ and $p_0$ we can define a smooth section of $\varphi^*P$ over the connected component containing $x_0$ by

$$\sigma(x) := \mathcal{P}_c^{\varphi^*\nabla}(p_0)\cdot \exp\left(2\pi i\int_c \vartheta\right)^{-1}. \tag{92}$$

Choosing $x_0$ and $p_0$ in each connected component of $A$ we obtain a smooth section of $\varphi^*P$ over all of $A$. If $\sigma'$ is a section obtained by different choices of the $x_0$'s and $p_0$'s, then $\sigma' = \sigma\cdot u$ where $u : A \to \mathrm{U}(1)$ is a locally constant function.
*Isomorphism classes of sections.* We further restrict the freedom in the choices of the $p_0$'s. Consider the pull-back diagram

$$\begin{array}{ccc} \varphi^*P & \xrightarrow{\Phi} & P \\ \downarrow & & \downarrow \\ A & \xrightarrow{\varphi} & X \end{array}$$

Equation (92) yields for any closed curve $c$ in $A$ starting and ending at $x_0$ that

$$\begin{aligned} \mathcal{P}^{\nabla}_{\varphi_* c}(\Phi(p_0)) &= \Phi(p_0)\cdot \exp\left(2\pi i\int_c \vartheta\right) \\ &= \Phi(p_0)\cdot f(\partial_\varphi(0,c)). \end{aligned}$$

For a closed curve $s$ in $X$ (instead of $A$) starting and ending at $\varphi(x_0)$ we have more generally

$$\mathcal{P}^{\nabla}_s(\Phi(p_0)) = \Phi(p_0)\cdot f(s,0).$$

Now, if $x_0$ and $x_0'$ lie in different connected components of $A$ but $\varphi(x_0)$ and $\varphi(x_0')$ lie in the same connected component of $X$, then we demand for any curve $s$ in $X$ starting at $\varphi(x_0)$ and ending at $\varphi(x_0')$ that

$$\mathcal{P}^{\nabla}_s(\Phi(p_0)) = \Phi(p_0')\cdot f(s, x_0 - x_0').$$

In this way, the choice of $p_0'$ is determined by the choice of $p_0$. Moreover, this relation does not depend on the choice of $s$. With this additional requirement the freedom to choose the $p_0$'s reduces to one choice for each maximal set of $x_0$'s which are mapped to the same connected component of $X$. Hence two sections $\sigma$ and $\sigma'$ constructed in this way are related by $\sigma' = \sigma \cdot (u \circ \varphi)$ where $u : X \to \mathrm{U}(1)$ is a locally constant function.

Said differently, a relative differential character $f \in \widehat{H}^2(\varphi; \mathbb{Z})$ determines an isomorphism class $[P, \nabla, \sigma]$ of $\mathrm{U}(1)$-bundles with connection $(P, \nabla)$ and section $\sigma$ along the map $\varphi$. Here $(P, \nabla, \sigma)$ is identified with $(P', \nabla', \sigma')$ if and only if there is a bundle isomorphism $\Psi : P \to P'$ such that $\nabla = \Psi^* \nabla'$ and $\Phi' \circ \sigma' = \Psi \circ \Phi \circ \sigma$. In particular, sections of the pull-back bundle are identified by bundle isomorphisms of $P$ (and not of the pull-back bundle $\varphi^* P$).
*Sections determine relative characters.* Conversely, let $P \to X$ be a $\mathrm{U}(1)$-bundle with connection $\nabla$ and $\sigma$ a section of $\varphi^* P$ over $A$. For any relative cycle of the form $(s, x - x')$ we define $f(s, x - x')$ by

$$\mathcal{P}_s^\nabla(\Phi(\sigma(x))) = \Phi(\sigma(x')) \cdot f(s, x - x').$$

Since $Z_1(\varphi; \mathbb{Z})$ is generated by cycles of this form, the differential character $f$ is uniquely determined. The definition of $f$ is invariant under bundle isomorphisms as defined above.
*Curvature and connection form.* To summarize, we have a 1-1 correspondence between relative differential characters $f \in \widehat{H}^2(\varphi; \mathbb{Z})$ and isomorphism classes $[P, \nabla, \sigma]$ of $\mathrm{U}(1)$-bundles with connection and section along $\varphi$. Under this correspondence, $-2\pi i \cdot \mathrm{curv}(f)$ is the curvature form of $(P, \nabla)$ and $-2\pi i \cdot \mathrm{cov}(f)$ is the connection 1-form of $\varphi^* \nabla$ with respect to $\sigma$.

*Remark 66. Relative differential cohomology.* The group $\widehat{H}^k(\varphi; \mathbb{Z})$ of relative differential characters may be considered as a relative differential cohomology group. Different versions of relative differential cohomology have appeared in the literature: In [40], de Rham-Federer currents on manifolds $X$ with boundary are used to describe differential cohomology relative to $A = \partial X$. In [65], relative differential cohomology is defined for the case of a submanifold $A \subset X$. In both these models, the curvature of a relative cohomology class vanishes upon restriction to the subset $A$.

However, the covariant derivatives of relative differential characters need not be closed. In this sense, the relative differential cohomology group defined by relative differential characters is more general (or is a larger group) than those described in [40] and [65].

Another version of relative differential cohomology – the relative Hopkins-Singer group $\check{H}^k(\varphi; \mathbb{Z})$ – is obtained from the mapping cone complex that arises from a modification of the Hopkins-Singer complex. This group is constructed in [8]. It turns out the be a subquotient of the group $\widehat{H}^k(\varphi; \mathbb{Z})$, see [8, Thm. 4.2].

Let $A \subset X$ be a smooth submanifold. Another notion of relative differential characters is discussed in [3], based on characters on the group of relative cycles $Z_*(X, A; \mathbb{Z})$. The corresponding group of relative characters turns out to be a subgroup of both $\widehat{H}^k(i_A;\mathbb{Z})$ and $\check{H}^k(i_A;\mathbb{Z})$. Here $i_A : A \to X$ denotes the embedding of the submanifold.

*Remark 67.* In [3] we use the formula (58) as a starting point to construct an external product between absolute and relative differential characters. Pull-back along a version of the diagonal map yields an internal product. This in turn provides the graded abelian group $\widehat{H}^*(\varphi;\mathbb{Z})$ with the structure of a right module over the differential cohomology ring $\widehat{H}^*(X;\mathbb{Z})$. By the identification of $\widehat{H}^*(X,A;\mathbb{Z})$ as a subgroup of $\widehat{H}^*(i_A;\mathbb{Z})$ and of $\check{H}^*(\varphi;\mathbb{Z})$ as a subquotient of $\widehat{H}^*(\varphi;\mathbb{Z})$, these graded groups carry induced module structures. These structures are compatible with the structure of the relative or mapping cone cohomology as a right module over the cohomology ring $H^*(X;\mathbb{Z})$.

## 8.2 Existence of Sections

Since differential cohomology is not a (generalized) cohomology theory, the question arises whether there are long exact sequences that relate the absolute and relative differential cohomology groups. Here we fit the maps from (91) into an exact sequence that characterizes those differential characters in $\widehat{H}^k(X;\mathbb{Z})$ that admit sections along $\varphi$.

**Theorem 68 (Exact Sequence).** *Let $\varphi : A \to X$ be a smooth map. Then the following sequences are exact:*

$$0 \longrightarrow \varphi^*\widehat{H}^{k-1}_{\mathrm{flat}}(X;\mathbb{Z}) \longrightarrow \widehat{H}^{k-1}(A;\mathbb{Z}) \xrightarrow{\breve{\imath}} \widehat{H}^k(\varphi;\mathbb{Z}) \xrightarrow{j\circ\varphi^*} \widehat{H}^k(X;\mathbb{Z}) \xrightarrow{\varphi^*\circ c} \varphi^*H^k(X;\mathbb{Z}) \longrightarrow 0 \tag{93}$$

*if $k \geq 2$ and*

$$0 \longrightarrow \widehat{H}^k(\varphi;\mathbb{Z}) \xrightarrow{\breve{p}} \widehat{H}^k(X;\mathbb{Z}) \xrightarrow{\varphi^*\circ c} \varphi^*H^k(X;\mathbb{Z}) \longrightarrow 0. \tag{94}$$

*if $k = 0$ or $k = 1$.*

*Remark 69.* Sequences (93) and (94) can be derived by homological algebraic methods. There are several ways to obtain differential cohomology as the cohomology of a chain complex. The smooth Deligne complex and the Hopkins-Singer complex that compute degree-$k$ differential cohomology both depend on $k$. Thus the cohomology groups in the long exact sequence obtained from the corresponding mapping cone complexes coincide with differential cohomology only in degree $k$.

The Hopkins-Singer complex can be modified so that all its cohomologies realize differential cohomology, see [8, p. 271]. The mapping cone construction then yields a long exact sequence where the absolute cohomology groups coincide with differential cohomology. But the corresponding relative groups for this modified complex are only subquotients of the groups of relative differential characters, [8, p. 278ff.].

Another long exact sequence for relative and absolute differential (generalized) cohomology is constructed in [65, Thm. 2.7]. Another way to define global trivializations of differential cohomology, based on the Hopkins-Singer complex, is described in [57].

Here we do not use any of these identifications of the groups of differential characters with the cohomologies of a chain complex, but give a direct proof.

*Proof (Proof of Theorem 68).* We only consider the case $k \geq 2$ because the case $k = 0$ is obvious and the case $k = 1$ is similar to but simpler than the case $k \geq 2$.

At several steps in the proof we use the fact that the group $\mathrm{U}(1)$ is divisible, hence that for every injective group homomorphism $G' \to G$ the induced homomorphism $\mathrm{Hom}(G, \mathrm{U}(1)) \to \mathrm{Hom}(G', \mathrm{U}(1))$ is surjective, see e.g. [55, pp. 32 and 372]. In other words, any homomorphism from a subgroup of $G$ to $\mathrm{U}(1)$ can be extended to a homomorphism from $G$ to $\mathrm{U}(1)$.

a) The map $\varphi^* \widehat{H}^{k-1}_{\mathrm{flat}}(X;\mathbb{Z}) \to \widehat{H}^{k-1}(A;\mathbb{Z})$ is the inclusion of a subgroup and hence injective.

b) We prove exactness at $\widehat{H}^{k-1}(A;\mathbb{Z})$. For $\bar{g} \in \widehat{H}^{k-1}_{\mathrm{flat}}(X;\mathbb{Z})$ and any $(s,t) \in Z_{k-1}(\varphi;\mathbb{Z})$ we have:

$$\begin{aligned} \breve{\imath}(\varphi^*\bar{g})(s,t) &= (\varphi^*\bar{g})(t) \\ &= \bar{g}(\varphi_* t) \\ &= \bar{g}(-\partial s) \\ &= \exp\Big(2\pi i \int_s -\mathrm{curv}(\bar{g})\Big) \\ &= 1\,, \end{aligned}$$

because $\mathrm{curv}(\bar{g}) = 0$. Hence $\varphi^* \widehat{H}^{k-1}_{\mathrm{flat}}(X;\mathbb{Z}) \subset \ker(\breve{\imath})$.

To show the converse inclusion we pick $g \in \widehat{H}^{k-1}(A;\mathbb{Z})$ with $\breve{\imath}(g) = 1$. Let $Q \subset Z_{k-2}(A;\mathbb{Z})$ be the subgroup of those $t \in Z_{k-2}(A;\mathbb{Z})$ for which there exists an $s \in C_{k-1}(X;\mathbb{Z})$ such that $(s,t) \in Z_{k-1}(\varphi;\mathbb{Z})$. The condition $\breve{\imath}(g) = 1$ is equivalent to $g$ being trivial on $Q$. We construct $\bar{g} \in \widehat{H}^{k-1}_{\mathrm{flat}}(X;\mathbb{Z})$ such that $\varphi^*\bar{g} = g$. Let $Q' \subset Z_{k-2}(X;\mathbb{Z})$ be the subgroup generated by $\varphi_* Z_{k-2}(A;\mathbb{Z})$ and $B_{k-2}(X;\mathbb{Z})$. We define a group homomorphism $\bar{g} : Q' \to \mathrm{U}(1)$ by setting

$$\bar{g}(\varphi_* x) := g(x)\,, \tag{95}$$

$$\bar{g}(\partial y) := 1\,. \tag{96}$$

The conditions are consistent since $\varphi_* Z_{k-2}(A;\mathbb{Z}) \cap B_{k-2}(X;\mathbb{Z}) = \varphi_* Q$. By (96), any extension to a group homomorphism $\bar{g} : Z_{k-2}(X;\mathbb{Z}) \to \mathrm{U}(1)$ yields a flat differential character on $X$. By (95), we have $\varphi^* \bar{g} = g$.

c) We prove exactness at $\widehat{H}^k(\varphi;\mathbb{Z})$. For every $z \in Z_{k-1}(X;\mathbb{Z})$, we have $\breve{p}(\breve{\imath}(g))(z) = \breve{\imath}(g)(z,0) = g(0) = 1$. Hence $\mathrm{im}(\breve{\imath}) \subset \ker(\breve{p})$.

Conversely, let $f \in \ker(\breve{p})$. Thus $f(z,0) = 1$ for every $z \in Z_{k-1}(X;\mathbb{Z})$. For cycles $(s,t),(s',t) \in Z_{k-1}(\varphi;\mathbb{Z})$, we have $\partial(s-s') = -\varphi_* t + \varphi_* t = 0$. Hence $f(s-s',0) = 1$ and thus $f(s,t) = f(s',t)$. Let $Q \subset Z_{k-2}(A;\mathbb{Z})$ be the subgroup defined in b). We define a group homomorphism $g : Q \to \mathrm{U}(1)$ by setting $g(t) := f(s,t)$.

Now $B_{k-2}(A;\mathbb{Z}) \subset Q$, since for $t = \partial y$; we have $(-\varphi_* y, t) = \partial_\varphi(0,-y) \in B_{k-1}(\varphi;\mathbb{Z}) \subset Z_{k-1}(\varphi;\mathbb{Z})$. We can extend $g$ as a group homomorphism $g : Z_{k-2}(A;\mathbb{Z}) \to \mathrm{U}(1)$. On $B_{k-2}(A;\mathbb{Z})$, we have

$$g(\partial y) = f(\partial_\varphi(0,-y)) = \exp\Big( -2\pi i \int_y \mathrm{cov}(f)\Big) .$$

Hence $g : Z_{k-2}(A;\mathbb{Z}) \to \mathrm{U}(1)$ is a differential character $g \in \widehat{H}^{k-1}(A;\mathbb{Z})$ with curvature $\mathrm{curv}(g) = -\mathrm{cov}(f)$. Since $f(s,t) = g(t)$ for every $(s,t) \in Z_k(\varphi;\mathbb{Z})$, we have $f = \breve{\imath}(g)$.

d) For the exactness at $\widehat{H}^k(X;\mathbb{Z})$ consider the following commutative diagram with exact columns:

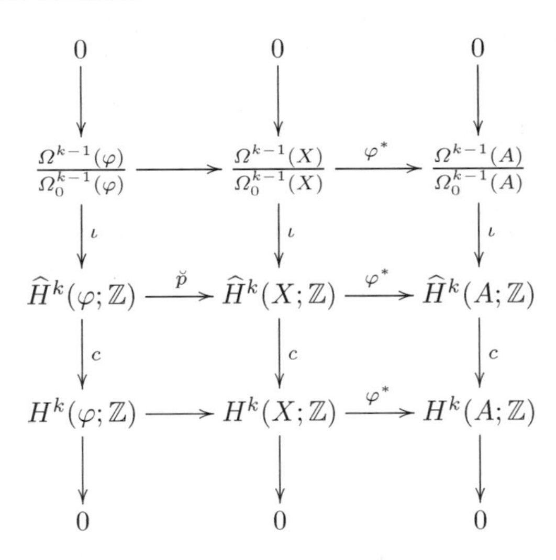

The bottom row is part of the long exact cohomology sequence obtained from the short exact sequence of chain complexes

$$0 \to C_*(X;\mathbb{Z}) \to C_*(\varphi;\mathbb{Z}) \to C_{*-1}(A;\mathbb{Z}) \to 0 .$$

Let $f \in \widehat{H}^k(\varphi;\mathbb{Z})$. From the commutativity of the diagram and the exactness of the bottom row we conclude $c(\varphi^* \breve{p}(f)) = 0$. Hence $\mathrm{im}(\breve{p}) \subset \ker(\varphi^* \circ c)$.

Conversely, let $h \in \ker(\varphi^* \circ c)$. We construct a section along $\varphi$. From the diagram we conclude that there exists a differential form $\chi \in \Omega^{k-1}(A)$ such that $\varphi^* h = \iota(\chi)$. Hence $\varphi^*\mathrm{curv}(h) = \mathrm{curv}(\varphi^* h) = \mathrm{curv}(\iota(\chi)) = d\chi$. Let $W \subset Z_{k-1}(\varphi;\mathbb{Z})$ be the subgroup generated by $B_{k-1}(\varphi;\mathbb{Z})$ and by pairs of the form $(s,t) = (z,0)$ with $z \in Z_{k-1}(X;\mathbb{Z})$. We define a group homomorphism $f : W \to \mathrm{U}(1)$ by setting:

$$f(\partial_\varphi(x,y)) := \exp\Big[2\pi i \cdot \Big(\int_x \mathrm{curv}(h) + \int_y \chi\Big)\Big], \tag{97}$$

$$f((z,0)) := h(z). \tag{98}$$

This definition is consistent, since for $(z,0) = \partial_\varphi(x,y) = (\partial x + \varphi_* y, -\partial y)$, we have

$$\begin{aligned} f((z,0)) &\overset{(98)}{=} h(\partial x + \varphi_* y) \\ &= \exp\Big(2\pi i \int_x \mathrm{curv}(h)\Big) \cdot \varphi^* h(y) \\ &= \exp\Big(2\pi i \int_x \mathrm{curv}(h)\Big) \cdot \iota(\chi)(y) \\ &= \exp\Big[2\pi i \cdot \Big(\int_x \mathrm{curv}(h) + \int_y \chi\Big)\Big] \\ &\overset{(97)}{=} f(\partial_\varphi(x,y)). \end{aligned}$$

We extend $f$ to a $\mathrm{U}(1)$-valued group homomorphism on $Z_{k-1}(\varphi;\mathbb{Z})$. By equation (97), this homomorphism satisfies (90). Thus $f \in \widehat{H}^k(\varphi;\mathbb{Z})$. Equation (98) implies that $\breve{p}(f) = h$.

e) Finally, the map $\varphi^* \circ c : \widehat{H}^k(X;\mathbb{Z}) \to \varphi^* H^k(X;\mathbb{Z})$ is surjective since $c$ is surjective by (30) and pull-back along $\varphi$ is surjective onto its image. □

**Corollary 70 (Long Exact Sequence).** *For $k \geq 2$ we have the following long exact sequence:*

$$\begin{aligned} \ldots &\longrightarrow H^{k-3}(A;\mathrm{U}(1)) \longrightarrow H^{k-2}(\varphi;\mathrm{U}(1)) \longrightarrow H^{k-2}(X;\mathrm{U}(1)) \longrightarrow \\ &\xrightarrow{j\circ\varphi^*} \widehat{H}^{k-1}(A;\mathbb{Z}) \xrightarrow{\breve{\imath}} \widehat{H}^k(\varphi;\mathbb{Z}) \xrightarrow{\breve{p}} \widehat{H}^k(X;\mathbb{Z}) \longrightarrow \\ &\xrightarrow{\varphi^*\circ c} H^k(A;\mathbb{Z}) \longrightarrow H^{k+1}(\varphi;\mathbb{Z}) \longrightarrow H^{k+1}(X;\mathbb{Z}) \longrightarrow \ldots \end{aligned} \tag{99}$$

*The sequence extends on the left and on the right as the mapping cone sequence for singular cohomology with coefficients* $\mathrm{U}(1)$ *and* $\mathbb{Z}$, *respectively.*

*Proof.* We use the identification $H^{k-2}(X;\mathrm{U}(1)) \xrightarrow{\cong} \widehat{H}^{k-1}_{\text{flat}}(X;\mathbb{Z})$ from diagram (30). In particular, the map $j : H^{k-2}(X;\mathrm{U}(1)) \to \widehat{H}^{k-1}(X;\mathbb{Z})$ is injective.

Exactness at the three middle terms is clear from Theorem 68. From the mapping cone sequence for cohomology with U(1)-coefficients, we conclude:

$$\begin{aligned}\ker\big[\, j\circ\varphi^* : H^{k-2}(X;\mathrm{U}(1)) \to \widehat{H}^{k-1}(A;\mathbb{Z})\,\big] \\ &= \ker\big[\,\varphi^* : H^{k-2}(X;\mathrm{U}(1)) \to H^{k-2}(A;\mathrm{U}(1))\,\big] \\ &= \mathrm{im}\big[\, H^{k-2}(\varphi;\mathrm{U}(1)) \to H^{k-2}(X;\mathrm{U}(1))\,\big]\,.\end{aligned}$$

This proves exactness at $H^{k-2}(X;\mathrm{U}(1))$.

From the mapping cone sequence for cohomology with integral coefficients and surjectivity of $c$, we conclude:

$$\begin{aligned}\ker\big[\, H^k(A;\mathbb{Z}) \to H^{k+1}(\varphi;\mathbb{Z})\,\big] &= \mathrm{im}\big[\,\varphi^* : H^k(X;\mathbb{Z}) \to H^k(A;\mathbb{Z})\,\big] \\ &= \mathrm{im}\big[\,\varphi^*\circ c : \widehat{H}^k(X;\mathbb{Z}) \to H^k(A;\mathbb{Z})\,\big]\,.\end{aligned}$$

This proves exactness at $H^k(A;\mathbb{Z})$. □

**Corollary 71.** *For $k \geq 2$ and the map $\varphi = \mathrm{id}_X$, the long exact sequence* (99) *yields the following five term exact sequence:*

$$0 \to H^{k-2}(X;\mathrm{U}(1)) \xrightarrow{j} \widehat{H}^{k-1}(X;\mathbb{Z}) \xrightarrow{\mathrm{curv}} \Omega^{k-1}(X) \xrightarrow{\iota} \widehat{H}^k(X;\mathbb{Z}) \xrightarrow{c} H^k(X;\mathbb{Z}) \to 0.$$

A differential character $h \in \widehat{H}^k(X;\mathbb{Z})$ is called *topologically trivial along* $\varphi$ if $\varphi^* c(h) = 0$. A differential character $h \in \widehat{H}^k(X;\mathbb{Z})$ is called *flat along* $\varphi$ if $\varphi^*\mathrm{curv}(h) = 0$.

As is well known, a U(1)-bundle is topologically trivial if and only if it admits sections. Topological triviality is detected by the first Chern class. Thus the pull-back bundle along a smooth map $\varphi$ is topologically trivial if and only if the original bundle admits sections along $\varphi$. A similar statement holds for differential characters of any degree:

**Corollary 72 (Properties of Sections).** *A differential character $h \in \widehat{H}^k(X;\mathbb{Z})$ admits sections along a smooth map $\varphi : A \to X$ if and only if it is topologically trivial along $\varphi$.*

*If $h$ admits parallel sections along $\varphi$, then $h$ is also flat along $\varphi$. Conversely, if $(\mathrm{curv}(h), 0) \in \Omega^k_0(\varphi)$ and $h$ is topologically trivial along $\varphi$, then it admits a parallel section.*

*Sections along $\varphi$ are uniquely determined by their covariant derivatives if $\varphi_* : H_{k-2}(A;\mathbb{Z}) \to H_{k-2}(X;\mathbb{Z})$ is injective. Explicitly, if $f_1, f_2 \in \widehat{H}^k(\varphi;\mathbb{Z})$ with $\breve{p}(f_1) = \breve{p}(f_2)$ and $\mathrm{cov}(f_1) = \mathrm{cov}(f_2)$, then $f_1 = f_2$.*

*Proof.* The first statement follows immediately from Theorem 68.

For the second, let $f \in \widehat{H}^k(\varphi;\mathbb{Z})$ with $\breve{p}(f) = h$ and $\mathrm{cov}(f) = 0$. Then $d_\varphi(\mathrm{curv}(f), \mathrm{cov}(f)) = 0$ implies $0 = \varphi^*\mathrm{curv}(f) - d\mathrm{cov}(f) = \varphi^*\mathrm{curv}(h)$. Conversely, by surjectivity of the map $(\mathrm{curv}, \mathrm{cov}) : \widehat{H}^k(\varphi;\mathbb{Z}) \to \Omega_0^k(X)$, we find a parallel section if $(\mathrm{curv}(h), 0) \in \Omega_0^k(\varphi)$. A necessary condition is $\varphi^*\mathrm{curv}(h) = 0$. That it is sufficient is shown in Corollary 76 below.

To show the last assertion, observe that $\varphi_* H_{k-2}(A;\mathbb{Z}) = H_{k-2}(X;\mathbb{Z})$ implies

$$\begin{aligned}
\varphi^* \widehat{H}^{k-1}_{\text{flat}}(X;\mathbb{Z}) &\cong \varphi^* H^{k-2}(X;\mathrm{U}(1)) \\
&= \varphi^* \operatorname{Hom}(H_{k-2}(X;\mathbb{Z}), \mathrm{U}(1)) \\
&= \operatorname{Hom}(\varphi_* H_{k-2}(A;\mathbb{Z}), \mathrm{U}(1)) \\
&= \operatorname{Hom}(H_{k-2}(A;\mathbb{Z}), \mathrm{U}(1)) \\
&= H^{k-2}(A;\mathrm{U}(1)) \\
&\cong \widehat{H}^{k-1}_{\text{flat}}(A;\mathbb{Z}) .
\end{aligned}$$

Now let $f_1, f_2 \in \widehat{H}^k(\varphi;\mathbb{Z})$ be sections of $h \in \widehat{H}^k(X;\mathbb{Z})$ with $\mathrm{cov}(f_1) = \mathrm{cov}(f_2)$. By Theorem 68, we have $f_1 - f_2 = \breve{\imath}(g)$ for some $g \in \widehat{H}^{k-1}(A;\mathbb{Z})$. Since $\mathrm{curv}(g) = -\mathrm{cov}(f_1 - f_2) = 0$, we have $g \in \widehat{H}^{k-1}_{\text{flat}}(A;\mathbb{Z}) = \varphi^* \widehat{H}^{k-1}_{\text{flat}}(X;\mathbb{Z})$. Hence $f_1 - f_2 = \breve{\imath}(g) = 0$ by Theorem 68. □

*Remark 73.* Any differential character $h \in \widehat{H}^k(X;\mathbb{Z})$ has local sections in the following sense: If $\varphi : A \to X$ is smooth where $A$ is contractible, then $H^k(A;\mathbb{Z}) = 0$. Hence $h$ is topologically trivial along $\varphi$ and therefore admits sections along $\varphi$.

*Example 74.* Let $G$ be a compact Lie group with Lie algebra $\mathfrak{g}$. An invariant polynomial, homogeneous of degree $k$, is a symmetric $\mathrm{Ad}_G$-invariant multilinear map $q : \mathfrak{g}^{\otimes k} \to \mathbb{R}$. The Chern-Weil construction associates to any principal $G$-bundle with connection $(P, \nabla)$ over $X$ a closed differential form $CW(q) = q(R^\nabla) \in \Omega^{2k}(X)$ by applying the polynomial $q$ to the curvature 2-form $R^\nabla$ of the connection $\nabla$. Consider those polynomials $q$ for which the Chern-Weil form $CW(q)$ has integral periods. Let $u \in H^{2k}(X;\mathbb{Z})$ be a universal characteristic class for principal $G$-bundles that coincides in $H^{2k}(X;\mathbb{R})$ with the de Rham class of $CW(q)$. The Cheeger-Simons construction [24, Thm 2.2] associates to this setting a differential character $\widehat{CW}(q,u) \in \widehat{H}^{2k}(X;\mathbb{Z})$ with curvature $\mathrm{curv}(\widehat{CW}(q,u)) = CW(q)$, the Chern-Weil form, and characteristic class $c(\widehat{CW}(q,u)) = u$, the fixed universal characteristic class. The construction is natural with respect to bundle maps.

Since the total space $EG$ of the universal principal $G$-bundle is contractible, universal characteristic classes vanish upon pull-back to the total space. By Theorem 68 the Cheeger-Simons character $\widehat{CW}(q,u)$ thus admits sections along the bundle projection $\pi : P \to X$. The so-called Cheeger-Chern-Simons construction of [4] yields a canonical section $\widehat{CCS}(q,u) \in \widehat{H}^{2k}(\pi;\mathbb{Z})$ with covariant derivative $\mathrm{cov}(\widehat{CCS}(q,u)) = CS(q) \in \Omega^{2k-1}(P)$, the Chern-Simons form. The construction is natural with respect to bundle maps.

## 8.3 Sections with Prescribed Covariant Derivative

In this section we discuss the extent to which we can prescribe the covariant derivative of sections of along a smooth map $\varphi : A \to X$. Exactness at $\widehat{H}^k(X;\mathbb{Z})$ of the long exact sequence (93) implies that a character $h \in \widehat{H}^k(X;\mathbb{Z})$ admits a section along a smooth map $\varphi : A \to X$ if any only if it is topologically trivial along $\varphi$, i.e. if and only if the pull-back $\varphi^*h$ is topologically trivial. In fact, in the proof of exactness at $\widehat{H}^k(X;\mathbb{Z})$ we have constructed a section with covariant derivative prescribed by the topological trivialization.

In [3] we show that the following diagram commutes:

$$\begin{array}{ccc} \widehat{H}^k(\varphi;\mathbb{Z}) & \xrightarrow{\breve{p}} & \widehat{H}^k(X;\mathbb{Z}) \\ \downarrow{\scriptstyle \mathrm{cov}} & & \downarrow{\scriptstyle \varphi^*} \\ \Omega^{k-1}(A) & \xrightarrow[\iota]{} & \widehat{H}^k(A;\mathbb{Z})\,. \end{array} \tag{100}$$

Summarizing these observations, we have the following characterization of the property of a character to admit sections with prescribed covariant derivative:

**Proposition 75 (Sections with Prescribed Covariant Derivative.).** *Let $\varphi : A \to X$ be a smooth map and $h \in \widehat{H}^k(X;\mathbb{Z})$. Then the following three statements are equivalent:*

1. *The character $h$ admits a section along $\varphi$ with covariant derivative $\chi$.*
2. *We have $\varphi^*h = \iota(\chi)$.*
3. *The character $h$ is topologically trivial along $\varphi$ and we have $d\chi = \varphi^*\mathrm{curv}(h)$.*

*Proof.* In the proof of Theorem 68, we have shown that from a topological trivialization $\varphi^*h = \iota(\chi)$ we can construct a section of $h$ along $\varphi$ with covariant derivative $\chi$. Thus the second statement implies the first.

Conversely, let $f \in \widehat{H}^k(\varphi;\mathbb{Z})$ be a section of $h$ with covariant derivative $\mathrm{cov}(f) = \chi$. Diagram (100) shows that $\varphi^*h = \iota(\mathrm{cov}(f))$. Hence the first statement implies the second.

The equivalence of the second with the third statement follows from commutativity of the upper right square of diagram (30): we have $\mathrm{curv}(\varphi^*h) = d\chi$ if and only if $\varphi^*h = \iota(\chi)$. □

From the equivalence of the second and third statement above, we further obtain:

**Corollary 76.** *Let $\varphi : A \to X$ be a smooth map. Let $\omega \in \Omega_0^k(X)$ and $\vartheta \in \Omega^{k-1}(A)$. Then we have $(\omega, \vartheta) \in \Omega_0^k(\varphi)$ if and only if $d\vartheta = \varphi^*\omega$. In other words, if the pair $(\omega, \vartheta)$ is $d_\varphi$-closed and $\omega$ has integral periods, then also $(\omega, \vartheta)$ has integral periods.*

*Proof.* By diagram (30) there exists a differential character $h \in \widehat{H}^k(X;\mathbb{Z})$ with $\mathrm{curv}(h) = \omega$. By Proposition 75, the condition $d\vartheta = \varphi^*\omega$ implies that $h$ admits a section $f \in \widehat{H}^k(\varphi;\mathbb{Z})$ along $\varphi$ with covariant derivative $\vartheta$. Hence the pair $(\omega, \vartheta) = (\mathrm{curv}(f), \mathrm{cov}(f))$ is $d_\varphi$-closed with integral periods. □

## *8.4 Relative Differential Characters and Fiber Integration*

Throughout this section, we consider the case that $A \subset X$ is a smooth subspace and $\varphi : A \to X$ the inclusion.

Let us consider the space $\widehat{H}^k(\mathrm{id}_X;\mathbb{Z})$ of differential characters with global sections. Let $(x, y) \in Z_k(\mathrm{id}_X;\mathbb{Z})$. Then we have $x = -\partial y$, hence $(x, y) = \partial(0, -y)$ and $Z_k(\mathrm{id}_X;\mathbb{Z}) = B_k(\mathrm{id}_X;\mathbb{Z})$. Therefore any relative differential character $f \in \widehat{H}^k(\mathrm{id}_X;\mathbb{Z})$ is of the form

$$f(c, -\partial c) = f(\partial(0, c)) = \exp\Big(2\pi i \int_c \mathrm{cov}(f)\Big).$$

Conversely, each $(k-1)$-form $\vartheta$ defines a relative differential character in $\widehat{H}^k(\mathrm{id}_X;\mathbb{Z})$ by

$$f(c, -\partial c) = \exp\Big(2\pi i \int_c \vartheta\Big).$$

Thus $\mathrm{cov} : \widehat{H}^k(\mathrm{id}_X;\mathbb{Z}) \to \Omega^{k-1}(X)$ is an isomorphism. Moreover, the diagram

$$\begin{array}{ccc} \widehat{H}^k(\mathrm{id}_X;\mathbb{Z}) & \xrightarrow[\cong]{\mathrm{cov}} & \Omega^{k-1}(X) \\ & \breve{p} \searrow \quad \swarrow \iota & \\ & \widehat{H}^k(X;\mathbb{Z}) & \end{array}$$

commutes. This is the special case of diagram (100) for $\varphi = \mathrm{id}_X$. The inverse of $\mathrm{cov} : \widehat{H}^k(\mathrm{id}_X;\mathbb{Z}) \to \Omega^{k-1}(X)$ is given by

$$\mathrm{cov}^{-1}(\vartheta) = \iota(\vartheta, 0)$$

since $(\mathrm{curv}, \mathrm{cov})(\iota(\vartheta, 0)) = d_{\mathrm{id}_X}(\vartheta, 0) = (d\vartheta, \vartheta)$. Consequently, we have

$$\breve{\imath}(g) = \iota_{\mathrm{id}_X}(-\mathrm{curv}(g), 0)$$

for any $g \in \widehat{H}^{k-1}(X;\mathbb{Z})$.

We may now reinterprete fiber integration for fibers $F$ with boundary as follows: Given $h \in H^k(E;\mathbb{Z})$, fiber integration along $F$ yields a form $\rho = (-1)^{k-\dim F} \fint_F \operatorname{curv}(h) \in \Omega^{k-\dim F}(X)$ in the notation of Proposition 54. Applying the isomorphism $\operatorname{cov}^{-1}$, we obtain a relative differential character $\widehat{\pi}_!^E h \in \widehat{H}^{k-\dim(F)+1}(\mathrm{id}_X;\mathbb{Z})$ with the property, that $\breve{p}(\widehat{\pi}_! h) = \iota(\rho) = \widehat{\pi}_!^{\partial E} h$. Hence we have constructed a fiber integration map

$$\widehat{\pi}_!^E = \operatorname{cov}^{-1} \circ \fint_F \circ (-1)^{k-\dim F} \operatorname{curv} : \quad H^k(E;\mathbb{Z}) \to H^{k-\dim(F)+1}(\mathrm{id}_X;\mathbb{Z}) \tag{101}$$

such that the diagram

$$\begin{array}{ccc} & \widehat{H}^k(E;\mathbb{Z}) & \\ {\scriptstyle \widehat{\pi}_!^E} \swarrow & & \searrow {\scriptstyle \widehat{\pi}_!^{\partial E}} \\ \widehat{H}^{k-\dim F+1}(\mathrm{id}_X;\mathbb{Z}) & \xrightarrow{\quad \breve{p} \quad} & \widehat{H}^{k-\dim \partial F}(X;\mathbb{Z}) \end{array} \tag{102}$$

commutes.

*Example 77.* Let the fibers of $\pi : E \to X$ be diffeomorphic to compact intervals and carry an orientation. Let $k = 2$ and let $P$ be a $\mathrm{U}(1)$-bundle with connection $\nabla$ corresponding to $h \in \widehat{H}^2(E;\mathbb{Z})$. In the notation of Examples 57 and 58 we have

$$\widehat{\pi}_!^{\partial E} P = (j^+)^* P \otimes (j^-)^* P^* = \operatorname{Hom}((j^-)^* P, (j^+)^* P)$$

where Hom stands for (unitary)[2] homomorphisms. Now $\widehat{\pi}_!^E P \in \widehat{H}^2(\mathrm{id}_X;\mathbb{Z})$ yields a global section $\sigma$ of $\widehat{\pi}_!^{\partial E} P = \operatorname{Hom}((j^-)^* P, (j^+)^* P)$, uniquely determined up to multiplication by an element in $\mathrm{U}(1)$ over each connected component of $X$. In the construction of $\sigma$ we choose $\sigma(x_0) \in (\widehat{\pi}_!^{\partial E} P)_{x_0} = \operatorname{Hom}(P_{j^-(x_0)}, P_{j^+(x_0)})$ as the parallel transport in $P$ along the fiber $E_{x_0}$ for some fixed $x_0$. Then one can check that $\sigma(x)$ is parallel transport in $P$ along the fiber $E_x$ for all $x$ in the connected component of $X$ containing $x_0$.

*Remark 78. Fiber integration for relative characters.* In [3] we construct fiber integration for relative differential characters: Let $\pi : E \to X$ be a fiber bundle with closed oriented fibers. Let $\varphi : A \to X$ be a smooth map and $\Phi : \varphi^* E \to E$ the induced bundle map. Let $k \geq \dim F + 2$. Fiber integration for relative characters is a homomorphism $\widehat{\pi}_! : \widehat{H}^k(\Phi;\mathbb{Z}) \to \widehat{H}^{k-\dim F}(\varphi;\mathbb{Z})$. It commutes with the maps $\breve{\imath}$ and $\breve{p}$ so that the diagram

[2] If we regard $\widehat{\pi}_!^{\partial E} P$ as a $\mathrm{U}(1)$-principal bundle, we have to take unitary homomorphisms. If we regard it as a complex line bundle, we have to take all $\mathbb{C}$-linear homomorphisms.

$$\begin{array}{ccccc}
\widehat{H}^{k-1}(\varphi^*E;\mathbb{Z}) & \xrightarrow{\check{\imath}} & \widehat{H}^k(\Phi;\mathbb{Z}) & \xrightarrow{\check{p}} & \widehat{H}^k(E;\mathbb{Z}) \\
\downarrow{\scriptstyle \widehat{\pi}_!} & & \downarrow{\scriptstyle \widehat{\pi}_!} & & \downarrow{\scriptstyle \widehat{\pi}_!} \\
\widehat{H}^{k-1-\dim F}(A;\mathbb{Z}) & \xrightarrow{\check{\imath}} & \widehat{H}^{k-\dim F}(\varphi;\mathbb{Z}) & \xrightarrow{\check{p}} & \widehat{H}^{k-\dim F}(X;\mathbb{Z}) .
\end{array}$$

commutes.

Fiber integration for relative characters has properties analogous to those for absolute characters, i.e. it commutes with curvature, covariant derivative, characteristic class and topological trivializations. Moreover, fiber integration in fiber products is compatible with cross products of characters, and we have an up-down formula.

# 9 Applications

We will now see how various constructions occurring in different contexts in the literature, such as higher-dimensional holonomy, parallel transport and transgression as well as chain field theories, can be described using the general calculus of absolute and relative differential characters developed in the preceding sections.

## *9.1 Higher Dimensional Holonomy and Parallel Transport*

In this section, we discuss holonomy and parallel transport of differential characters along compact oriented smooth manifolds. Holonomy of smooth Deligne classes has been discussed in [19, Sec. 3]. Surface holonomy was considered as classical action for a quantum field theory in [35, 33]. An approach to holonomy along surfaces with boundary using $D$-branes is described in [33, Sec. 6].

For a $\mathrm{U}(1)$-bundle with connection $(P,\nabla)$ on $X$, holonomy around a closed loop is defined geometrically by parallel transport. Parallel transport along a path $\gamma:[0,1]\to X$ in the associated complex line bundle takes values in the line $L^*_{\gamma(0)}\otimes L_{\gamma(1)}$. Holonomy along a closed path $\gamma:[0,1]\to X$ is the element in $\mathrm{U}(1)$ that corresponds to the value of the parallel transport in $L^*_{\gamma(0)}\otimes L_{\gamma(0)}\cong\mathbb{C}$.

*Higher dimensional holonomy.* In abritrary degree $k$, let $h\in\widehat{H}^k(X;\mathbb{Z})$ be a differential character. In view of Example 19, we may think of the map $h$ as defining holonomy around orientable closed manifolds of dimension $k-1$. More explicitly, for a smooth map $\varphi:\Sigma\to X$ from an oriented closed $(k-1)$-manifold $\Sigma$, we set

$$\mathrm{Hol}^h(\varphi) := \varphi^* h([\Sigma]) = \varphi^* h([\Sigma]_{\partial S_k}) = h(\varphi_*[\Sigma]_{\partial S_k}). \tag{103}$$

Holonomy is invariant under thin cobordism in the sense of [17]: for a cobordism $\Phi : W \to X$ from $\varphi : \Sigma \to X$ to $\varphi' : \Sigma' \to X$, we have $\varphi'_*[\Sigma']_{\partial S_k} - \varphi_*[\Sigma]_{\partial S_k} = \partial\Phi_*[W]_{S_k}$. If the cobordism is thin, then $\Phi_* c \in S_k(X;\mathbb{Z})$ for any fundamental cycle $c$ of $W$. Thus $\mathrm{Hol}^h(\varphi') = h(\varphi'_*[\Sigma']_{\partial S_k}) = h(\varphi_*[\Sigma]_{\partial S_k}) = \mathrm{Hol}^h(\varphi)$.

Higher dimensional parallel transport will be defined analogously by evaluating differential characters along oriented smooth $(k-1)$-manifolds with boundary. The result will be an element in a complex line attached to the boundary. For surfaces such constructions are well known from Chern-Simons theory, see e.g. [56, Sec. 2] and [29, Sec. 2].

*Construction of the line bundle $\mathcal{L}^h$.* Let $h \in \widehat{H}^k(X;\mathbb{Z})$ and let $W$ be a compact oriented $(k-1)$-manifold $W$ with boundary $\partial W = \Sigma$. Let $\varphi : \Sigma \to X$ be a smooth map which extends to a map defined on $W$. In other words, it lies in the image of the restriction map $r : C^\infty(W,X) \to C^\infty(\partial W, X)$, $\Phi \mapsto \Phi|_{\partial W}$. For a smooth map $\Phi : W \to X$ we set $-\Phi : \overline{W} \to X$ for the same map from the manifold with reversed orientation. On the set $C^\infty(W,X) \times \mathbb{C}$, we consider the equivalence relation

$$(\Phi, c) \sim (\Phi', c') :\Leftrightarrow\ r(\Phi) = r(\Phi') \text{ and } c = \mathrm{Hol}^h(\Phi' \cup_\varphi -\Phi) \cdot c'.^3 \tag{104}$$

For $\varphi \in r(C^\infty(W,X))$, this defines a complex line

$$\mathcal{L}^h_\varphi := \{(\Phi, c) \mid \Phi \in r^{-1}(\varphi), c \in \mathbb{C}\}/\sim\ . \tag{105}$$

Varying the map $\varphi$, we obtain a complex line bundle $\mathcal{L}^h \to r(C^\infty(W,X))$. Holonomy defines a Hermitian metric on $\mathcal{L}^h$ by

$$\langle [\Phi_1, c_1], [\Phi_2, c_2] \rangle := \mathrm{Hol}^h(\Phi_1 \cup_\varphi -\Phi_2) \cdot c_1 \cdot \overline{c_2}\,. \tag{106}$$

This is well defined, since for any two representents $(\Phi_1, c_1) \sim (\Phi'_1, c'_1)$ we have $c'_1 = \mathrm{Hol}^h(\Phi_1 \cup_\varphi -\Phi'_1) \cdot c_1$ and thus

$$\begin{aligned}
\langle [\Phi'_1, c'_1], [\Phi_2, c_2] \rangle &= \mathrm{Hol}^h(\Phi'_1 \cup_\varphi -\Phi_2) \cdot c'_1 \cdot \overline{c_2} \\
&= \mathrm{Hol}^h(\Phi_1 \cup_\varphi -\Phi'_1) \cdot \mathrm{Hol}^h(\Phi'_1 \cup_\varphi -\Phi_2) \cdot c_1 \cdot \overline{c_2} \\
&= \mathrm{Hol}^h(\Phi_1 \cup_\varphi -\Phi_2) \cdot c_1 \cdot \overline{c_2} \\
&= \langle [\Phi_1, c_1], [\Phi_2, c_2] \rangle\,.
\end{aligned}$$

*Higher dimensional parallel transport.* Parallel transport along $\Phi : W \to X$ is defined by

$$\mathrm{PT}^h(\Phi) := [\Phi, 1] \in \mathcal{L}^h_{r(\Phi)}. \tag{107}$$

[3] In general, $\Phi' \cup_\varphi -\Phi$ is not smooth as a map defined on the *manifold* $W \cup_{\partial W} \overline{W}$ but it defines a smooth singular cycle if the fundamental cycle of $W \cup_{\partial W} \overline{W}$ is chosen appropriately.

The map $\mathrm{PT}^h : C^\infty(W,X) \to \mathcal{L}^h$, $\Phi \mapsto [\Phi, 1]$, is a section of $\mathcal{L}^h$ along the restriction map $r$. Moreover, $[\Phi, 1]$ has unit length. Thus parallel transport yields a section of the U(1)-bundle associated with the Hermitian line bundle $\mathcal{L}^h$.

*The connection $\nabla^h$ on $\mathcal{L}^h$.* We construct a connection $\nabla^h$ on the bundle $\mathcal{L}^h$ by describing its parallel transport (not to be confused with the higher dimensional parallel transport constructed above): Choose a path $\gamma : [0,1] \to r(C^\infty(W,X))$ and a lift $\Gamma : [0,1] \to C^\infty(W,X)$ with $r \circ \Gamma = \gamma$. Define $F : [0,1] \times W \to X$ by $F(t,w) := \Gamma(t)(w)$.

The path $\Gamma$ yields a lift of the path $\gamma$ to the total space $\mathcal{L}^h$, defined by

$$\overline{\Gamma} : [0,1] \to \mathcal{L}^h, \quad \overline{\Gamma}(t) := [\Gamma(t), 1].$$

We define parallel transport along the path $\gamma$ to be the homomorphism

$$\mathcal{P}_\gamma^{\nabla^h} : \mathcal{L}^h_{\gamma(0)} \to \mathcal{L}^h_{\gamma(t)}, \; \overline{\Gamma}(0) \mapsto \exp\Big( -2\pi i \int_{[0,t]\times W} F^*\mathrm{curv}(h)\Big) \cdot \overline{\Gamma}(t). \quad (108)$$

*Identification of the holonomy of $\nabla^h$.* Now we compute the holonomy of this connection. Let $\gamma : [0,1] \to r(C^\infty(W,X))$ be a closed curve, i.e. $\gamma(0) = \gamma(1) = \varphi \in r(C^\infty(W,X))$. Then the lift $\Gamma : [0,1] \to C^\infty(W,X)$ need not be closed. But for any $w \in \partial W$, we have:

$$F(0,w) = \Gamma(0)(w) = \gamma(0)(w) = \varphi(w) = \gamma(1)(w) = F(1,w).$$

Hence $F|_{[0,1]\times\partial W}$ descends to a map $f : S^1 \times \partial W \to X$.

By definition, holonomy along $\gamma$ in the bundle $(\mathcal{L}^h, \nabla^h)$ is the complex number $\mathrm{Hol}^{\nabla^h}(\gamma) \in \mathbb{C}^*$ defined by

$$\mathcal{P}_\gamma^{\nabla^h}(\overline{\Gamma}(0)) = \mathrm{Hol}^{\nabla^h}(\gamma) \cdot \overline{\Gamma}(0)\,.$$

By (104), we may write

$$\overline{\Gamma}(0) = \big(\mathrm{Hol}^h(\Gamma(1)\cup_\varphi - \Gamma(0))\big)^{-1} \cdot [\Gamma(1), 1] = \big(\mathrm{Hol}^h(\Gamma(1)\cup_\varphi - \Gamma(0))\big)^{-1} \cdot \overline{\Gamma}(1).$$

Thus we obtain for the parallel transport along the closed curve $\gamma$:

$$\begin{aligned}
\mathcal{P}_\gamma^{\nabla^h}(\overline{\Gamma}(0)) &= \exp\Big( -2\pi i \int_{[0,1]\times W} F^*\mathrm{curv}(h)\Big) \cdot \overline{\Gamma}(1) \\
&= \exp\Big( -2\pi i \int_{F_*([0,1]\times W)} \mathrm{curv}(h)\Big) \cdot \overline{\Gamma}(1) \\
&= h\big( -\partial F_*([0,1]\times W)\big) \cdot \overline{\Gamma}(1) \\
&= h\big(F_*([0,1]\times \partial W) - F_*(\{1\}\times W \sqcup \{0\} \times \overline{W})\big) \cdot \overline{\Gamma}(1) \\
&= h\big(F_*([0,1]\times\partial W)\big) \cdot h\big(\Gamma(1)_*W - \Gamma(0)_*W)\big)^{-1} \cdot \overline{\Gamma}(1)
\end{aligned}$$

$$
\begin{aligned}
&= f^*h([S^1 \times \partial W]) \cdot \mathrm{Hol}^h(\Gamma(1) \cup_\varphi -\Gamma(0))^{-1} \cdot \overline{\Gamma}(1) \\
&= f^*h([S^1 \times \partial W]) \cdot \mathrm{Hol}^h(\Gamma(1) \cup_\varphi -\Gamma(0))^{-1} \cdot \overline{\Gamma}(1) \\
&= \mathrm{Hol}^h(f) \cdot \overline{\Gamma}(0)\,.
\end{aligned}
$$

Consequently,

$$\mathrm{Hol}^{\nabla^h}(\gamma) = \mathrm{Hol}^h(f) \in \mathrm{U}(1) \subset \mathbb{C}^*\,. \tag{109}$$

Thus the holonomy of $\nabla^h$ along the path $\gamma$ coincides with the higher dimensional holonomy along the map $f : S^1 \times \partial W \to X$.

In particular, we have defined a unitary connection $\nabla^h$ on the line bundle $\mathcal{L}^h \to r(C^\infty(W,X))$ with holonomy given by the holonomy of the differential character $h \in \widehat{H}^k(X;\mathbb{Z})$.

*Computation of the connection 1-form.* The line bundle $\mathcal{L}^h \to r(C^\infty(W,X))$ with unitary connection $\nabla^h$ and section $\mathrm{PT}^h$ along the restriction map $r : C^\infty(W,X) \to C^\infty(\partial W, X)$ yields a relative character $[\mathcal{L}^h, \nabla^h, \mathrm{PT}^h] \in \widehat{H}^2(r;\mathbb{Z})$. To complete the picture of the equivalence class $[\mathcal{L}^h, \nabla^h, \mathrm{PT}^h]$ as a differential character, it remains to compute the 1-form $\mathrm{cov}([\mathcal{L}^h, \nabla^h, \mathrm{PT}^h]) \in \Omega^1(C^\infty(W,X))$. By Example 65, this corresponds to the connection 1-form of $r^*\nabla^h$ with respect to the section $\mathrm{PT}^h$. We now compute this 1-form.

Let $\Gamma : [0,1] \to C^\infty(W,X)$ be a path as above and $\overline{\Gamma}$ the corresponding lift of the path $\gamma = r \circ \Gamma$ to the total space $r^*\mathcal{L}^h$. The connection 1-form $\vartheta^{r^*\nabla^h}$ of $r^*\nabla^h$ is determined by parallel transport along the path $\Gamma$ through the equation

$$\mathcal{P}_\gamma^{r^*\nabla^h} : r^*\mathcal{L}^h_{\gamma(0)} \to r^*\mathcal{L}^h_{\gamma(t)}, \quad \overline{\Gamma}(0) \mapsto \exp\Big( - \int_0^t \vartheta^{r^*\nabla^h}(\overline{\Gamma}')(s) ds \Big) \cdot \overline{\Gamma}(t)\,.$$

Comparing with (108), we obtain

$$
\begin{aligned}
\exp\Big( \int_0^t (\vartheta^{r^*\nabla^h}(\overline{\Gamma}'))(s)\, ds \Big) &= \exp\Big( 2\pi i \int_{[0,t]\times W} F^*\mathrm{curv}(h) \Big) \\
&= \exp\Big( 2\pi i \int_{[0,t]} \fint_W F^*\mathrm{curv}(h) \Big) \\
&= \exp\Big( 2\pi i \int_{[0,t]} \Gamma^*\Big( \fint_W \mathrm{ev}_W^*\mathrm{curv}(h) \Big) \Big). 
\end{aligned} \tag{110}
$$

Here $\mathrm{ev}_W : C^\infty(W,X) \times W \to X$, $(\Phi, w) \mapsto \Phi(w)$, denotes the evaluation map and $\fint_W$ the fiber integration in the trivial bundle $[0,1] \times W \to [0,1]$. Since (110) holds for any $t \in [0,1]$, we have $\vartheta^{r^*\nabla^h}(\overline{\Gamma}')(s) = 2\pi i \cdot \Gamma^*(\fint_W \mathrm{ev}_W^*\mathrm{curv}(h))_s(\frac{\partial}{\partial s})$. This determines the connection 1-form of $r^*\nabla^h$ with respect to the section $\overline{\Gamma}$ along the path $\gamma : [0,1] \to r(C^\infty(W,X))$. By Example 65, we conclude

$$\mathrm{cov}([\mathcal{L}^h, \nabla^h, \mathrm{PT}^h]) = -\int_W \mathrm{ev}_W^* \mathrm{curv}(h) \,. \tag{111}$$

The transgression maps defined in the following sections use fiber integration to generalize the construction of the line bundle with connection $(\mathcal{L}^h, \nabla^h)$ and the section $\mathrm{PT}^h$ along the restriction map.

## *9.2 Higher Dimensional Transgression*

In this section, we define transgression of differential characters of arbitrary degree along oriented closed manifolds. The classical case studied in the literature is transgression along $S^1$ for degree-2 and degree-3 differential cohomology. Our construction generalizes these classical cases to transgression along oriented closed manifolds of arbitrary finite dimension. It turns out that the holonomy defined in Sect. 9.1 is a special case of this transgression.

Transgression along $S^1$ has been discussed in several models of differential cohomology, e.g. for Deligne cohomology in [9] and for bundle gerbes in [66], [67]. Transgression along higher dimensional oriented closed manifolds has been constructed in [38].

Let $\Sigma$ be a compact smooth manifold without boundary, and let $X$ be any smooth manifold. Then the space $C^\infty(\Sigma, X)$ of smooth maps from $\Sigma$ to $X$ is again a smooth space as explained in Sect. 2. The best-known space of this type is the *free loop space* $\mathcal{L}(X) := C^\infty(S^1, X)$ of smooth maps from the circle $S^1$ to $X$.

The evaluation map $\mathrm{ev}_\Sigma$ is defined in the obvious way:

$$\mathrm{ev}_\Sigma : C^\infty(\Sigma, X) \times \Sigma \to X, \quad (\varphi, s) \mapsto \varphi(s).$$

We consider the pull-back $\mathrm{ev}_\Sigma^* : \widehat{H}^k(X;\mathbb{Z}) \to \widehat{H}^k(C^\infty(\Sigma, X) \times \Sigma;\mathbb{Z})$. If $\Sigma$ is oriented, then we can integrate characters in $\widehat{H}^k(C^\infty(\Sigma, X) \times \Sigma;\mathbb{Z})$ over the fiber of the trivial fiber bundle $C^\infty(\Sigma, X) \times \Sigma \xrightarrow{\pi} C^\infty(\Sigma, X)$.

**Definition 79 (Transgression along Closed Manifold).** Let $\Sigma$ be a compact oriented smooth manifold without boundary, and let $X$ be any smooth manifold. *Transgression along* $\Sigma$ is the map

$$\tau_\Sigma : \widehat{H}^*(X;\mathbb{Z}) \to \widehat{H}^{*-\dim \Sigma}(C^\infty(\Sigma, X);\mathbb{Z}) \,, \quad h \mapsto \widehat{\pi}_!(\mathrm{ev}_\Sigma^* h) \,. \tag{112}$$

In particular, for $\Sigma = S^1$ we have

$$\tau_{S^1} : \widehat{H}^*(X;\mathbb{Z}) \to \widehat{H}^{*-1}(\mathcal{L}(X);\mathbb{Z}) \,, \quad h \mapsto \widehat{\pi}_!(\mathrm{ev}_{S^1}^* h) \,.$$

*Example 80.* For $k = 2$, transgression along $\Sigma = S^1$ associates to a U(1)-bundle on $X$ its holonomy map $\mathcal{L}(X) \to \mathrm{U}(1)$.

For $k = 3$, transgression along $\Sigma = S^1$ has been discussed in quantum field theory to construct the anomaly bundle over loop space [35, 9, 20]. In this case, the image of the transgression map has been characterized [66, 67].

*Example 81.* Let $h \in \widehat{H}^k(X;\mathbb{Z})$. Let $\dim \Sigma = k-1$. Transgression along $\Sigma$ yields a differential character $\tau_\Sigma h \in \widehat{H}^1(C^\infty(\Sigma,X);\mathbb{Z})$, which by Example 18 corresponds to a U(1)-valued function on the mapping space $C^\infty(\Sigma,X)$. We verify that this function coincides with the holonomy of $h$ as defined in Sect. 9.1. For any fixed $\varphi \in C^\infty(\Sigma,X)$, we have the pull-back diagram:

$$\begin{array}{ccccc} \{\varphi\}\times\Sigma & \xrightarrow{\imath_\varphi\times\mathrm{id}} & C^\infty(\Sigma,X)\times\Sigma & \xrightarrow{\mathrm{ev}_\Sigma} & X \\ \downarrow{\scriptstyle \widehat{\pi}_!} & & \downarrow{\scriptstyle \widehat{\pi}_!} & & \\ \{\varphi\} & \xrightarrow{\imath_\varphi} & C^\infty(\Sigma,X)\,. & & \end{array}$$

(with the composite top arrow $\widetilde{\varphi}: \{\varphi\}\times\Sigma \to X$)

Thus by naturality of fiber integration, we have:

$$\begin{aligned} \mathrm{Hol}^h(\varphi) &\overset{(103)}{=} (\varphi^*h)([\Sigma]) \\ &= (\widehat{\pi}_!\widetilde{\varphi}^*h) \\ &= (\widehat{\pi}_!(\imath_\varphi\times\mathrm{id})^*\mathrm{ev}_\Sigma^*h)) \\ &\overset{(60)}{=} \imath_\varphi^*\underbrace{(\widehat{\pi}_!\mathrm{ev}_\Sigma^*h)}_{=\tau_\Sigma h} \\ &\overset{(112)}{=} (\tau_\Sigma h)(\varphi)\,. \end{aligned} \tag{113}$$

To evaluate the U(1)-valued function $\tau_\Sigma h$ on the map $\varphi$, we could have used Definition 40 instead of the pull-back diagram. But the argument above can be generalized to compute the holonomy of the character $\tau_\Sigma h$ for transgression of any degree: Let $h \in \widehat{H}^k(X;\mathbb{Z})$, and let $\Sigma_2$ be an oriented closed manifold. Let $\Sigma_1$ be an oriented closed manifold of dimension $\dim(\Sigma_1) = k - \dim(\Sigma_2) - 1$, and let $\varphi : \Sigma_1 \to C^\infty(\Sigma_2,X)$ be a smooth map. By (113) we have:

$$\mathrm{Hol}^{\tau_{\Sigma_2}h}(\varphi) = (\tau_{\Sigma_1}(\tau_{\Sigma_2}h))(\varphi). \tag{114}$$

We generalize this equation, replacing holonomy by transgression: Let $\Sigma_1$ and $\Sigma_2$ be compact oriented smooth manifolds without boundary. The evaluation in the first entry yields a canonical identification

$$\mathrm{ev}_1 : C^\infty(\Sigma_1\times\Sigma_2,X) \xrightarrow{\cong} C^\infty(\Sigma_1,C^\infty(\Sigma_2,X)), \quad f \mapsto (t\mapsto f(t,\cdot)).$$

Using functoriality and naturality of fiber integration, we conclude that higher dimensional transgression is functorial and graded commutative:

**Proposition 82 (Functoriality of Transgression).** *Let $\Sigma_1$ and $\Sigma_2$ be compact oriented smooth manifolds without boundary. Let $h \in \widehat{H}^k(X;\mathbb{Z})$. Then we have:*

$$\tau_{\Sigma_1\times\Sigma_2}h = \mathrm{ev}_1^*(\tau_{\Sigma_1}\circ\tau_{\Sigma_2})h = (-1)^{\dim\Sigma_1\cdot\dim\Sigma_2}\tau_{\Sigma_2\times\Sigma_1}h. \tag{115}$$

*Proof.* The canonical diffeomorphism $\Sigma_1\times\Sigma_2 \xrightarrow{\cong} \Sigma_2\times\Sigma_1$ yields a canonical identification $C^\infty(\Sigma_2\times\Sigma_1,X) \xrightarrow{\cong} C^\infty(\Sigma_1\times\Sigma_2,X)$. The fiber orientation in the trivial fiber bundles with fiber $\Sigma_2\times\Sigma_1$ is $(-1)^{\dim\Sigma_1\cdot\dim\Sigma_2}$ times the one in the bundles with fiber $\Sigma_1\times\Sigma_2$. According to Proposition 49, we obtain $\tau_{\Sigma_1\times\Sigma_2}h = (-1)^{\dim\Sigma_1\cdot\dim\Sigma_2}\tau_{\Sigma_2\times\Sigma_1}$.

The evaluation maps fit into the commutative diagram:

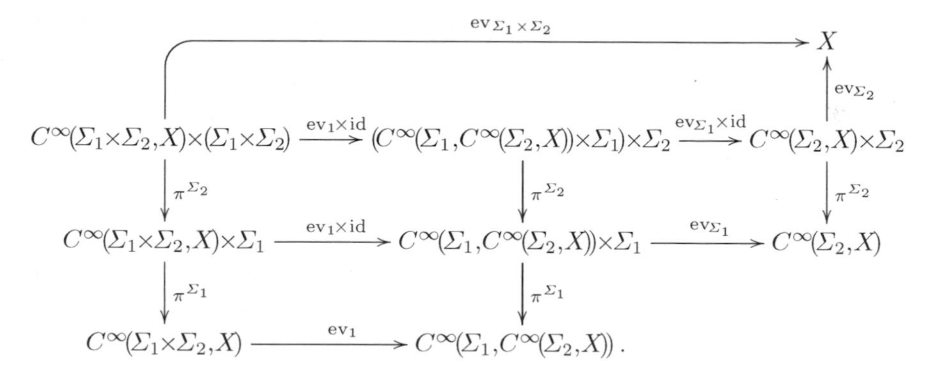


Here $\pi^{\Sigma_i}$ denotes the various projections in trivial bundles with fiber $\Sigma_i$ and $\pi^{\Sigma_1\times\Sigma_2} = \pi^{\Sigma_1}\circ\pi^{\Sigma_2}$ denotes the projection in the trivial bundle with fiber $\Sigma_1\times\Sigma_2$. We decompose $\mathrm{ev}_{\Sigma_1\times\Sigma_2}$ as in the top row of the above diagram:

$$\mathrm{ev}_{\Sigma_1\times\Sigma_2} = \mathrm{ev}_{\Sigma_2}\circ(\mathrm{ev}_{\Sigma_1}\times\mathrm{id}_{\Sigma_2})\circ(\mathrm{ev}_1\times\mathrm{id}_{\Sigma_1\times\Sigma_2}).$$

Using naturality of fiber integration, we obtain:

$$\begin{aligned}
\tau_{\Sigma_1\times\Sigma_2} &= \widehat{\pi}_!^{\Sigma_1\times\Sigma_2}(\mathrm{ev}^*_{\Sigma_1\times\Sigma_2}h)\\
&\overset{(75)}{=} \widehat{\pi}_!^{\Sigma_1}\pi_!^{\Sigma_2}(\mathrm{ev}_1\times\mathrm{id}_{\Sigma_1\times\Sigma_2})^*(\mathrm{ev}_{\Sigma_1}\times\mathrm{id}_{\Sigma_2})^*\mathrm{ev}^*_{\Sigma_2}h\\
&\overset{(60)}{=} \widehat{\pi}_!^{\Sigma_1}(\mathrm{ev}_1\times\mathrm{id}_{\Sigma_1})^*\widehat{\pi}_!^{\Sigma_2}(\mathrm{ev}_{\Sigma_1}\times\mathrm{id}_{\Sigma_2})^*\mathrm{ev}^*_{\Sigma_2}h\\
&\overset{(60)}{=} \mathrm{ev}_1^*\underbrace{(\widehat{\pi}_!^{\Sigma_1}\circ\mathrm{ev}^*_{\Sigma_1})}_{=\tau_{\Sigma_1}}\underbrace{(\widehat{\pi}_!^{\Sigma_2}\circ\mathrm{ev}^*_{\Sigma_2})}_{=\tau_{\Sigma_2}}h\\
&= \mathrm{ev}_1^*(\tau_{\Sigma_1}\circ\tau_{\Sigma_2})h.
\end{aligned}$$

□

Holonomy of differential characters is additive with respect to topological sums (i.e. disjoint union of oriented closed manifolds): for $h \in \widehat{H}^k(X;\mathbb{Z})$ and

$\varphi : \Sigma_1 \sqcup \Sigma_2 \to X$, we have

$$\begin{aligned}\mathrm{Hol}^h(\varphi) &= h(\varphi_*[\Sigma_1 \sqcup \Sigma_2]_{\partial S_k}) \\ &= h(\varphi_{1*}[\Sigma_1]_{\partial S_k} + \varphi_{2*}[\Sigma_2]_{\partial S_k}) \\ &= \mathrm{Hol}^h(\varphi_1) \cdot \mathrm{Hol}^h(\varphi_2).\end{aligned}$$

Here $\varphi_i$ denotes the restriction of $\varphi : \Sigma_1 \sqcup \Sigma_2 \to X$ to $\Sigma_i$ for $i = 1, 2$.

Likewise, transgression along oriented closed manifolds is additive with respect to topological sums: Denote by $r_i : C^\infty(\Sigma_1 \sqcup \Sigma_2, X) \to C^\infty(\Sigma_i; X)$, $\varphi \mapsto \varphi_i$, $i = 1, 2$, the restriction maps. Then we have:

**Proposition 83 (Additivity of Transgression).** *Let $\Sigma_1$ and $\Sigma_2$ be oriented closed manifolds. Let $h \in \widehat{H}^k(X;\mathbb{Z})$. Then we have:*

$$\tau_{\Sigma_1 \sqcup \Sigma_2} h = r_1^*(\tau_{\Sigma_1} h) + r_2^*(\tau_{\Sigma_2} h)\,. \tag{116}$$

*Proof.* For $i = 1, 2$ set:

$$\begin{aligned}E &:= C^\infty(\Sigma_1 \sqcup \Sigma_2, X) \times (\Sigma_1 \sqcup \Sigma_2) \\ D_i &:= C^\infty(\Sigma_1 \sqcup \Sigma_2, X) \times \Sigma_i \\ E_i &:= C^\infty(\Sigma_i, X) \times \Sigma_i\,.\end{aligned}$$

The canonical inclusions $\Sigma_i \hookrightarrow (\Sigma_1 \sqcup \Sigma_2)$ yield inclusions $j_i : D_i \hookrightarrow E$. From the restriction maps $r_i$ and the evaluation maps, we obtain the commutative diagram:

(117)

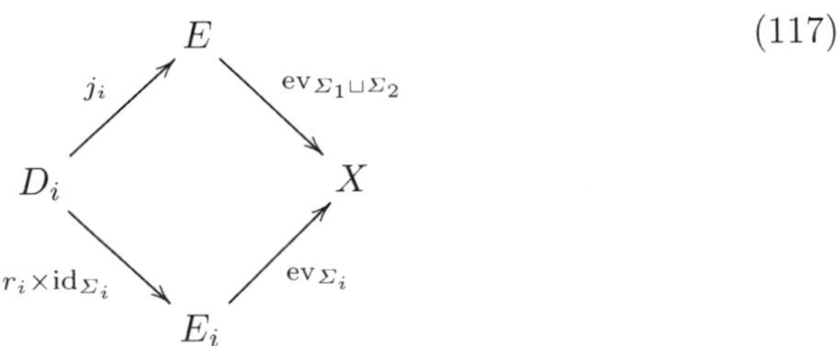


Let $z \in Z_{k-1-\dim \Sigma_i}(C^\infty(\Sigma_1 \sqcup \Sigma_2, X);\mathbb{Z})$ be a singular cycle. Choose the geometric cycle $\zeta(z) \in \mathcal{Z}_{k-1-\dim \Sigma_i}(C^\infty(\Sigma_1 \sqcup \Sigma_2, X))$ and the singular chain $a(z) \in C_{k-\dim \Sigma_i}(C^\infty(\Sigma_1 \sqcup \Sigma_2, X);\mathbb{Z})$ such that $[z - \partial a(z)]_{\partial S_{k-\dim \Sigma_i}} = [\zeta(z)]_{\partial S_{k-\dim \Sigma_i}}$. Moreover, choose $\zeta(r_{i*}z) = r_{i*}\zeta(z)$ and $a(r_{i*}z) = r_{i*}a(z)$. Then we have:

$$[\mathrm{PB}_E \zeta(z)]_{\partial S_k} = j_{1*}[\mathrm{PB}_{D_1}\zeta(z)]_{\partial S_k} + j_{2*}[\mathrm{PB}_{D_2}\zeta(z)]_{\partial S_k}\,.$$

Applying the evaluation map $\mathrm{ev}_{\Sigma_1 \sqcup \Sigma_2}$ and using (117), we obtain:

$$(\mathrm{ev}_{\Sigma_1 \sqcup \Sigma_2})_*[\mathrm{PB}_E \zeta(z)]_{\partial S_k} = (\mathrm{ev}_{\Sigma_1 \sqcup \Sigma_2})_*\Big(\sum_{i=1}^{2} j_{i*}[\mathrm{PB}_{D_i}\zeta(z)]_{\partial S_k}\Big)$$

$$\overset{(117)}{=} \sum_{i=1}^{2} \mathrm{ev}_{\Sigma_i *}((r_i \times \mathrm{id}_{\Sigma_i})_*[\mathrm{PB}_{D_i}\zeta(z)]_{\partial S_k})$$

$$= \sum_{i=1}^{2} \mathrm{ev}_{\Sigma_i *}[\mathrm{PB}_{E_i}\zeta(r_{i*}z)]_{\partial S_k} \,. \tag{118}$$

In the last equality we have used (11) for the pull-back diagram:

$$\begin{array}{ccc} D_i & \xrightarrow{r_i \times \mathrm{id}_{\Sigma_i}} & E_i \\ \downarrow & & \downarrow \\ C^\infty(\Sigma_1 \sqcup \Sigma_2, X) & \xrightarrow[r_i]{} & C^\infty(\Sigma_i, X) \,. \end{array}$$

Now we compute the transgression along $\Sigma_1 \sqcup \Sigma_2$:

$$\begin{aligned}
(\tau_{\Sigma_1 \sqcup \Sigma_2} h)(z) &= (\widehat{\pi}_!(\mathrm{ev}_{\Sigma_1 \sqcup \Sigma_2})^* h)(z) \\
&\overset{(66)}{=} (\mathrm{ev}_{\Sigma_1 \sqcup \Sigma_2})^* h([\mathrm{PB}_E \zeta(z)]_{\partial S_k}) \\
&\qquad \cdot \exp\Big(2\pi i \int_{a(z)} \fint_{\Sigma_1 \sqcup \Sigma_2} \mathrm{curv}(\mathrm{ev}^*_{\Sigma_1 \sqcup \Sigma_2} h)\Big) \\
&= h\Big((\mathrm{ev}_{\Sigma_1 \sqcup \Sigma_2})_*[\mathrm{PB}_E \zeta(z)]_{\partial S_k}\Big) \\
&\qquad \cdot \exp\Big(2\pi i \int_{a(z)} \sum_{i=1}^{2} \fint_{\Sigma_i} \jmath_i^* (\mathrm{ev}_{\Sigma_1 \sqcup \Sigma_2})^* \mathrm{curv}(h)\Big) \\
&\overset{(118),(117)}{=} h\Big(\sum_{i=1}^{2} (\mathrm{ev}_{\Sigma_i})_*[\mathrm{PB}_{E_i}\zeta(r_{i*}z)]_{\partial S_k}\Big) \\
&\qquad \cdot \exp\Big(2\pi i \int_{a(z)} \sum_{i=1}^{2} \fint_{\Sigma_i} (r_i \times \mathrm{id}_{\Sigma_i})^* (\mathrm{ev}_{\Sigma_i})^* \mathrm{curv}(h)\Big) \\
&\overset{(60)}{=} h\Big(\sum_{i=1}^{2} (\mathrm{ev}_{\Sigma_i})_*[\mathrm{PB}_{E_i}\zeta(r_{i*}z)]_{\partial S_k}\Big) \\
&\qquad \cdot \exp\Big(2\pi i \sum_{i=1}^{2} \int_{a(r_{i*}z)} \fint_{\Sigma_i} (\mathrm{ev}_{\Sigma_i})^* \mathrm{curv}(h)\Big) \\
&= \tau_{\Sigma_1} h(r_{1*}z) + \tau_{\Sigma_2} h(r_{2*}z) \\
&= (r_1^* \tau_{\Sigma_1} h + r_2^* \tau_{\Sigma_2} h)(z) \,.
\end{aligned}$$

□

*Example 84.* Let $W$ be a compact oriented closed $(k-1)$-manifold with boundary $\partial W = \Sigma$. In Sect. 9.1, we have constructed a Hermitian line bundle with unitary connection $(\mathcal{L}^h, \nabla^h)$ on $r(C^\infty(W, X))$. By Example 19, this corresponds to a degree-2 differential character on $r(C^\infty(W, X)$. To show that $[\mathcal{L}^h, \nabla^h] = \tau_S h \in \widehat{H}^2((r(C^\infty(W,X));\mathbb{Z})$ it suffices to compare the holonomies, since holonomy classifies line bundles with connection up to isomorphism.[4] Let $\gamma : S^1 \to r(C^\infty(W,X))$ be a closed path, and let $f : S^1 \times \Sigma \to X$ be the induced map as in Sect. 9.1. We then have:

$$\mathrm{Hol}^{\nabla^h}(\gamma) \overset{(109)}{=} \mathrm{Hol}^h(f) \overset{(113)}{=} (\tau_{S^1\times\Sigma}h)(f) \overset{(115)}{=} \tau_{S^1}(\tau_\Sigma h)(\gamma) \overset{(114)}{=} \mathrm{Hol}^{\tau_\Sigma h}(\gamma)\,.$$

Thus $[\mathcal{L}^h, \nabla^h] = \tau_\Sigma h$.

*Remark 85. Transgression and Topological Quantum Field Theories.* Topological quantum field theories in the sense of Atiyah [1] are symmetric monoidal functors from a cobordism category to the category of complex vector spaces. In particular, they associate to topological sums of closed oriented manifolds the tensor products of the vector spaces associated to the summands. Transgression of differential characters has similar functorial properties in the sense that it is additive with respect to topological sums.

Topological quantum field theories associate to an oriented compact manifold with boundary an element in the vector space associated to the boundary. Similarly, transgression along oriented manifolds with boundary yields a section along the restriction map of the differential character obtained by transgression along the boundary. Transgression along manifolds with boundary will be constructed in the following section.

## 9.3 Transgression along Manifolds with Boundary

Let $W$ be a compact oriented smooth manifold with boundary $\partial W$. Restriction to the boundary defines a map $r : C^\infty(W,X) \to C^\infty(\partial W,X)$, $r(\varphi) = \varphi|_{\partial W}$. We consider the trivial bundles

$$\begin{aligned} E &= C^\infty(W,X) \times W \to C^\infty(W,X)\,, \\ \partial E &= C^\infty(W,X) \times \partial W \to C^\infty(W,X) \end{aligned}$$

and the evaluation map

$$\mathrm{ev}_W : C^\infty(W,X) \times W \to X, \quad (\varphi, w) \mapsto \varphi(w)\,.$$

In analogy to the transgression along oriented closed manifolds, we define:

**Definition 86 (Transgression along Manifolds with Boundary).** Let $W$ be a compact oriented smooth manifold with boundary $\partial W$ and let $X$ be a smooth manifold. Fiber integration for fibers with boundary yields the following two *transgression maps* along $W$ and $\partial W$:

[4] Here we do not distinguish notationally between $\tau_\Sigma h$ as differential character on $C^\infty(S,X)$ and its restriction to $r(C^\infty(W,X)) \subset C^\infty(S,X)$.

$$\begin{aligned}\tau^E : \widehat{H}^k(X;\mathbb{Z}) &\to \widehat{H}^{k-\dim W+1}(\mathrm{id}_{C^\infty(W,X)};\mathbb{Z}),\\ h &\mapsto \widehat{\pi}_!^E \mathrm{ev}_W^* h,\\ \tau^{\partial E} : \widehat{H}^k(X;\mathbb{Z}) &\to \widehat{H}^{k-\dim \partial W}(C^\infty(W,X);\mathbb{Z}),\\ h &\mapsto \widehat{\pi}_!^{\partial E} \mathrm{ev}_W^* h.\end{aligned}$$

*Example 87.* We consider the special case $k = 2$ and $W = I = [0,1]$. The space $\mathcal{P}(X) := C^\infty(I,X)$ is called the *path space* of $X$. In this case, the trivial bundle $\partial E = C^\infty(I,X) \times \partial I = \mathcal{P}(X) \times \{0,1\} \to \mathcal{P}(X)$ is a twofold covering. By Example 19, any differential character $h \in \widehat{H}^2(X;\mathbb{Z})$ corresponds to (the isomorphism class of) a U(1)-bundle with connection $(P,\nabla)$ on $X$. Transgression along $\partial I$ yields a U(1)-bundle with connection $\tau^{\partial E}(P,\nabla)$ on the path space $\mathcal{P}(X)$. Its fiber over a path $\gamma \in \mathcal{P}(X)$ is given by $P^*_{\gamma(0)} \otimes P_{\gamma(1)}$. Transgression along $I$ yields a section $\sigma$ of this bundle along the restriction map $r : \mathcal{P}(X) \to C^\infty(\{0,1\},X) = X \times X$, $\gamma \mapsto (\gamma(0),\gamma(1))$. As we have seen in Example 77, $\sigma(\gamma)$ can be chosen to be the parallel transporter along $\gamma \in \mathcal{P}(X)$.

In the following we consider the relations between the three transgression maps $\tau_{\partial W}$, $\tau^{\partial E}$ and $\tau^E$. We first note that $\tau^E h$ is a section of $\tau^{\partial E} h$:

$$\breve{p}(\tau^E h) = \breve{p}(\widehat{\pi}_!^E \mathrm{ev}_W^* h) \overset{(102)}{=} \widehat{\pi}_!^{\partial E} \mathrm{ev}_W^* h = \tau^{\partial E} h. \tag{119}$$

In particular we have:

$$\mathrm{curv}(\tau^E h) = \mathrm{curv}(\tau_{\partial E} h) \overset{(62)}{=} \fint_{\partial W} \mathrm{ev}_W^* \mathrm{curv}(h). \tag{120}$$

For the covariant derivative of $\tau^E h \in \widehat{H}^{k-\dim W}(\mathrm{id}_{C^\infty(W,X)};\mathbb{Z})$, we have:

$$\mathrm{cov}(\tau^E h) \overset{(101)}{=} (-1)^{k-\dim W} \fint_W \mathrm{ev}_W^* \mathrm{curv}(h). \tag{121}$$

We note further that $\tau^{\partial E}$ is not the same as $\tau_{\partial W}$ defined in Sect. 9.2 (with $\Sigma = \partial W$) since the former takes values in differential characters on $C^\infty(W,X)$ rather than on $C^\infty(\partial W,X)$. But they are related by the restriction map $r : C^\infty(W,X) \to C^\infty(\partial W,X)$, $\varphi \mapsto \varphi|_{\partial W}$. We have the pull-back diagram:

$$\begin{array}{ccc} \partial E = C^\infty(W,X) \times \partial W & \xrightarrow{R} & C^\infty(\partial W,X) \times \partial W \\ \downarrow & & \downarrow \\ C^\infty(W,X) & \xrightarrow{r} & C^\infty(\partial W,X). \end{array} \tag{122}$$

By (60), fiber integration is natural with respect to pull-back along smooth maps, hence $\widehat{\pi}_!^{\partial E} \circ R^* = r^* \circ \widehat{\pi}_!^E$. This yields

$$\tau^{\partial E} h = \widehat{\pi}_!^{\partial E}(\mathrm{ev}_W^* h) = \widehat{\pi}_!^{\partial E}(R^* \mathrm{ev}_{\partial W}^* h) = r^* \widehat{\pi}_!(\mathrm{ev}_{\partial W}^* h) = r^*(\tau_{\partial W} h). \quad (123)$$

Thus the three transgression maps fit into the following commutative diagram:

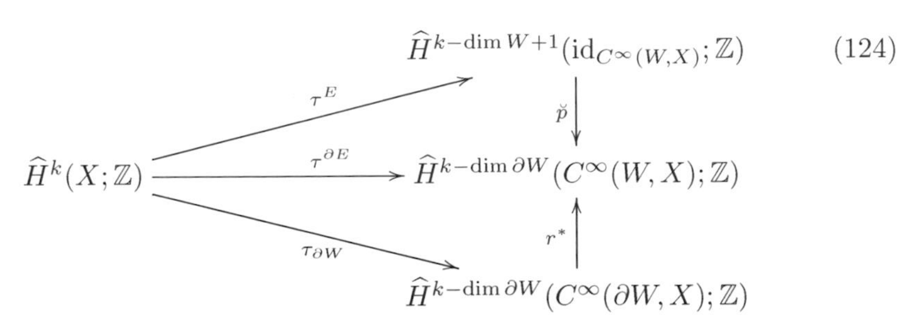


In particular, $\tau_{\partial W} h \in \widehat{H}^{k-\dim \partial W}(C^\infty(\partial W, X);\mathbb{Z})$ is topologically trivial along $r$: for the pull-back along $r$ of the characteristic class, we find:

$$r^* c(\tau_{\partial W} h) = c(r^* \tau_{\partial W} h) \overset{(123)}{=} c(\tau^{\partial E} h) \overset{(119)}{=} c(\breve{p}(\tau^E h)) \overset{(93)}{=} 0.$$

By Corollary 72, we conclude that $\tau_{\partial W} h$ has sections along the restriction map. Thus there exist relative characters $f \in \widehat{H}^{k-\dim \partial W}(r;\mathbb{Z})$ with $\breve{p}(f) = \tau_{\partial W} h$. It would be nice to extend the transgression maps to a construction of such a section. In some cases, it is possible to presribe its covariant derivative. In more special cases, this uniquely determines the section.

*Sections for $\tau_{\partial W} h$ with prescribed covariant derivative.* We want to construct a section $f$ of the character $\tau_{\partial W} h$ along the restriction map $r$ with prescribed covariant derivative. Assume that $r$ induces the trivial map $r^* : H^{k-\dim \partial W-1}(C^\infty(\partial W, X);\mathrm{U}(1)) \to H^{k-\dim \partial W-1}(C^\infty(W, X);\mathrm{U}(1))$. This holds for instance if $W = I$ and $X$ is connected, since in this case the path space $C^\infty(W, X) = \mathcal{P}(X)$ is contractible. We start with a pair of forms $(\mathrm{curv}(\tau_{\partial W} h), \chi) \in \Omega_0^{k-\dim \partial W}(r)$ in the mapping cone de Rham complex. Since the map $(\mathrm{curv}, \mathrm{cov}) : \widehat{H}^{k-\dim \partial W}(\mathrm{id}_{C^\infty(\partial W,X)};\mathbb{Z}) \to \Omega_0^{k-\dim \partial W}(r)$ is surjective, we find a relative differential character $f_0 \in H^{k-\dim \partial W}(r;\mathbb{Z})$ with $(\mathrm{curv}(f_0), \mathrm{cov}(f_0)) = (\mathrm{curv}(\tau_{\partial W} h), \chi)$. Now take any section $f_1$ of $\tau_{\partial W} h$. Since $\mathrm{curv}(f_0) = \mathrm{curv}(\tau_{\partial W} h) = \mathrm{curv}(f_1)$, we have $\breve{p}(f_1) - \breve{p}(f_0) = j(u)$ for some $u \in H^{k-\dim \partial W-1}(C^\infty(\partial W, X);\mathrm{U}(1))$. By the mapping cone sequence for cohomology with $\mathrm{U}(1)$-coefficients and the assumption on the restriction map, we find $\bar{u} \in H^{k-\dim \partial W-1}(\mathrm{id}_{C^\infty(\partial W,X)};\mathrm{U}(1)) = \{0\}$ which maps to $u$. Now put $f := f_0 + j(\bar{u})$. Then we have $\breve{p}(f) = \breve{p}(f_0) + j(u) = \breve{p}(f_1) = \tau_{\partial W} h$. Moreover, $\mathrm{cov}(f) = \mathrm{cov}(f_0) = \chi$. Thus we have found a section $f$ of $\tau_{\partial W} h$ along the restriction map $r$ with prescribed covariant derivative $\mathrm{cov}(f) = \chi$. By Corollary 72, section $f$ is uniquely determined by its covariant derivative $\chi$, if the map $r^* : H^{k-\dim \partial W-1}(C^\infty(\partial W, X); \mathrm{U}(1)) \to H^{k-\dim \partial W-1}(C^\infty(W, X); \mathrm{U}(1))$ is surjective. Together with the above assumption for $r_*$ to be the trivial map, we thus obtain the requirement that $H^{k-\dim \partial W-1}(C^\infty(W, X); \mathrm{U}(1)) = \{0\}$. We have proved:

**Corollary 88 (Transgression w. Prescribed Covariant Derivative I).** *Let $X$ be a smooth manifold, and let $h \in \widehat{H}^k(X; \mathbb{Z})$. Let $W$ be an oriented manifold with boundary. Assume $H^{k-\dim \partial W-1}(C^\infty(W, X); \mathrm{U}(1)) = \{0\}$. Let $\chi \in \Omega^{k-\dim \partial W-1}(C^\infty(W, X))$ be a differential form such that $(\mathrm{curv}(\tau_{\partial W} h), \chi) \in \Omega_0^{k-\dim \partial W}(r)$. Then the transgression maps $\tau_{\partial W}$, $\tau^{\partial E}$ and $\tau^E$ defined in Sections 9.2 and 9.3 uniquely determine a relative differential character $\tau^\chi_{W,\partial W} h \in \widehat{H}^{k-\dim \partial W}(r; \mathbb{Z})$ satisfying*

$$\begin{aligned} \breve{p}(\tau^\chi_{W,\partial W} h) &= \tau_{\partial W} h \\ \mathrm{cov}(\tau^\chi_{W,\partial W} h) &= \chi \,. \end{aligned}$$

A distinguished form $\chi \in \Omega^{k-\dim \partial W-1}(C^\infty(W, X))$ is obtained by integrating $\mathrm{ev}_W^* \mathrm{curv}(h)$ over the fiber of the trivial bundle $C^\infty(W, X) \times W \to C^\infty(W, X)$. In the remainder of this section we take this form as prescribed covariant derivative:

*Sections for $\tau_{\partial W} h$ with covariant derivative determined by transgression.* Transgression along $W$ yields the form $\mathrm{cov}(\tau^E h) \in \Omega^{k-\dim \partial W-1}(C^\infty(W, X))$ as a natural candidate for the covariant derivative of a section $\tau_{\partial W} h$ along the restriction map. The pair $(\mathrm{curv}(\tau_{\partial W} h), \mathrm{cov}(\tau^E h))$ is $d_r$-closed since

$$r^* \mathrm{curv}(\tau_{\partial W} h) = \mathrm{curv}(r^* \tau_{\partial W} h) \overset{(123)}{=} \mathrm{curv}(\tau^{\partial E} h) \overset{(119)}{=} d\mathrm{cov}(\tau^E h) \,.$$

By Corollary 76 we thus have

$$(\mathrm{curv}(\tau_{\partial W} h), \mathrm{cov}(\tau^E h)) \in \Omega_0^{k-\dim \partial W}(r; \mathbb{Z}) \,.$$

This yields:

**Corollary 89 (Transgression w. Prescribed Covariant Derivative II).** *Let $X$ be a smooth manifold, and let $h \in \widehat{H}^k(X; \mathbb{Z})$. Let $W$ be an oriented manifold with boundary. Assume $H^{k-\dim \partial W-1}(C^\infty(W, X); \mathrm{U}(1)) = \{0\}$. Then the transgression maps $\tau_{\partial W}$, $\tau^{\partial E}$ and $\tau^E$ defined in Sections 9.2 and 9.3 uniquely determine a relative differential character $\tau_{W,\partial W} h \in \widehat{H}^{k-\dim \partial W}(r; \mathbb{Z})$ satisfying*

$$\begin{aligned} \breve{p}(\tau_{W,\partial W} h) &= \tau_{\partial W} h \\ \mathrm{cov}(\tau_{W,\partial W} h) &= \mathrm{cov}(\tau^E h) \overset{(121)}{=} (-1)^{k-\dim W} \fint_W \mathrm{ev}_W^* \mathrm{curv}(h) \,. \end{aligned}$$

*Example 90.* Let $k = \dim W$. In this case, the assumption on $r^*$ is automatically satisfied. Given a differential character $h \in \widehat{H}^k(X;\mathbb{Z})$, we obtain the relative character $\tau_{W,\partial W} h \in \widehat{H}^1(r;\mathbb{Z})$, in other words a U(1)-valued function $\tau_{\partial W} h = \mathrm{Hol}^h$ on $C^\infty(\partial W, X)$ together with a real-valued function $\mathrm{cov}(\tau^E h)$ on $C^\infty(W,X)$. The condition $\breve{p}(\tau_{W,\partial W}) = \tau_{\partial W} h$ says that $\mathrm{Hol}^h \circ r = \exp \circ 2\pi i \cdot \mathrm{cov}(\tau^E h)$. In the special case $k = 2$, this is the well-known fact that the holonomy along a contractible loop is given by the integral of the curvature over a spanning disk.

*Example 91.* Let $k = \dim W + 1$. In Sect. 9.1 we have constructed a Hermitian line bundle with connection $(\mathcal{L}^h, \nabla^h)$ on $r(C^\infty(W,X))$ together with a section $\mathrm{PT}^h$ along the restriction map $r$. By Example 65, this determines a relative differential character $[\mathcal{L}^h, \nabla^h, \mathrm{PT}^h] \in \widehat{H}^2(r;\mathbb{Z})$. By Example 84, we have $\breve{p}([\mathcal{L}^h, \nabla^h, \mathrm{PT}^h]) = [\mathcal{L}^h, \nabla^h] = \tau_{\partial W} h$. Moreover, by (111), we have $\mathrm{cov}([\mathcal{L}^h, \nabla^h, \mathrm{PT}^h]) = -\fint_W \mathrm{ev}_W^* \mathrm{curv}(h) = \mathrm{cov}(\tau^E h)$. Under the assumption of Corollary 89, we conclude $[\mathcal{L}^h, \nabla^h, \mathrm{PT}^h] = \tau_{W,\partial W} h$.

## 9.4 Chain Field Theories

Topological quantum field theories in the sense of Atiyah [1] are symmetric monoidal functors from a cobordism category to the category of complex vector spaces. This concept of topological field theories has been modified in several directions, e.g. by replacing the source or target category.

Chain field theories in the sense of [64] are a modification of topological field theories where the source category is replaced by a category with smooth cycles as objects and chains as morphisms. Chain field theories are closely related to differential characters. Using the notion of thin chains, we generalize [17, Thm. 3.5] from 2-dimensional thin invariant field theories to chain field theories of arbitrary dimension: chain field theories are invariant under thin 2-morphisms.

We briefly recall the notion of chain field theories: The objects of the category $\mathtt{Chain}^{n+1}(X)$ are smooth singular $n$-cycles in $X$. A morphism from $x$ to $x'$ is an $(n+1)$-chain $a$ such that $\partial a = x' - x$. Taking the additive group structure of $Z_n(X;\mathbb{Z})$ and $C_{n+1}(X;\mathbb{Z})$ as the tensor product turns $\mathtt{Chain}^{n+1}(X)$ into a strict monoidal category, more precisely a strict symmetric monoidal groupoid, see [64, Prop. 1.1].

Let $z \in Z_{n+1}(X;\mathbb{Z})$ and let $x$ be any object in the category $\mathtt{Chain}^{n+1}(X)$. Then we have $\partial z = 0 = x - x$. This yields a 1-1 correspondence of the automorphism group of any object of $\mathtt{Chain}^{n+1}(X)$ with the group $Z_{n+1}(X;\mathbb{Z})$ of smooth singular $(n+1)$-cycles in $X$.

Let $a, a' \in C_{n+1}(X;\mathbb{Z})$ and $x, x' \in Z_n(X;\mathbb{Z})$ with $\partial a = \partial a' = x' - x$. Then the chains $a, a'$ yield morphisms $x \xrightarrow{a} x'$, $x \xrightarrow{a'} x'$ in $\mathtt{Chain}^{n+1}(X)$ between the same objects $x, x'$. A chain $b \in C_{n+2}(X;\mathbb{Z})$ satisfying $\partial b = a' - a$ is called a 2-*morphism* from the morphism $x \xrightarrow{a} x'$ to the morphism $x \xrightarrow{a'} x'$.[5] We write $a \overset{b}{\Rightarrow} a'$ for a 2-morphism from $x \xrightarrow{a} x'$ to $x \xrightarrow{a'} x'$. If $b \in C_{n+2}(X;\mathbb{Z})$ is thin in the sense of Definition 4, i.e. $b \in S_{n+2}(X;\mathbb{Z})$, then we call $a \overset{b}{\Rightarrow} a'$ a *thin 2-morphism.*

Denote by $\mathbb{C}$-`Lines` the category whose objects are Hermitian lines and whose morphisms are isometries. A *chain field theory* on $X$ is defined to be a functor of symmetric monoidal tensor categories $E : \mathtt{Chain}^{n+1}(X) \to \mathbb{C}\text{-}\mathtt{Lines}$ with an additional smoothness condition. To formulate this condition, note that $E$ maps the automorphism group $Z_{n+1}(X;\mathbb{Z})$ of the monoidal unit 0 of $\mathtt{Chain}^{n+1}(X)$ to the automorphism group $\mathrm{U}(1)$ of the monoidal unit $\mathbb{C}$ of $\mathbb{C}$-`Lines`. Hence we obtain a homomorphism $Z_{n+1}(X;\mathbb{Z}) \to \mathrm{U}(1)$. The smoothness condition for the functor $E$ is the requirement that there exists a closed differential form $\omega \in \Omega^{n+2}(X)$ such that for any chain $b \in C_{n+2}(X;\mathbb{Z})$, we have

$$E(0 \xrightarrow{\partial b} 0) = \exp\Big(2\pi i \int_b \omega\Big) \in \mathrm{U}(1)\,. \tag{125}$$

Thus a chain field theory $E$ induces a homomorphism $Z_{n+1}(X;\mathbb{Z}) \to \mathrm{U}(1)$, $z \mapsto E(z)(1)$. By the smoothness condition (125), this yields a differential character in $\widehat{H}^{n+1}(X;\mathbb{Z})$ with curvature $\omega$. Moreover, chain field theories are classified up to equivalence by the differential characters obtained in this manner, see [64, Thm. 2.1].

For any Hermitian line $L$, the group of isometric automorphisms of $L$ is canonically identified with $\mathrm{U}(1)$. Thus let $E$ be a chain field theory, $x \in Z_n(X;\mathbb{Z})$ an object, and $z \in Z_{n+1}(X;\mathbb{Z})$ an automorphism of $x$. Then the isometry $E(x \xrightarrow{z} x)$ of the Hermitian line $E(x)$ is given as

$$E(x \xrightarrow{z} x) = \big(E(0 \xrightarrow{z} 0)(1)\big) \cdot \mathrm{id}_{E(x)}\,. \tag{126}$$

By [17, p. 434], chain field theories in the sense of [64] generalize thin invariant field theories in the sense of [17]. By [17, Thm. 3.5], thin invariant field theories are invariant under thin cobordism of morphisms. In the context of chain field theories, we obtain the analogous result:

**Proposition 92 (Thin Invariance).** *Chain field theories are invariant under thin 2-morphisms: Let* $E : \mathtt{Chain}^{n+1}(X) \to \mathbb{C}\text{-}\mathtt{Lines}$ *be a chain field theory. Let* $x, x' \in Z_n(X;\mathbb{Z})$ *be objects and* $x \xrightarrow{a} x'$, $x \xrightarrow{a'} x'$ *morphisms in* $\mathtt{Chain}^{n+1}(X)$. *Let* $b \in S_{k+2}(X;\mathbb{Z})$ *with* $\partial b = a' - a$ *and* $a \overset{b}{\Rightarrow} a'$ *the corresponding thin 2-morphism. Then we have*

$$E(x \xrightarrow{a} x') = E(x \xrightarrow{a'} x').$$

[5] In [64, p. 91], this is called a chain deformation.

*Proof.* The composition of the morphism $x \xrightarrow{a'} x'$ with the inverse of $x \xrightarrow{a} x'$ yields an automorphism of $x'$. For the corresponding automorphism of $E(x')$, we have:

$$\begin{aligned} E(x \xrightarrow{a'} x') \circ (E(x \xrightarrow{a} x'))^{-1} &= E((x \xrightarrow{a'} x') \circ (x' \xrightarrow{-a} x)) \\ &= E(x' \xrightarrow{a'-a} x') \\ &\overset{(126)}{=} (E(0 \xrightarrow{a'-a} 0)(1)) \cdot \mathrm{id}_{x'} \\ &\overset{(125)}{=} \exp\Big(2\pi i \underbrace{\int_b \omega}_{=0}\Big) \cdot \mathrm{id}_{x'} \\ &= \mathrm{id}_{x'} \,. \end{aligned}$$

Thus $E(x \xrightarrow{a'} x') = E(x \xrightarrow{a} x')$. □

## References


1. M. Atiyah: *Topological quantum field theories.* Publ. Math. Inst. Hautes Etudes Sci. **68** (1989), 285–299
2. J. W. Barrett: *Holonomy and path structures in general relativity and Yang-Mills theory.* Internat. J. Theoret. Phys. **30** (1991), 1171–1215
3. C. Becker: *Part II: Relative differential cohomology*, this volume
4. C. Becker: *Cheeger-Chern-Simons theory and differential String classes.* arXiv:1404.0716, 2014
5. S. Bloch: *The dilogarithm and extensions of Lie algebras.* Algebraic K-theory, Lecture Notes in Math. **854** Springer, Berlin 1981, 1–23
6. A. Borel, F. Hirzebruch: *Characteristic classes and homogeneous spaces I.* Amer. J. Math. **80** (1958), 458–538
7. R. Bott, L.W. Tu: *Differential Forms in Algebraic Topology.* Springer 1982
8. M. Bright, P. Turner: *Relative differential characters.* Comm. Anal. Geom. **14** (2006), 269–282
9. J.-L. Brylinski: *Loop Spaces, Characteristic Classes and Geometric Quantization.* Progress in Mathematics, Birkhäuser, Boston 1993
10. U. Bunke: *Differential cohomology.* arXiv:1208.3961, 2012
11. U. Bunke, D. Gepner: *Differential function spectra, the differential Becker-Gottlieb transfer, and applications to differential algebraic K-theory* arXiv:1306.0247, 2013
12. U. Bunke, M. Kreck, T. Schick: *A geometric description of differential cohomology.* Ann. Math. Blaise Pascal **17** (2010), 1–16
13. U. Bunke, T. Nikolaus, M. Völkl: *Differential cohomology theories as sheaves of spectra.* arXiv:1311.3188, 2013
14. U. Bunke, T. Schick: *Smooth K-theory.* Astérisque **328** (2009), 43–135
15. U. Bunke, T. Schick: *Uniqueness of smooth extensions of generalized cohomology theories.* J. Topol. **3** (2010), 110–156
16. U. Bunke, G. Tamme: *Multiplicative differential algebraic K-theory and applications.* arXiv:1311.1421, 2013
17. U. Bunke, P. Turner, S. Willerton: *Gerbes and homotopy quantum field theories* Algebr. Geom. Topol. **4** (2004), 407–437

18. A. Caetano, R. F. Picken: *An axiomatic definition of holonomy.* Internat. J. Math. **5** (1994), 835–848
19. A. L. Carey, S. Johnson, M. K. Murray: *Holonomy on D-branes.* J. Geom. Phys. **52** (2004), 186–216
20. A. L. Carey, M. K. Murray: *Faddeev's anomaly and bundle gerbes.* Lett. Math. Phys. **37** (1996), 29–36
21. A. L. Carey, S. Johnson, M. K. Murray, D. Stevenson, B.-L. Wang: *Bundle gerbes for Chern-Simons and Wess-Zumino-Witten theories.* Comm. Math. Phys. **259** (2005), 577–613
22. A. L. Carey, J. Mickelsson, M. K. Murray: *Bundle gerbes applied to quantum field theory.* Rev. Math. Phys. **12** (2000), 65–90
23. J. Cheeger: *Multiplication of differential characters.* Symposia Mathematica, Vol. XI, 441–445. Academic Press, London 1973
24. J. Cheeger, J. Simons: *Differential Characters and Geometric Invariants.* Geometry and topology, Lecture Notes in Math. **1167**, Springer, Berlin 1985, 50–80
25. S.Chern: *On the characteristic classes of complex sphere bundles and algebraic varieties.* Amer. J. Math. **75** (1953), 565–597
26. A. Dold: *Lectures on algebraic topology.* Springer, Berlin 1995
27. J.L. Dupont, R. Ljungmann: *Integration of simplicial forms and Deligne cohomology.* Math. Scand. **97** (2005), 11–39
28. C.-O. Ewald: *A de Rham isomorphism in singular cohomology and Stokes theorem for stratifolds.* Int. J. Geom. Methods Mod. Phys. **2** (2005), 63–81
29. D. Freed: *Classical Chern-Simons theory. I.* Adv. Math. **113** (1995), 237–303
30. D. Freed: *Classical Chern-Simons theory. II.* Houston J. Math. **28** (2002), 293–310
31. D. Freed, M. Hopkins: *On Ramond-Ramond fields and K-theory.* J. High Energy Phys. **5** (2000), paper no. 05(2000)044, 14 p.
32. D. Freed, J. Lott: *An index theorem in differential K-theory.* Geom. Topol. **14** (2010), 903–966
33. J. Fuchs, T. Nikolaus, C. Schweigert, K. Waldorf: *Bundle gerbes and surface holonomy.* European Congress of Mathematics, 167–195, Eur. Math. Soc., Zürich 2010
34. P. Gajer: *Higher holonomies, geometric loop groups and smooth Deligne cohomology.* in: Advances in geometry, 195–235, Progr. Math. **172**, Birkhäuser, Boston 1999
35. K. Gawȩdzki: *Topological actions in two-dimensional quantum field theories.* In: *Non-perturbative quantum field theory*, 101–141, NATO Adv. Sci. Inst. Ser. B Phys. **185**, Plenum, New York 1988
36. K. Gawȩdzki, N. Reis: *WZW branes and gerbes.* Rev. Math. Phys. **14** (2002), 1281–1334
37. K. Gomi, Y. Terashima: *A fiber integration formula for the smooth Deligne cohomology.* Internat. Math. Res. Notices **13** (2000), 699–708
38. K. Gomi, Y. Terashima: *Higher-dimensional parallel transports.* Math. Res. Lett. **8** (2001), 25–33
39. W. Greub, S. Halperin, R. Vanstone: *Connections, curvature, and cohomology. Vol. I: De Rham cohomology of manifolds and vector bundles.* Academic Press, New York-London 1972
40. R. Harvey, B. Lawson: *Lefschetz-Pontrjagin duality for differential characters.* An. Acad. Brasil. Ciênc. **73** (2001), 145–159
41. R. Harvey, B. Lawson: *From sparks to grundles – differential characters.* Comm. Anal. Geom. **14** (2006), 25–58
42. R. Harvey, B. Lawson: *D-bar sparks.* Proc. Lond. Math. Soc. **97** (2008), 1–30
43. R. Harvey, B. Lawson, J. Zweck: *The de Rham-Federer theory of differential characters and character duality.* Amer. J. Math. **125** (2003), 791–847
44. N. Hitchin: *Lectures on special Lagrangian submanifolds.* In: Winter School on Mirror Symmetry, Vector Bundles and Lagrangian Submanifolds, 151–182, Amer. Math. Soc., Providence 2001

45. M. J. Hopkins, I. M. Singer: *Quadratic functions in geometry, topology, and M-theory.* J. Diff. Geom. **70** (2005), 329–452
46. P. Iglesias-Zemmour: *Diffeology.* Amer. Math. Soc., Providence 2013
47. M. Kreck: *Differentiable Algebraic Topology. From Stratifolds to Exotic Spheres.* Amer. Math. Soc., Providence 2010
48. A. Kriegl, P. W. Michor: *The Convenient Setting of Global Analysis.* Amer. Math. Soc., Providence 1997
49. R. Ljungmann: *Secondary invariants for families of bundles.* PhD thesis, Aarhus 2006
50. R. Mąka, P. Urbański: *Differential forms on differential spaces.* Demonstratio Math. **27** (1994), 99–108
51. M.K. Murray: *Bundle gerbes.* J. London Math. Soc. **54** (1996), 403–416
52. M.K. Murray, D. Stevenson: *Bundle gerbes: stable isomorphism and local theory.* J. London Math. Soc. **62** (2000), 925–937
53. M.K. Murray, D. Stevenson: *Higgs field, bundle gerbes and String structures.* Comm. Math. Phys. **243** (2003), 541–555
54. J. McCleary: *A user's guide to spectral sequences.* 2. ed., Cambridge University Press, Cambridge 2001
55. V. V. Prasolov: *Elements of Homology Theory.* Amer. Math. Soc., Providence 2007
56. T.R. Ramadas, I.M. Singer, J. Weitsman: *Some comments on Chern-Simons gauge theory.* Comm. Math. Phys. **126** (1989), 409–420
57. C. Redden: *Trivializations of differential cocycles.* `arXiv:1201.2919`, 2012, to appear in J. Homotopy Relat. Struct.
58. U. Schreiber, C. Schweigert, K. Waldorf: *Unoriented WZW models and holonomy of bundle gerbes.* Comm. Math. Phys. **274** (2007), 31–64
59. J. P. Serre: *Homologie singulière des espaces fibrés.* Ann. Math. **54** (1951), 425–505
60. J. Simons, D. Sullivan: *Axiomatic characterization of ordinary differential cohomology.* J. Topol. **1** (2008), 45–56
61. J. Simons, D. Sullivan: *Structured vector bundles define differential $K$-theory.* in: Quanta of maths, 579–599, Amer. Math. Soc., Providence 2010
62. R. Sikorski: *Differential modules.* Colloq. Math. **24** (1972), 45–79
63. D. Stevenson: *Bundle 2-gerbes.* Proc. London Math. Soc. **88** (2004), 405–435
64. P. Turner: *A functorial approach to differential characters.* Algebr. Geom. Topol. **4** (2004), 81–93
65. M. Upmeier: *Products in Generalized Differential Cohomology.* `arXiv:1112.4173v2`, 2011
66. K. Waldorf: *Transgression to Loop Spaces and its Inverse, I: Diffeological Bundles and Fusion Maps.* Cah. Topol. Géom. Différ. Catég. **53** (2012), 162–210
67. K. Waldorf: *Transgression to Loop Spaces and its Inverse, II: Gerbes and Fusion Bundles with Connection.* `arXiv:1004.0031`, 2010
68. K. Waldorf: *String connections and Chern-Simons theory.* Trans. Amer. Math. Soc. **365** (2013), 4393–4432

# Relative differential cohomology

Christian Becker

**Abstract** We study two notions of relative differential cohomology, using the model of differential characters. The two notions arise from the two options to construct relative homology, either by cycles of a quotient complex or of a mapping cone complex. We discuss the relation of the two notions of relative differential cohomology to each other. We discuss long exact sequences for both notions, thereby clarifying their relation to absolute differential cohomology. We construct the external and internal product of relative and absolute characters and show that relative differential cohomology is a right module over the absolute differential cohomology ring. Finally we construct fiber integration and transgression for relative differential characters.

## 1 Introduction

Differential cohomology is a refinement of integral cohomology by differential forms. The first model for differential cohomology is the graded group $\widehat{H}^*(X;\mathbb{Z})$ of differential characters, defined by J. Cheeger and J. Simons in [16]. Differential characters of degree $k$ are certain homomorphisms $h : Z_{k-1}(X;\mathbb{Z}) \to \mathrm{U}(1)$ on the abelian group of smooth singular $(k-1)$-cycles on $X$. By now there exist lots of different models for differential cohomology, formulated in terms of smooth Deligne (hyper-)cohomology [8], [9], [14], de Rham-Feder currents [21], [24], [22], [23], differential cocycles [26], [11] or simplicial forms [17]. In low degrees there exist more special models like gerbes, Hitchin gerbes [25], and bundle gerbes [32] for $k = 3$ and bundle 2-gerbes [38] for $k = 4$. Axiomatic definitions have been provided in [36] and [1] for differential refinements of ordinary cohomology and in [12], [13] for

Christian Becker
Universität Potsdam, Institut für Mathematik, Am Neuen Palais 10, 14469 Potsdam, Germany, e-mail: becker@math.uni-potsdam.de

differential refinements of generalized cohomology theories. Constructions of generalized differential cohomology theories have appeared in [26] and [9], [10]. As a particular case of generalized cohomology, there are several models of differential $K$-theory [30], [20], [12], [37]. Most of these treatments do not cover relative differential cohomology.

In analogy to the case for absolute cohomology, we may define relative differential cohomology as a refinement of relative integral cohomology by differential forms. Relative differential cohomology groups have been considered in several contexts like differential characters [21], [7], [45], differential cocyles [45], [7], Čech cocycles [45]. They have also appeared in more special models like relative gerbes [34], and for differential extensions of generalized cohomology theories [42]. Relative algebraic differential characters have been studied in [4]. Relative differential cohomology groups are closely related to trivializations of differential cohomology as considered in [44], [33]. Physical applications of relative differential cohomology groups have been sketched in [27] and [35]. Applications to Chern-Simons theory are discussed in [44] and [2].

It seems that a systematic discussion of notions and models for relative differential cohomology including e.g. uniqueness, module structures, long exact sequences etc. is still missing. In the present paper we treat the case of relative differential (ordinary) cohomology. As in the preceding paper [1] we work with the group $\widehat{H}^*(X;\mathbb{Z})$ of differential characters as a model for differential cohomology. The definition and elementary properties of differential characters are easily transferred from absolute to relative cohomology.

We consider the following situation: let $X$ be a smooth manifold and $i_A : A \to X$ the embedding of a smooth submanifold. There are two ways to define the relative singular homology: either as the homology of the (smooth singular) mapping cone complex $C_*(i_A;\mathbb{Z})$ or as the homology of the quotient complex $C_*(X,A;\mathbb{Z}) := C_*(X;\mathbb{Z})/\mathrm{im}(i_{A*})$. There arise two different notions of relative cycles and hence two notions of relative differential characters.

The first option was treated in [7] and will be reviewed in Sect. 3.1 below. The characters on $Z_{k-1}(i_A;\mathbb{Z})$ thus obtained are called *relative differential characters*. We adopt this notion, although it would also be appropriate to call them *mapping cone characters*. We denote the corresponding group of relative differential characters by $\widehat{H}^k(i_A;\mathbb{Z})$. It is a differential refinement of the relative cohomology $H^k(i_A;\mathbb{Z}) \cong H^k(X,A;\mathbb{Z})$. In fact, the notion of relative differential characters is established in [7] not just for embeddings but for any smooth maps $\varphi : A \to X$. This way one obtains a differential refinement $\widehat{H}^k(\varphi;\mathbb{Z})$ of the mapping cone cohomology $H^k(\varphi;\mathbb{Z})$. These characters are treated also in [1, Chap. 8], where we derive a long exact sequences that relates the groups of relative and absolute differential characters. As a particularly nice example, in Sect. 3.2 we show that a bundle gerbe with connection $\mathcal{G}$ over $X$, represented by a submersion $\pi : Y \to X$, represents a relative differential character $h_{\mathcal{G}} \in \widehat{H}^3(\pi;\mathbb{Z})$. In this case, the covariant derivative of the character $h_{\mathcal{G}}$ coincides with the curving of the bundle gerbe $\mathcal{G}$.

The second option has appeared in [21] for the special case of the inclusion of the boundary $i_{\partial M} : \partial M \to M$ of a smooth manifold with boundary. We treat this version of relative differential cohomology as groups of characters on $Z_{k-1}(X, A; \mathbb{Z})$ in detail in Sect. 3.3. Here $A \subset X$ is an arbitrary smooth submanifold. We denote the corresponding group of differential characters by $\widehat{H}^k(X, A; \mathbb{Z})$. It yields another differential refinement of the relative cohomology $H^k(X, A; \mathbb{Z})$. We show that the group $\widehat{H}^k(X, A; \mathbb{Z})$ corresponds to the subgroup of *parallel* characters in $\widehat{H}^k(i_A; \mathbb{Z})$. In this sense, $\widehat{H}^k(i_A; \mathbb{Z})$ is finer as a refinement of $H^k(X, A; \mathbb{Z})$ than $\widehat{H}^k(X, A; \mathbb{Z})$. We derive a long exact sequence that relates the group $\widehat{H}^k(X, A; \mathbb{Z})$ to absolute differential cohomology groups on $X$ and $A$.

We clarifiy the relation of the two notions of relative differential cohomology above to another notion that has appeared in the literature, namely the relative Hopkins-Singer groups $\check{H}^k(\varphi; \mathbb{Z})$ for a smooth map $\varphi : A \to X$ and $\check{H}^k(i_A; \mathbb{Z})$ for the embedding $i_A : A \to X$ of a smooth submanifold. These groups have been constructed in [7]. It is shown there that $\check{H}^k(\varphi; \mathbb{Z})$ is a subquotient of $\widehat{H}^k(\varphi; \mathbb{Z})$. In Sect. 3.4 we show that $\widehat{H}^k(X, A; \mathbb{Z})$ is a subgroup of $\check{H}^k(i_A; \mathbb{Z})$.

In Sect. 4 we discuss internal and external products in differential cohomology. The internal product and ring structure on $\widehat{H}^*(X; \mathbb{Z})$ was constructed first in [16]. Uniqueness of the ring structure is proved in [36] and [1]. The proof in [1, Chap. 6] starts from an axiomatic definition of internal and external products and ends up with a new formula for the latter. In that sense it is constructive. In the present paper we give a new proof of the key lemma in the uniqueness proof from [1, Chap. 6]. This new proof starts from the original definition in [16] and ends up with the formula in [1, Chap. 6]. Further, we use the methods from [1, Chap. 6] to construct a product of absolute and relative differential characters. This provides the graded group $\widehat{H}^*(\varphi; \mathbb{Z})$ of relative differential characters with the structure of a module over the ring $\widehat{H}^*(X; \mathbb{Z})$ of absolute differential characters. The module structure is natural with respect to pull-back and the structure maps (curvature, covariant derivative, characteristic class and topological trivializations) are multiplicative.

Last but not least, in Sect. 5 we construct fiber integration of relative differential characters and transgression maps as we did for absolute differential characters in [1, Chap. 7–9]. To some extent thus, the present work is a "relativization" of the results obtained in [1] on the absolute differential cohomology ring $\widehat{H}^*(X; \mathbb{Z})$ of a smooth manifold $X$. In fact, the "relativization" is a generalization of those results from absolute to relative differential cohomology. The results for absolute differential cohomology are reproduced as a special case. We show that fiber integration in fiber products is compatible with cross products of characters, and we derive the up-down formula from this. We generalize in two ways a result from [1] on integration over fibers that bound: For integration of relative differential characters over fibers that bound, we find topological trivializations of the integrated characters as in [1]. For integration of absolute characters in fiber bundles that bound along a smooth map into the base – a notion we introduce in Sect. 5.2 – we show that the integrated characters admit sections along that map with covariant derivatives prescribed by fiber integration.

The methods in [1] use representations of smooth homology classes by certain geometric cycles, namely Kreck's stratifolds [28]. In the present paper we need to adapt these representations to mapping cone cohomology classes. This is done in Sect. 2 below. It provides the necessary prerequisites from relative (or mapping cone) stratifold homology that are needed in the rest of the paper.

**Acknowledgements** It is a great pleasure to thank Christian Bär and Matthias Kreck for very helpful discussions. Moreover, the author thanks *Sonderforschungsbereich 647* funded by *Deutsche Forschungsgemeinschaft* for financial support.

# 2 Stratifold Homology

In this chapter we construct relative stratifold homology as a geometric homology isomorphic to mapping cone homology of a smooth map $\varphi : A \to X$. We first discuss the concept of thin chains from [1]. This yields the notion of refined fundamental classes of closed oriented smooth manifolds or stratifolds. We introduce the notion of geometric relative cycle and their refined fundamental classes. We prove that the bordism theory of relative stratifolds in $(X, A)$ is isomorphic to the smooth singular mapping cone homology of a smooth map $\varphi : A \to X$. Finally, we adapt the construction of the pull-back operation of geometric cycles and of the transfer map for smooth singular cycles in the base of a fiber bundle $\pi : E \to X$ with closed oriented fibers from [1, Chap. 3] to mapping cone homology.

## *2.1 Thin Chains*

We briefly recall the concept of thin chains defined in [1, Chap. 3] and certain equivalence relations on singular chains and cycles, respectively. Let $C_k(X; \mathbb{Z})$ denote the abelian group of smooth singular $k$-chains in a smooth manifold $X$. Let $Z_k(X; \mathbb{Z})$ and $B_k(X; \mathbb{Z})$ denote the subgroups of smooth singular $k$-cycles and $k$-boundaries, respectively. A *thin* $k$-chain is a chain $s \in C_k(X; \mathbb{Z})$ such that for every $k$-form $\omega \in \Omega^k(X)$, we have $\int_s \omega = 0$. This happens for instance if $s$ is supported in a $(k-1)$-dimensional submanifold. Thin chains are preserved by the boundary operator and thus form a subcomplex $S_*(X;\mathbb{Z}) \subset C_*(X;\mathbb{Z})$.

### 2.1.1 Degenerate Chains

Let $\Delta^k := \{\sum_{i=0}^k t_i e_i \mid \sum_{i=0}^k t_i = 1\} \subset R^{k+1}$ be the standard $k$-simplex. Let $l_j : \Delta^{k+1} \to \Delta^k$ be the $j$-th degeneracy map. A smooth singular $(k+1)$-simplex $\sigma : \Delta^{k+1} \to X$ is called *degenerate*, it if is of the form $\sigma = \sigma' \circ l_j$ for some $k$-simplex $\sigma'$ and $j \in \{0, \dots, k\}$. Let $D_k(X;\mathbb{Z}) \subset C_k(X;\mathbb{Z})$ be the submodule generated by degenerate simplexes. Elements of $D_*(X;\mathbb{Z})$ are called *degenerate chains*. It is easy to see that degenerate chains are preserved by the boundary operator $\partial$ of the singular chain complex. Thus $(D_*(X;\mathbb{Z}), \partial)$ is a subcomplex which is well-known to have vanishing homology [41]. Hence the homology of the quotient complex $C_*(X;\mathbb{Z})/D_*(X;\mathbb{Z})$ is canonically isomorphic to the smooth singular homology $H_*(X;\mathbb{Z})$.

Degenerate chains are special examples of thin chains, i.e. $D_*(X;\mathbb{Z}) \subset S_*(X;\mathbb{Z})$, since differential forms vanish upon pull-back by degeneracy maps. In particular, any degenerate cycle $z \in Z_*(X;\mathbb{Z}) \cap D_*(X;\mathbb{Z})$ is the boundary of a thin chain: since $[z] = 0 \in H_*(D_*(X;\mathbb{Z}))$, we find a chain $c \in D_*(X;\mathbb{Z}) \subset S_*(X;\mathbb{Z})$ such that $\partial c = z$. This might not be the case for arbitrary thin cycles.

### 2.1.2 The Mapping Cone Complex

Let $\varphi : A \to X$ be a smooth map. We will denote by $C_k(\varphi;\mathbb{Z}) := C_k(X;\mathbb{Z}) \times C_{k-1}(A;\mathbb{Z})$ the group of $k$-chains in the *mapping cone complex* of $\varphi$. The differential $\partial_\varphi : C_k(\varphi;\mathbb{Z}) \to C_{k-1}(\varphi;\mathbb{Z})$ of the mapping cone complex is defined as $\partial_\varphi(s,t) := (\partial s + \varphi_* t, -\partial t)$. We denote by $Z_k(\varphi;\mathbb{Z})$ and $B_k(\varphi;\mathbb{Z})$ the $k$-cycles and $k$-boundaries of this complex. Moreover, set $S_k(\varphi;\mathbb{Z}) := S_k(X;\mathbb{Z}) \times S_{k-1}(A;\mathbb{Z})$ for the space of thin chains in the mapping cone complex.

The homology of the mapping cone complex is denoted by $H_*(\varphi;\mathbb{Z})$. The short exact sequence of chain complexes

$$0 \longrightarrow C_*(X;\mathbb{Z}) \longrightarrow C_*(\varphi;\mathbb{Z}) \longrightarrow C_{*-1}(A;\mathbb{Z}) \longrightarrow 0$$

induces a long exact sequence of homology groups:

$$\dots \longrightarrow H_*(X;\mathbb{Z}) \longrightarrow H_*(\varphi;\mathbb{Z}) \longrightarrow H_{*-1}(A;\mathbb{Z}) \xrightarrow{\varphi_*} H_{*-1}(X;\mathbb{Z}) \longrightarrow \dots .$$

The connecting homomorphism coincides with the map on homology induced by $\varphi$.

The mapping cone cochain complex $(C^*(\varphi;\mathbb{Z}),\delta_\varphi)$ associated with the cochain map $\varphi^* : C^*(X;\mathbb{Z}) \to C^*(A;\mathbb{Z})$ coincides with the dual complex to $(C_*(\varphi;\mathbb{Z}),\partial_\varphi)$. The cohomology of this complex is denoted by $H^*(\varphi;\mathbb{Z})$ and will be referred to as the *mapping cone cohomology.* We obtain the corresponding long exact sequence:

$$\ldots \longrightarrow H^{*-1}(A;\mathbb{Z}) \longrightarrow H^*(\varphi;\mathbb{Z}) \longrightarrow H^*(X;\mathbb{Z}) \xrightarrow{\varphi^*} H^*(A;\mathbb{Z}) \longrightarrow \ldots .$$

In case $\varphi = i_A : A \hookrightarrow X$ is the inclusion of a subset, we have a natural chain map $q : C_*(i_A;\mathbb{Z}) \to C_*(X,A;\mathbb{Z})$, $(v,w) \mapsto v + \mathrm{im}(i_{A*})$. Here $C_k(X,A;\mathbb{Z}) := C_k(X;\mathbb{Z})/i_{A*}(C_k(A;\mathbb{Z}))$ is the *relative chain complex.* The long exact sequences together with the five lemma provide identifications $H_*(\varphi;\mathbb{Z}) \cong H_*(X,A;\mathbb{Z})$ and $H^*(\varphi;\mathbb{Z}) \cong H^*(X,A;\mathbb{Z})$.

Let $(\Omega^*(\varphi),d_\varphi)$ be the relative or mapping cone de Rham complex as defined in [6, p. 78]. Thus $\Omega^k(\varphi) := \Omega^k(X) \times \Omega^{k-1}(A)$ with the differential $d_\varphi(\omega,\vartheta) := (d\omega, \varphi^*\omega - d\vartheta)$. We denote the cohomology of this complex by $H^*_{\mathrm{dR}}(\varphi)$ and call it the *mapping cone de Rham cohomology.*

Integration of a pair $(\omega,\vartheta) \in \Omega^k(\varphi)$ over a chain $(a,b) \in C_k(\varphi;\mathbb{Z})$ is defined in the obvious manner:

$$\int_{(a,b)} (\omega,\vartheta) := \int_a \omega + \int_b \vartheta .$$

Thus pairs of differential forms $(\omega,\varphi) \in \Omega^k(\varphi)$ can be considered as differential cochains in $C^k(\varphi;\mathbb{R})$. Moreover, by the mapping cone Stokes theorem

$$\begin{aligned}\int_{\partial_\varphi(a,b)} (\omega,\vartheta) &= \int_{(\partial a + \varphi_* b, -\partial b)} (\omega,\vartheta) \\ &= \int_{(a,b)} (d\omega, \varphi^*\omega - d\vartheta) \\ &= \int_{(a,b)} d_\varphi(\omega,\vartheta) \end{aligned} \tag{1}$$

the inclusion $(\Omega^*(\varphi);d_\varphi) \hookrightarrow (C^*(\varphi;\mathbb{R}),\delta_\varphi)$ is a chain map.

The short exact sequence of de Rham complexes

$$0 \longrightarrow \Omega^{*-1}(A) \longrightarrow \Omega^*(\varphi) \longrightarrow \Omega^*(X) \longrightarrow 0$$

gives rise to the long exact sequence

$$\ldots \longrightarrow H^{k-1}_{\mathrm{dR}}(A) \longrightarrow H^k_{\mathrm{dR}}(\varphi) \longrightarrow H^k_{\mathrm{dR}}(X) \longrightarrow H^k_{\mathrm{dR}}(A) \longrightarrow \ldots$$

in de Rham cohomology. The de Rham theorem together with the five Lemma yields the identification $H^*_{\mathrm{dR}}(\varphi) \cong H^*(\varphi;\mathbb{R})$.

### 2.1.3 Equivalence Classes

Let $c \in C_k(X;\mathbb{Z})$ be a smooth singular $k$-chain in $X$. We consider its equivalence class modulo thin chains, i.e. its image in $C_k(X;\mathbb{Z})/S_k(X;\mathbb{Z})$. We denote this class by $[c]_{S_k}$. Similarly, for a smooth singular $k$-cycle $z \in Z_k(X;\mathbb{Z})$ we consider its equivalence class modulo boundaries of thin chains, i.e. the image in the quotient $Z_k(X;\mathbb{Z})/\partial S_{k+1}(X;\mathbb{Z})$. We denote this class by $[z]_{\partial S_{k+1}}$. Finally, for a cycle $(s,t) \in Z_k(\varphi;\mathbb{Z})$ of the mapping cone complex we consider its equivalence class $[s,t]_{\partial_\varphi S_{k+1}} \in Z_k(\varphi;\mathbb{Z})/\partial S_{k+1}(\varphi;\mathbb{Z})$ modulo boundaries of thin chains.

These equivalence classes show up rather naturally when considering fundamental classes of oriented closed manifolds or $p$-stratifolds, as shall be explained in the next section.

Moreover, by definition of thin chains, integration of differential forms over smooth singular cycles descends to the equivalence classes modulo boundaries of thin chains. Thus we have well-defined integration maps

$$\begin{aligned} Z_k(\varphi;\mathbb{Z})/\partial_\varphi S_{k+1}(\varphi;\mathbb{Z}) \times \Omega^k(\varphi) &\to \mathbb{R}\,, \\ ([s,t]_{\partial_\varphi S_{k+1}}, (\omega,\vartheta)) &\mapsto \int_{[s,t]_{\partial_\varphi S_{k+1}}} (\omega,\vartheta) := \int_{(s,t)} (\omega,\vartheta)\,, \end{aligned}$$

and similary for absolute cycles and differential forms, see [1, Chap. 3].

## *2.2 Refined Fundamental Classes*

Let $M$ be a closed oriented $k$-dimensional smooth manifold. Triangulation yields a fundamental cycle $z \in Z_k(M;\mathbb{Z})$. Any two such cycles differ by a boundary $\partial a \in B_k(M;\mathbb{Z})$. For dimensional reasons, we have $C_{k+1}(M;\mathbb{Z}) = S_{k+1}(M;\mathbb{Z})$. Thus the fundamental class of $M$ may be regarded as equivalence class in $Z_k(M;\mathbb{Z})/\partial S_{k+1}(M;\mathbb{Z})$. We denote this class by $[M]_{\partial S_{k+1}}$. A smooth map $f : M \to X$ yields an induced class $f_*[M]_{\partial S_{k+1}} \in Z_k(X;\mathbb{Z})/\partial S_{k+1}(X;\mathbb{Z})$. We refer to $[M]_{\partial S_{k+1}}$ (resp. $f_*[M]_{\partial S_{k+1}}$) as the *refined fundamental class* of $M$ (in $X$).

Now let $M$ be a compact oriented smooth $k$-dimensional manifold with boundary and $i_{\partial M} : \partial M \to M$ the inclusion of the boundary. By triangulation we obtain a smooth singular chain $x \in C_k(M;\mathbb{Z})$ together with a smooth singular cycle $y \in Z_{k-1}(\partial M;\mathbb{Z})$ such that $\partial x = -y$. Thus the pair $(x,y) \in C_k(M;\mathbb{Z}) \times C_{k-1}(\partial M;\mathbb{Z})$ is a cycle in $Z_k(i_{\partial M};\mathbb{Z})$. Moreover, $y$ is a fundamental cycle of $\partial M$.

For any two such chains $x, x' \in C_k(M;\mathbb{Z})$, obtained from triangulations of $M$, we find a chain $a \in C_{k+1}(M;\mathbb{Z}) = S_{k+1}(M;\mathbb{Z})$ such that $x - x' - \partial a$ is supported in $\partial M$. Since $M$ is supposed to be $k$-dimensional, we have $x - x' - \partial a =: b \in C_k(\partial M;\mathbb{Z}) = S_k(\partial M;\mathbb{Z})$. Thus $M$ comes together with a well-defined equivalence class in $C_k(M;\mathbb{Z})/S_k(M;\mathbb{Z})$. We denote this class by $[M]_{S_k}$.

We may also collect the data on $\partial M$ into the equivalence class: any two pairs $(x,y)$ and $(x',y')$, obtained as above from triangulations, differ by the relative boundary $\partial(a,b) = (\partial a + b, -\partial b)$ of a pair $(a,b) \in C_{k+1}(M;\mathbb{Z}) \times C_k(\partial M;\mathbb{Z}) = C_{k+1}(i_{\partial M};\mathbb{Z})$. For dimensional reasons, we then have $C_{k+1}(\varphi;\mathbb{Z}) = S_{k+1}(\varphi;\mathbb{Z})$. Thus the pair $(M,\partial M)$ comes together with a well-defined equivalence class in $Z_k(i_{\partial M};\mathbb{Z})/\partial S_{k+1}(i_{\partial M};\mathbb{Z})$. We denote this class by $[M,\partial M]_{\partial S_{k+1}}$.

A commutative diagram of smooth maps

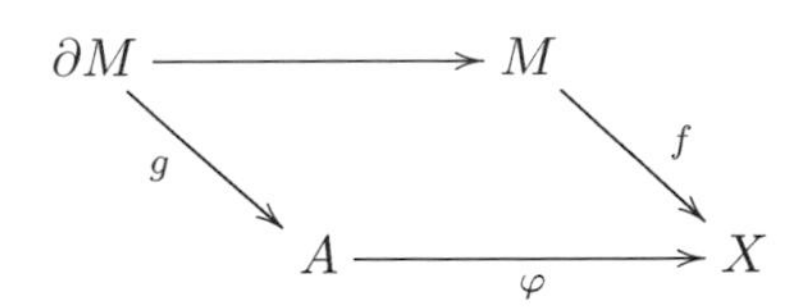


yields an induced class $(f,g)_*[M,\partial M]_{\partial S_{k+1}} \in Z_k(\varphi;\mathbb{Z})/\partial_\varphi S_{k+1}(\varphi;\mathbb{Z})$. We refer to $[M,\partial M]_{\partial S_{k+1}}$ (resp. $(f,g)_*[M,\partial M]_{\partial S_{k+1}}$) as the *refined fundamental class* of $(M,\partial M)$ (in $(X,A)$).

Restriction of the triangulation to the boundary maps the refined fundamental class $[M,\partial M]_{\partial S_{k+1}} \in Z_k(i_{\partial M};\mathbb{Z})/\partial S_{k+1}(i_{\partial M};\mathbb{Z})$ of $(M,\partial M)$ to the refined fundamental class $[\partial M]_{\partial S_k} \in Z_{k-1}(\partial M;\mathbb{Z})/\partial S_k(\partial M;\mathbb{Z})$ of the boundary (and similarly for the classes in $(X,A)$).

*Remark 1.* In the same way as explained for a closed oriented smooth manifold, we can associate to a closed oriented $p$-stratifold $M$ a refined fundamental class $[M]_{\partial S_{k+1}} \in Z_k(M;\mathbb{Z})/\partial S_{k+1}(M;\mathbb{Z})$ obtained from a triangulation of the top-dimensional stratum. Similarly, to a compact oriented $p$-stratifold $M$ with closed boundary $N = \partial M$ we associate an equivalence class $[M,N]_{\partial S_{k+1}} \in Z_k(i_{\partial M};\mathbb{Z})/\partial S_{k+1}(i_{\partial M};\mathbb{Z})$. Restriction to the boundary maps the refined fundamental class $[M,N]_{\partial S_{k+1}}$ to the refined fundamental class $[\partial M]_{\partial S_k}$ of the boundary.

## 2.3 Geometric Cycles

We use a notion of geometric cycles similar to the one in [1, Chap. 4]. In contrast to the cycles and boundaries approach formulated there, in the present work we use the original construction from [28] of geometric or stratifold homology $\mathcal{H}_*(X)$ as the bordism theory of stratifolds in $X$.

The concept of geometric cycles is motivated by the aim to represent singular homology classes in a smooth manifold $X$ by smooth submanifolds. By work of Thom [40] this is not allways possible: in general, there are homology classes not representable as fundamental classes of submanifolds, see e.g. [3].

Replacing smooth manifolds by certain types of singular manifolds, it is possible to represent all homology classes by certain geometric spaces. This is achieved either by pseudomanifolds in the sense of Baas and Sullivan, or by stratifolds in the sense of Kreck [28]. Here we use stratifolds, as we did in [1].

Let $n \in \mathbb{N}_0$. The abelian semigroup $\mathcal{Z}_k(X)$ of *geometric $k$-cycles* is the set of smooth maps $f : M \to X$, where $M$ is a $k$-dimensional oriented compact $p$-stratifold without boundary (see [28, pp. 35 and 43] for the definition of stratifolds). The semigroup structure is defined by disjoint union. For $n < 0$ put $\mathcal{Z}_n(X) := \{0\}$. For a smooth map $f : X \to Y$ we define $f_* : \mathcal{Z}_n(X) \to \mathcal{Z}_n(Y)$ by concatenation, i.e. $f_*(M \xrightarrow{g} X) := M \xrightarrow{f\circ g} Y$.

For an oriented stratifold $S$ we denote by $\overline{S}$ the same stratifold $S$ with reversed orientation. A bordism between geometric $k$-cycles $f : S \to X$ and $f' : S' \to X$ in $X$ is a smooth map $F : W \to X$ from a $(k+1)$-dimensional compact oriented $p$-stratifold with boundary $\partial W = \overline{S} \sqcup S'$ such that $F|_S = f$ and $F|_{S'} = f'$. Geometric $k$-cycles $S \xrightarrow{f} X$ and $S' \xrightarrow{f'} X$ are called *bordant* if there exists a bordism between them. This defines an equivalence relation on $\mathcal{Z}_k(X)$. For transience note that stratifolds with boundary can be glued along their boundary.

The bordism class of a geometric $k$-cycle $S \xrightarrow{f} X$ is denoted by $[S \xrightarrow{f} X]$. The *$k$-th stratifold homology of* $X$ is the set of bordism classes

$$\mathcal{H}_k(X) := \{[S \xrightarrow{f} X] \,|\, f \in \mathcal{Z}_k(X)\}$$

of geometric $k$-cycles in $X$. Orientation reversal defines an involution $\overline{\phantom{x}}$ on $\mathcal{Z}_k(X)$ which maps $S \xrightarrow{f} X$ to $\overline{S} \xrightarrow{f} X$. The involution is compatible with the bordism relation and thus yields a well-defined involution on $\mathcal{H}_k(X)$. The stratifold homology $\mathcal{H}_k(X)$ is an abelian group where the inverse of $[S \xrightarrow{f} X]$ is given by $[\overline{S} \xrightarrow{f} X]$. A null-bordism of $S \sqcup \overline{S} \xrightarrow{f\sqcup f} X$ is given by $S \times [0,1] \xrightarrow{F} X$, $F(s,t) := f(s)$.

The stratifold homology group $\mathcal{H}_k(X)$ can equivalently be defined as the quotient of the semigroup $\mathcal{Z}_k(X)$ of geometric cycles by a sub semigroup of geometric boundaries, as in [1, Chap. 3]. In the present work we use the bordism theory formulation, since this seems more suitable for generalization to relative homology.

A $k$-dimensional closed $p$-stratifold $M$ has a fundamental class $[M] \in H_k(M;\mathbb{Z})$, see [28, p. 186]. More precisely, triangulation of the top dimensional stratum yields a smooth singular cycle in $Z_k(M;\mathbb{Z})$. Any two such cycles differ by the boundary of a smooth singular chain in $C_{k+1}(M;\mathbb{Z}) = S_{k+1}(M;\mathbb{Z})$. Thus we have a well-defined equivalence class $[M]_{\partial S_{k+1}} \in Z_k(M;\mathbb{Z})/\partial S_{k+1}(M;\mathbb{Z})$. As in the previous section, we call it the *refined fundamental class* of $M$. Composition with smooth maps yields a well-defined semigroup homomorphism $\mathcal{Z}_k(X) \to Z_k(X;\mathbb{Z})/\partial S_{k+1}(X;\mathbb{Z})$ mapping the geometric cycle $M \xrightarrow{f} X$ to the equivalence class $f_*[M]_{\partial S_{k+1}}$. It descends to a group isomorphism $\mathcal{H}_k(X) \to H_k(X;\mathbb{Z})$, $[M \xrightarrow{f} X] \mapsto [f_*[M]_{\partial S_{k+1}}]$, see [28, p. 186].

Differential forms in $\Omega^*(X)$ can be pulled back to a stratifold $S$ along a smooth map $f : S \to X$. Integration of differential forms over (refined fundamental classes of) compact oriented stratifolds is well-defined and the Stokes theorem holds [18]. For a geometric cycle $\zeta \in \mathcal{Z}_k(X)$, represented by $S \xrightarrow{f} X$, and a differential form $\omega \in \Omega^k(X)$, we write:

$$\int_{[\zeta]_{\partial S_{k+1}}} \omega = \int_{[S]_{\partial S_{k+1}}} f^*\omega = \int_S f^*\omega \,.$$

## 2.4 Relative Stratifold Homology

In this section we introduce relative stratifold homology by adapting the well-known definition of relative bordism groups to stratifolds. More precisely we modify the classical notion in order to represent the mapping cone homology $H_*(\varphi;\mathbb{Z})$ of a smooth map as a bordism theory of stratifolds. The standard construction yields a long exact sequence that relates the absolute and relative stratifold homology groups.

### 2.4.1 Geometric Relative Cycles

Let $k \geq 0$. Let $S$ be a $k$-dimensional compact oriented regular $p$-stratifold with boundary $\partial S = \overline{T}$, and let $\partial T = 0$. By a smooth map $(S,T) \xrightarrow{(f,g)} (X,A)$ we understand a pair of smooth maps $f : S \to X$ and $g : T \to A$ such that the diagram

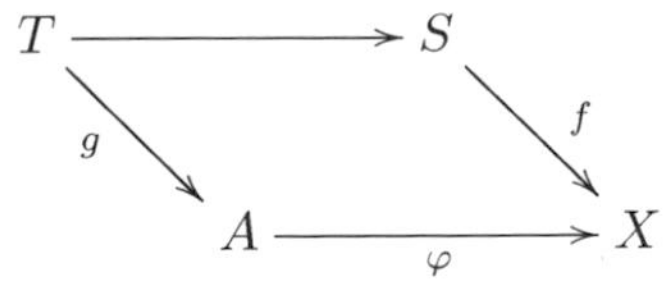


commutes. We define the abelian semigroup

$$\mathcal{Z}_k(\varphi) := \{(S,T) \xrightarrow{(f,g)} (X,A) \,|\, \partial S = \overline{T}, \partial T = \emptyset\}$$

of *geometric relative cycles*, where the semigroup structure is given by disjoint union. For $k = 0$, we have $\mathcal{Z}_0(\varphi) = \mathcal{Z}_0(X)$.

A *bordism* from $(S_0,T_0) \xrightarrow{(f_0,g_0)} (X,A)$ to $(S_1,T_1) \xrightarrow{(f_1,g_1)} (X,A)$ is a smooth map $(W,M) \xrightarrow{(F,G)} (X,A)$ with the following properties:

- $W$ is a $(k+1)$-dimensional compact oriented regular $p$-stratifold with boundary $\partial W$ and $M$ is a $k$-dimensional compact oriented regular $p$-stratifold with boundary $\partial M$.
- The boundary $\partial W$ is the union of compact oriented stratifolds (diffeomorphic to) $\overline{S_0}$, $S_1$ and $\overline{M}$. Moreover, $\partial M = \partial S_1 \sqcup \partial \overline{S_0}$ and $S_i \cap M = \partial S_i$ for $i = 0, 1$.
- On the boundary components $S_i$, $i = 0, 1$, we have $F|_{S_i} = f_i$. Similarly, on $T_i$ we have $G|_{T_i} = g_i$.

Geometric relative $k$-cycles in $(X, A)$ are called *bordant* if there exists a bordism between them. This defines an equivalence relation on $\mathcal{Z}_k(\varphi)$. For transience of the relation, we note that bordisms can be glued along parts of the boundary, as explained in [28, Sec. A.2]. The bordism class of a geometric relative $k$-cycle $(f, g) \in \mathcal{Z}_k(\varphi)$ is denoted by $[f, g]$.

For a pair of compact oriented $p$-stratifolds $(S, T)$ as above, we denote by $(\overline{S}, \overline{T})$ the same stratifolds with reversed orientation. Orientation reversal defines an involution on geometric relative cycles which maps $(S, T) \xrightarrow{(f,g)} (X, A)$ to $(\overline{S}, \overline{T}) \xrightarrow{(f,g)} (X, A)$. The involution is compatible with the bordism relation in the sense that orientation reversal on bordisms mapping $(W, M) \xrightarrow{(F,G)} (X, A)$ to $(\overline{W}, \overline{M}) \xrightarrow{(F,G)} (X, A)$ induces orientation reversal on the cycles related by the bordisms. In other words, we have a well-defined involution on bordism classes $\overline{\phantom{x}} : [f, g] \mapsto \overline{[f, g]} := [\overline{(f, g)}]$.

We define the *$k$-th relative stratifold homology group* $\mathcal{H}_k(\varphi)$ as the set of all bordism classes $[f, g]$ of geometric relative $k$-cycles $(f, g) \in \mathcal{Z}_k(\varphi)$. Given a geometric relative $k$-cycle $(S, T) \xrightarrow{(f,g)} (X, A)$, we find a null-bordism $(W, M) \xrightarrow{(F,G)} (X, A)$ of $(f, g) \sqcup \overline{(f, g)}$ by setting $W := S \times [0, 1]$, $M := \partial S \times [0, 1]$ and $F(s, t) := f(s)$, $G(s, t) := g(s)$. The semigroup structure on $\mathcal{Z}_k(\varphi)$ thus yields an abelian group structure on $\mathcal{H}_k(\varphi)$, where the inverse of a bordism class $[f, g]$ is given by $\overline{[f, g]}$. In the following we will write $[f, g] + [f', g']$ instead of $[(f, g) \sqcup (f', g')]$ and correspondingly $-[f, g]$ instead of $\overline{[f, g]}$.

### 2.4.2 Long Exact Sequence

Any smooth map $\varphi : A \to X$ induces a semigroup homomorphism $\varphi_* : \mathcal{Z}_k(A) \to \mathcal{Z}_k(X)$, $f \mapsto \varphi \circ f$. Moreover, we have the canonical semigroup homomorphisms $i : \mathcal{Z}_k(X) \to \mathcal{Z}_k(\varphi)$, $f \mapsto (f, \emptyset)$, and $p : \mathcal{Z}_k(\varphi) \to \mathcal{Z}_{k-1}(A)$, $(f, g) \mapsto g$. To simplify notation, we write geometric cycles in $X$ as $\zeta \in \mathcal{Z}_k(X)$ and cycles relative to $\varphi$ as pairs $(\zeta, \tau) \in \mathcal{Z}_k(\varphi)$. The bordism class of a geometric relative $k$-cycle $(\zeta, \tau) \in \mathcal{Z}_k(\varphi)$ is denoted by $[\zeta, \tau]$. Instead of the empty map $\emptyset$ we write $0$ for the neutral element in the semigroup $\mathcal{Z}_{k-1}(A)$ (and similarly for the other semigroups of geometric cycles). The maps just defined now read $\varphi_* : \zeta \mapsto \varphi_* \zeta$ and $i : \zeta \mapsto (\zeta, 0)$ and $p : (\zeta, \tau) \mapsto \tau$.

These semigroup homomorphisms are bordism invariant and hence descend to group homomorphisms on stratifold homology. They fit into the following long exact sequence:

**Proposition 2 (Long Exact Sequence).** *Let $\varphi : A \to X$ be a smooth map and $k \geq 0$. Then we have the following exact sequence relating absolute and relative stratifold homology groups:*

$$\ldots \longrightarrow \mathcal{H}_k(X) \xrightarrow{i_*} \mathcal{H}_k(\varphi) \xrightarrow{p_*} \mathcal{H}_{k-1}(A) \xrightarrow{\varphi_*} \ldots \longrightarrow \mathcal{H}_0(\varphi) \longrightarrow 0\,.$$

*Proof.* Conceptually, the proof of exactness of the sequence is the same as for oriented smooth bordism homology, see e.g. [39, Chap. 21]:

*Exactness at $\mathcal{H}_k(X)$:* Let $g : T \to A$ be a geometric cycle in $\mathcal{Z}_k(A)$. Then we have $i(\varphi_* g) = (\varphi \circ g, \emptyset) : (T, \emptyset) \to (X, A)$. We set $W := T \times [0,1]$ and $F : W \to X$, $F(x,t) := \varphi(g(x))$. Then we have $\partial W = \overline{T} \times \{0\} \sqcup T \times \{1\}$. Moreover, we set $M := T \times \{1\}$ and $G : M \to A$, $G(x,1) := g(x)$. This defines a null bordism $(W, M) \xrightarrow{(F,G)} (X, A)$ of $i(g) = (\varphi \circ g, \emptyset)$. Thus the composition $\mathcal{H}_k(A) \xrightarrow{\varphi_*} \mathcal{H}_k(X) \xrightarrow{i_*} \mathcal{H}_k(\varphi)$ is the trivial map.

Now let $f : S \to X$ be a geometric cycle in $X$ such that $i_*([S \xrightarrow{f} X]) = 0 \in \mathcal{H}_k(\varphi)$. Choose a null bordism $(W, M) \xrightarrow{(F,G)} (X, A)$ of $i(f) = (f, \emptyset) : (S, \emptyset) \to (X, A)$. Then we have $\partial M = \partial \overline{S} \sqcup \emptyset = \emptyset$ and $S \cap M = \partial S = \emptyset$, thus $\partial W = \overline{S} \sqcup M$. Moreover, $F|_S = f$ and $F|_M = \varphi \circ G$. Thus we have a geometric cycle $G : M \to A$, and a bordism $F : W \to X$ from $f$ to $\varphi \circ g$. This shows $[S \xrightarrow{f} X] = \varphi_*[M \xrightarrow{g} A]$.

*Exactness at $\mathcal{H}_k(\varphi)$:* By definition, the image of the composition $\mathcal{Z}_k(X) \xrightarrow{i} \mathcal{Z}_k(\varphi) \xrightarrow{p} \mathcal{Z}_{k-1}(A)$ is the empty map, which represents the trivial bordism class.

Now let $[(S, T) \xrightarrow{(f,g)} (X, A)] \in \mathcal{H}_k(\varphi)$ be a relative bordism class with $p_*[f, g] = [g] = 0 \in \mathcal{H}_{k-1}(A)$. Then we find a $k$-dimensional compact oriented stratifold $Q$ with boundary $\partial Q = T$ and a smooth map $G : Q \to A$ such that $G|_{\partial Q} = g$. Glueing $S$ and $\overline{Q}$ along $T = \partial Q = \partial S$, we obtain a $k$-dimensional oriented closed stratifold $N := S \cup_T \overline{Q}$. We extend the maps $f$ and $\varphi \circ G$ to a smooth map $r : N \to X$.

It remains to construct a bordism $(W, M) \xrightarrow{(F,G)} (X, A)$ from the cycle $(S, T) \xrightarrow{(f,g)} (X, A)$ to $(N, \emptyset) \xrightarrow{(r,\emptyset)} (X, A)$. To this end, set $W := N \times [0,1]$ and $F : W \to X$, $F(n,t) := r(n)$. Thus $\partial W = \overline{N} \times \{0\} \sqcup N \times \{1\}$. We set $M := Q \times \{0\}$. This yields a smooth map $(W, M) \xrightarrow{(F,G)} (X, A)$.

We have $\partial W = \overline{S} \times \{0\} \cup M \cup N \times \{1\}$. Moreover $\partial M = \partial Q \times \{0\} = \partial S \times \{0\} = T$ and $S \times \{0\} \cap M = \partial S = T$ and $N \times \{1\} \cap N = \emptyset$. By construction, we have $F|_{S \times \{0\}} = f$ and $F|_{N \times \{1\}} = r$. Thus we have constructed a bordism $(W, M) \xrightarrow{(F,G)} (X, A)$ from $(S, T) \xrightarrow{(f,g)} (X, A)$ to $(N, \emptyset) \xrightarrow{(r,\emptyset)} (X, A)$. In other words, we have shown that $[f, g] = i_*[r, \emptyset]$.

*Exactness at $\mathcal{H}_{k-1}(A)$:* Let $(S,T) \xrightarrow{(f,g)} (X,A) \in \mathcal{Z}_k(\varphi)$ be a geometric relative cycle. Then we have $\partial S = T$ and $f|_{\partial S} = \varphi \circ g$. Thus $S \xrightarrow{f} X$ is a null bordism of $\varphi_*(p(f,g)) = \varphi \circ g : T \to X$. In other words, the composition $\mathcal{H}_k(\varphi) \xrightarrow{p_*} \mathcal{H}_{k-1}(A) \xrightarrow{\varphi_*} \mathcal{H}_{k-1}(X)$ is trivial.

Let $T \xrightarrow{g} A$ be a geometric cycle in $\mathcal{Z}_{k-1}(A)$ with $\varphi_*[g] = [T \xrightarrow{\varphi \circ g} X] = 0 \in \mathcal{H}_{k-1}(X)$. Choose a null bordism $S \xrightarrow{F} X$ of $T \xrightarrow{\varphi \circ g} X$. Then $(S,T) \xrightarrow{(F,g)} (X,A)$ is a relative geometric cycle in $\mathcal{Z}_k(\varphi)$ and $p(F,g) = g$.

*Exactness at $\mathcal{H}_0(\varphi)$:* The map $i_* : \mathcal{H}_0(X) \to \mathcal{H}_0(\varphi)$ is induced by the isomorphism $j : \mathcal{Z}_0(X) \xrightarrow{\cong} \mathcal{Z}_0(\varphi)$. Hence it is surjective. □

### 2.4.3 Relative Stratifold Homology and Mapping Cone Homology

In [1] we used geometric cycles in $\mathcal{Z}_k(X)$ to represent singular homology classes in $X$. A geometric cycle $\zeta \in \mathcal{Z}_k(X)$, given by a smooth map $M \xrightarrow{f} X$, yields an equivalence class $f_*[M]_{\partial S_{k+1}} \in Z_k(X;\mathbb{Z})/\partial S_{k+1}(X;\mathbb{Z})$. By a slight abuse of notation, we denote this class as $[\zeta]_{\partial S_{k+1}}$ and refer to it as the *refined fundamental class* of $\zeta$. The map $\mathcal{Z}_k(X) \to Z_k(X;\mathbb{Z})/\partial S_{k+1}(X;\mathbb{Z})$, $\zeta \mapsto [\zeta]_{\partial S_{k+1}}$, is a semigroup homomorphism and commutes with the boundary operators. By [28, p. 186], the induced map $\mathcal{H}_k(X) \to H_k(X;\mathbb{Z})$, $[\zeta] \mapsto f_*[M]$, is a group isomorphism. Similarly, a geometric relative cycle $(\zeta,\tau) \in \mathcal{Z}_k(\varphi)$, given by a smooth map $(S,T) \xrightarrow{(f,g)} (X,A)$, has a *refined fundamental class* $[\zeta,\tau]_{\partial_\varphi S_{k+1}} := (f,g)_*[S,T]_{\partial S_{k+1}} \in Z_k(\varphi;\mathbb{Z})/\partial_\varphi S_{k+1}(\varphi;\mathbb{Z})$.

Taking refined fundamental classes commutes with maps $i : \mathcal{Z}_k(X) \to \mathcal{Z}_k(\varphi)$ and $p : \mathcal{Z}_k(\varphi) \to \mathcal{Z}_{k-1}(A)$ defined above: Restriction to the boundary maps the refined fundamental class $[\zeta,\tau]_{\partial_\varphi S_{k+1}} \in Z_k(\varphi;\mathbb{Z})/\partial_\varphi S_{k+1}(\varphi;\mathbb{Z})$ to the refined fundamental class $[\tau]_{\partial S_k} \in Z_{k-1}(A;\mathbb{Z})/\partial S_k(A;\mathbb{Z})$ of the boundary. Similarly, under the map $i : \mathcal{Z}_k(X) \to \mathcal{Z}_k(\varphi)$ the refined fundamental class $[\zeta]_{\partial S_{k+1}} \in \mathbb{Z}_k(X;\mathbb{Z})/\partial S_{k+1}(X;\mathbb{Z})$ of a geometric cycle is mapped to the refined fundamental class $[\zeta,\emptyset]_{\partial_\varphi S_{k+1}} \in \mathbb{Z}_k(\varphi;\mathbb{Z})/\partial_\varphi S_{k+1}(\varphi;\mathbb{Z})$ of the corresponding relative cycle.

Let $(\zeta,\tau),(\zeta',\tau') \in \mathcal{Z}_k(\varphi)$ be geometric cycles, represented by smooth maps $(S,T) \xrightarrow{(f,g)} (X,A)$ and $(S',T') \xrightarrow{(f',g')} (X,A)$. Let $(W,M) \xrightarrow{(F,G)} (X,A)$ be a bordism from $(S,T) \xrightarrow{(f,g)} (X,A)$ to $(S',T') \xrightarrow{(f',g')} (X,A)$. Choose a triangulation of $W$ and the induced triangulations of $S,S',M \subset \partial W$. We thus obtain a chain $(w,m) \in C_{k+1}(W;\mathbb{Z}) \times C_k(M;\mathbb{Z})$. Denote the corresponding fundamental cycles of $(S,T)$ and $(S',T')$ by $(s,t)$ and $(s',t')$, respectively. By definition of the bordism relation, we find

$$(f',g')_*(s',t') - (f,g)_*(s,t) = \partial_\varphi\big((F,G)_*(w,m)\big)\,.$$

This yields for the refined fundamental classes:

$$[\zeta',\tau']_{\partial_\varphi S_{k+1}} - [\zeta,\tau]_{\partial_\varphi S_{k+1}} = \partial_\varphi\big((F,G)_*[W,M]_{S_{k+1}}\big)\,. \tag{2}$$

Thus the fundamental classes coincide: $(f',g')_*[S',T'] = (f,g)_*[S,T] \in H_k(\varphi;\mathbb{Z})$.

The refined fundamental class map $\mathcal{Z}_k(\varphi) \to Z_k(\varphi;\mathbb{Z})/\partial_\varphi S_{k+1}(\varphi;\mathbb{Z})$, $(\zeta,\tau) \mapsto [\zeta,\tau]_{\partial_\varphi S_{k+1}} := (f,g)_*[S,T]_{\partial S_{k+1}}$, thus descends to a group homomorphism $\mathcal{H}_k(\varphi) \to H_k(\varphi;\mathbb{Z})$, $[\zeta,\tau] \mapsto \big[[\zeta,\tau]_{\partial_\varphi S_{k+1}}\big] = (f,g)_*[S,T]$. Here $[S,T] \in H_k(S,T;\mathbb{Z})$ denotes the usual fundamental class of the stratifold $S$ with boundary $T$ and $\big[[\zeta,\tau]_{\partial_\varphi S_{k+1}}\big]$ denotes the image of the refined fundamental class $[\zeta,\tau]_{\partial_\varphi S_{k+1}} \in Z_k(\varphi;\mathbb{Z})/\partial_\varphi S_{k+1}(\varphi;\mathbb{Z})$ in the mapping cone homology $H_k(\varphi;\mathbb{Z}) = Z_k(\varphi;\mathbb{Z})/\partial_\varphi C_{k+1}(\varphi;\mathbb{Z})$.

In fact, this map is a group isomorphism. Thus geometric relative cycles represent homology classes of the mapping cone:

**Theorem 3 (Relative Stratifold Homology).** *Let $\varphi : A \to X$ be a smooth map and $k \geq 0$. Then the map $\mathcal{H}_k(\varphi) \to H_k(\varphi;\mathbb{Z})$, $[\zeta,\tau] \mapsto \big[[\zeta,\tau]_{\partial_\varphi S_{k+1}}\big]$, is a group isomorphism.*

*Proof.* The refined fundamental classes of absolute and relative geometric cycles yield a commutative diagram

$$\begin{array}{ccccccccc}
\mathcal{Z}_k(A) & \xrightarrow{\varphi_*} & \mathcal{Z}_k(X) & \xrightarrow{i} & \mathcal{Z}_k(\varphi) & \xrightarrow{p} & \mathcal{Z}_{k-1}(A) & \xrightarrow{\varphi_*} & \mathcal{Z}_{k-1}(X) \\
\downarrow & & \downarrow & & \downarrow & & \downarrow & & \downarrow \\
\frac{Z_k(A;\mathbb{Z})}{\partial S_{k+1}(A;\mathbb{Z})} & \longrightarrow & \frac{Z_k(X;\mathbb{Z})}{\partial S_{k+1}(X;\mathbb{Z})} & \longrightarrow & \frac{Z_k(\varphi;\mathbb{Z})}{\partial_\varphi S_{k+1}(\varphi;\mathbb{Z})} & \longrightarrow & \frac{Z_{k-1}(A;\mathbb{Z})}{\partial S_k(A;\mathbb{Z})} & \longrightarrow & \frac{Z_{k-1}(X;\mathbb{Z})}{\partial S_k(X;\mathbb{Z})}\,.
\end{array} \tag{3}$$

In the induced diagram on homology, the two left as well as the two right vertical maps are group isomorphisms by [28, p. 186]. By the five lemma, so is the middle vertical map. □

### 2.4.4 Integration of Differential Forms

As above let $\Omega^*(\varphi)$ be the mapping cone de Rham complex of a smooth map $\varphi : A \to X$. Integration of differential forms in $\Omega^*(\varphi)$ over refined fundamental classes of geometric relative cycles is well-defined. For $(\omega,\vartheta) \in \Omega^k(\varphi)$ and $(\zeta,\tau) \in \mathcal{Z}_k(\varphi)$, represented by $(S,T) \xrightarrow{(f,g)} (X,A)$, we write:

$$\int_{[\zeta,\tau]_{\partial_\varphi S_{k+1}}} (\omega,\vartheta) = \int_{[S,T]_{\partial S_{k+1}}} (f,g)^*(\omega,\vartheta) = \int_{(S,T)} (f,g)^*(\omega,\vartheta)\,.$$

## 2.5 The Cross Product

For geometric cycles $\zeta \in \mathcal{Z}_k(X)$ and $\zeta' \in \mathcal{Z}_{k'}(X')$ the cartesian product of the corresponding stratifolds defines a cross product on stratifold homology, see [28, Chap. 10] and [1, Chap. 6]: if $\zeta$ is represented by $M \xrightarrow{f} X$ and $\zeta'$ is represented by $M' \xrightarrow{f'} X'$ then the cross product $\zeta \times \zeta'$ is the stratifold represented by $M \times M' \xrightarrow{f \times f'} X \times X'$. This cartesian product of stratifolds is compatible with bordism: if $W \xrightarrow{F} X$ is a bordism from $\zeta_0 \to \zeta_1$, then $W \times S' \xrightarrow{F \times f'} X \times X'$ is a bordism from $\zeta_0 \times \zeta'$ to $\zeta_1 \times \zeta'$, and similarly for bordisms of the second factor. Thus the cartesian product descends to a product of stratifold bordism groups. This coincides with the homology cross product under the isomorphism $\mathcal{H}_*(X) \xrightarrow{\cong} H_*(X;\mathbb{Z})$.

### 2.5.1 Cross Products of Geometric Cycles

Analogously, we define the cross product of a geometric relative cycle $(\zeta, \tau) \in \mathcal{Z}_k(\varphi)$, represented by $(S,T) \xrightarrow{(f,g)} (X,A)$, with a geometric cycle $\zeta' \in \mathcal{Z}_{k'}(X')$, represented by $S' \xrightarrow{f'} X'$: the stratifold

$$(S,T) \times S' \xrightarrow{(f,g)\times f'} (X,A) \times X'$$

represents a geometric relative cycle $(\zeta, \tau) \times \zeta' \in \mathcal{Z}_{k+k'}(\varphi \times \mathrm{id}_{X'})$.

The cartesian product of stratifolds is compatible with the bordism relation: A bordism $(W,M) \xrightarrow{(F,G)} (X,A)$ from $(\zeta_0,\tau_0)$ to $(\zeta_1,\tau_1)$, yields a bordism $(W,M)\times S' \xrightarrow{(F,G)\times f'} (X,A)\times X'$ from $(\zeta_0,\tau_0)\times\zeta'$ to $(\zeta_1,\tau_1)\times\zeta'$. Likewise, a bordism $W' \xrightarrow{F'} X'$ from $\zeta_0'$ to $\zeta_1'$ yields a bordism $(S,T)\times W' \xrightarrow{(f,g)\times F'} (X,A)\times X'$ from $(\zeta,\tau)\times\zeta_0'$ to $(\zeta,\tau)\times\zeta_1'$. Thus the cartesian product of stratifolds descends to a cross product

$$\times : \mathcal{H}_*(\varphi) \otimes \mathcal{H}_*(X') \to \mathcal{H}_*(\varphi \times \mathrm{id}_{X'})$$

on stratifold homology.

Choosing triangulations of the stratifolds involved and refining them to triangulations of the various cartesian products, it is easy to see that the cross product on stratifold bordism groups coincides with the ordinary homology cross product. In the same way, the cross product $\times : \mathcal{H}_*(\varphi) \otimes \mathcal{H}_*(X') \to \mathcal{H}_*(\varphi \times \mathrm{id}_{X'})$ on stratifold bordism groups is identified with the homology cross product $\times : H_*(\varphi;\mathbb{Z}) \otimes H_*(X';\mathbb{Z}) \to H_*(\varphi \times \mathrm{id}_{X'}\ Z)$.

## 2.6 The Pull-Back Operation

Let $\pi : E \to X$ be a fiber bundle with closed oriented fibers. Let $\varphi : A \to X$ be a smooth map and $\Phi : \varphi^*E \to E$ the induced fiber bundle map in the pull-back diagram

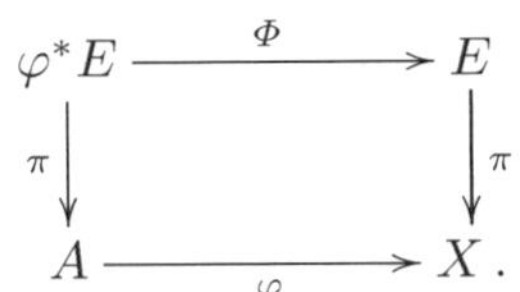

We adapt the pull-back operation $\mathrm{PB}_E$ on geometric cycles in the base of a fiber bundle from [1, Chap. 4]. In the notation as above, we define the pull-back operation $\mathrm{PB}_E : \mathcal{Z}_{k-\dim F}(X) \to \mathcal{Z}_k(E)$ by mapping the geometric cycle $M \xrightarrow{f} X$ to $f^*E \xrightarrow{F} E$. Here $F : f^*E \to E$ denotes the induced bundle map on the total space of the pull-back bundle $\pi : f^*E \to M$. Similarly, we may define a pull-back operation $\mathrm{PB}_{E,\varphi^*E} : \mathcal{Z}_{k-\dim F}(\varphi) \to \mathcal{Z}_k(\Phi)$ by mapping the relative cycle $(S,T) \xrightarrow{(f,g)} (X,A)$ to $(f^*E, g^*(\varphi^*E)) \xrightarrow{(F,G)} (E,\varphi^*E)$. Here $G : g^*(\varphi^*E) \to \varphi^*E$ is the bundle map in the pull-back diagram induced by $g : T \to A$ and the bundle $\varphi^*E \to A$.

These maps fit into the following commutative diagram of pull-back bundles:

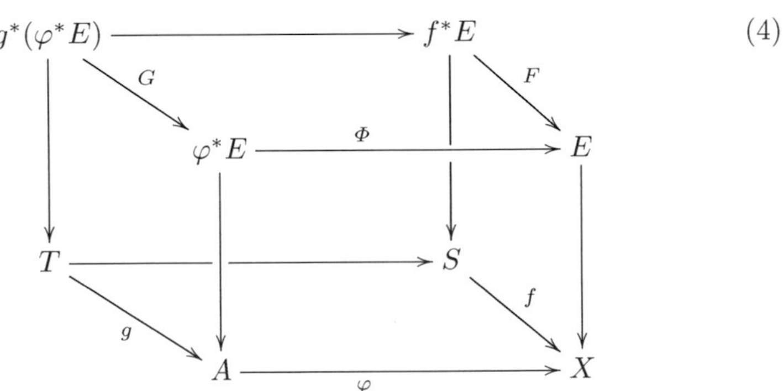

(4)

Since $T = \partial S$ and $\varphi \circ g = f|_{\partial S}$, we have in particular $g^*(\varphi^*E) = (f|_{\partial S})^*E = \partial(f^*E)$ and $\Phi \circ G = F$. Thus $(f^*E, g^*(\varphi^*E)) \xrightarrow{(F,G)} (E,\varphi^*E)$ indeed defines a geometric cycle in $\mathcal{Z}_k(\Phi)$.

The pull-back operation $\mathrm{PB}_E$ on geometric relative cycles is compatible with the maps $i : \mathcal{Z}_{k-\dim F}(X) \to \mathcal{Z}_{k-\dim F}(\varphi)$ and $p : \mathcal{Z}_{k-\dim F}(\varphi) \to \mathcal{Z}_{k-\dim F-1}(A)$ and the pull-back operations on geometric cycles in $X$ and $A$, respectively. Thus we have a commutative diagram of pull-back operations:

$$\begin{array}{ccccc} \mathcal{Z}_k(E) & \xrightarrow{i} & \mathcal{Z}_k(\Phi) & \xrightarrow{p} & \mathcal{Z}_{k-1}(\varphi^*E) \\ \uparrow \mathrm{PB}_E & & \uparrow \mathrm{PB}_{E,\varphi^*E} & & \uparrow \mathrm{PB}_{\varphi^*E} \\ \mathcal{Z}_{k-\dim F}(X) & \xrightarrow[i]{} & \mathcal{Z}_{k-\dim F}(\varphi) & \xrightarrow[p]{} & \mathcal{Z}_{k-\dim F-1}(A) \,. \end{array} \tag{5}$$

Now let $\pi : (E, \partial E) \to X$ be a fiber bundle with compact oriented fibers with boundary. In particular, the total space $E$ is a manifold with boundary $\partial E$ and retriction to the boundary yields a fiber bundle $\pi : \partial E \to X$ with closed oriented fibers. Denote by $i_{\partial E} : \partial E \to E$ the inclusion of the boundary. We introduce a pull-back operation from absolute cycles to relative cycles as follows: In [1, Chap. 4] we defined the pull-back operations $\mathrm{PB}_E : \mathcal{Z}_{k-\dim F}(X) \to \mathcal{C}_k(E)$ and $\mathrm{PB}_{\partial E} : \mathcal{Z}_{k-\dim \partial F}(X) \to \mathcal{Z}_k(\partial E)$.

We generalize the former to a pull-back operation on geometric chains: Let $\beta \in \mathcal{C}_k(X)$ be a geometric chain, represented by a smooth map $W \xrightarrow{f} X$ from a (not necesarily closed) oriented $k$-dimensional stratifold to $X$. Then we may pull-back the bundle $E$ to $W$. The total space is still a compact oriented stratifold, and we have an induced smooth map $F : f^*E \to E$. Thus we define the pull-back operation

$$\mathrm{PB}_E : \mathcal{C}_{k-\dim F}(X) \to \mathcal{C}_k(E)$$

by mapping the geometric chain $W \xrightarrow{f} X$ to the geometric chain $f^*E \xrightarrow{F} E$.

We define the pull-back operation from absolute to relative cycles

$$\mathrm{PB}_{E,\partial E} : \mathcal{Z}_{k-\dim F}(X) \to \mathcal{Z}_k(i_{\partial E})$$

by mapping a geometric cycle $S \xrightarrow{f} X$ to the geometric relative cycle $(f^*E, f^*\partial E) \xrightarrow{F} (E, \partial E)$ if $k - \dim F$ is odd and to $(\overline{f^*E}, f^*\partial E) \xrightarrow{F} (E, \partial E)$ if $k - \dim F$ is even. In the following, we write $((-1)^{k-\dim F} f^*E, f^*\partial E)$ to account for both cases. We then have $\partial((-1)^{k-\dim F+1} f^*E) = \overline{f^*\partial E}$ and $\partial(f^*\partial E) = \emptyset$. Thus $\mathrm{PB}_{E,\partial E}(\zeta)$ is indeed a geometric relative cycle in $\mathcal{Z}_k(i_{\partial E})$.

Now let $\zeta$, $\zeta' \in \mathcal{Z}_{k-\dim F}(X)$ be geometric cycles, represented by smooth maps $S \xrightarrow{f} X$ and $S' \xrightarrow{f'} X$, and let $W \xrightarrow{F} X$ be a bordism from $\zeta$ to $\zeta'$. Then the induced bundle map $F^*E \xrightarrow{\mathbf{F}} E$ yields a bordism

$$(\mathbf{W}, \mathbf{M}) := ((-1)^{k-\dim F+2} F^*E, F^*\partial E) \xrightarrow{\mathbf{F}} (E, \partial E)$$

from $\mathrm{PB}_{E,\partial E}(\zeta')$ to $\mathrm{PB}_{E,\partial E}(\zeta)$. Namely, we have

$$\begin{aligned} \partial(\mathbf{W}) &= \partial((-1)^{k-\dim F+2} F^*E)) \\ &= (-1)^{k-\dim F+2} F|_{\partial W} \cup \overline{F^*\partial E} \\ &= (-1)^{k-\dim F+1} (\overline{f'^*E} \cup f^*E) \cup \overline{F^*\partial E} \end{aligned}$$

and

$$\begin{aligned}\partial(\mathbf{M}) &= \partial(F^*\partial E)\\ &= (-1)^{k-\dim F+1}(\partial\overline{f'^*E} \sqcup \partial f^*E).\end{aligned}$$

### 2.6.1 Compatibility with Fiber Integration of Differential Forms

As above let $\Omega^*(\Phi)$ be the mapping cone de Rham complex for the induced bundle map $\Phi : \varphi^*E \to E$. Fiber integration on the relative de Rham complex is defined componentwise: for $(\omega,\vartheta) \in \Omega^k(\Phi)$ put

$$\fint_F (\omega,\vartheta) := \left(\fint_F \omega \,,\, \fint_F \vartheta\right).$$

This obviously defines a map $\fint_F : \Omega^k(\Phi) \to \Omega^{k-\dim F}(\varphi)$.

Fiber integration of differential forms is natural with respect to pull-back along the induced bundle maps in the pull-back diagram (4). In other words, for a smooth map $(Y,B) \xrightarrow{(f,g)} (X,A)$ and the corresponding map $(f^*E, g^*(\varphi^*E)) \xrightarrow{(F,G)} (E,\varphi^*E)$, we have:

$$\fint_F (F,G)^*(\omega,\vartheta) = (f,g)^* \fint_F (\omega,\vartheta). \tag{6}$$

Moreover, fiber integration is compatible with the mapping cone de Rham differentials:

$$\begin{aligned}d_\varphi \fint_F (\omega,\vartheta) &= \left(d\fint_F \omega \,,\, \varphi^*\fint_F \omega + d\fint_F \vartheta\right)\\ &= \left(\fint_F d\omega \,,\, \fint_F \Phi^*\omega + d\vartheta\right)\\ &= \fint_F d_\Phi(\omega,\vartheta). \end{aligned}\tag{7}$$

Thus fiber integration of differential forms descends to a well-defined homomorphism $H^k_{\mathrm{dR}}(\Phi) \to H^{k-\dim F}_{\mathrm{dR}}(\varphi)$ of the mapping cone de Rham cohomologies.

The pull-back operation $\mathrm{PB}_E$ is compatible with fiber integration of differential forms in the following sense: Let $(\zeta,\tau) \in \mathcal{Z}_{k-\dim F}(\varphi)$ be a geometric relative cycle, represented by $(S,T) \xrightarrow{(f,g)} (X,A)$, and $(\omega,\vartheta) \in \Omega^k(\Phi)$. Then we have:

$$\int_{[\mathrm{PB}_{E,\varphi^*E}(\zeta,\tau)]_{\partial_\Phi S_{k+1}}} (\omega,\vartheta) = \int_{(f^*E,g^*(\varphi^*E))} (F,G)^*(\omega,\vartheta)$$

$$\stackrel{(6)}{=} \int_{(S,T)} (f,g)^* \fint_F (\omega, \vartheta)$$
$$= \int_{[\zeta,\tau]_{\partial_\varphi S_{k-\dim F+1}}} \fint_F (\omega, \vartheta) . \tag{8}$$

Now let $\pi : (E, \partial E) \to X$ be a fiber bundle with compact oriented fibers with boundary. Then the pull-back operation $\mathrm{PB}_{E,\partial E}$ also satsfies a compatibility property for fiber integration of differential forms. This time the orientations have to be taken into account more carefully than in the case of closed fibers. Namely, for a pair of differential forms $(\omega, \vartheta) \in \Omega^k(i_{\partial E})$ and a geometric cycle $\zeta \in \mathcal{Z}_{k-\dim F}(X)$, represented by a smooth map $S \xrightarrow{f} X$, we have:

$$\begin{aligned}
\int_{[\mathrm{PB}_{E,\partial E}(\zeta)]_{\partial S_{k+1}}} (\omega, \vartheta) &= \int_{(f^*E, f^*\partial E)} F^*((-1)^{k-\dim F+1}\omega, \vartheta) \\
&= \int_S \fint_F (-1)^{k-\dim F+1} F^*\omega + \fint_{\partial F} F^*\vartheta \\
&= \int_S f^*\Big((-1)^{k-\dim F+1} \fint_F \omega + \fint_{\partial F} \vartheta\Big) \\
&= \int_{[\zeta]_{\partial S_{k-\dim F+1}}} \Big((-1)^{k-\dim F+1} \fint_F \omega + \fint_{\partial F} \vartheta\Big) \qquad (9)
\end{aligned}$$

Note that the sign on the second summand comes from the orientation convention on the geometric cycle $f^*\partial E \xrightarrow{F} \partial E$ in the pull-back operation $\mathrm{PB}_{E,\partial E}$.

### 2.6.2 Compatibility with Bordism and Refined Fundamental Classes

The pull-back operation $\mathrm{PB}_E$ is compatible with the bordism relation in the following sense: Let $(\zeta, \tau), (\zeta', \tau') \in \mathcal{Z}_{k-\dim F}(\varphi)$ be geometric relative cycles, represented by $(S,T) \xrightarrow{(f,g)} (X,A)$ and $(S',T') \xrightarrow{(f',g')} (X,A)$, respectively. Let $(W,M) \xrightarrow{(F,G)} (X,A)$ be a bordism from $(\zeta,\tau)$ to $(\zeta',\tau')$. Then the induced bundle map $(F^*E, G^*(\varphi^*E)) \xrightarrow{(\mathbf{F},\mathbf{G})} (E, \varphi^*E)$ defines a bordism from $\mathrm{PB}_{E,\varphi^*E}(\zeta,\tau)$ to $\mathrm{PB}_{E,\varphi^*E}(\zeta',\tau')$. Consequently the pull-back operation yields a homomorphism on relative stratifold homology $\mathcal{H}_{k-\dim F}(\varphi) \to \mathcal{H}_k(\Phi)$, $[(\zeta,\tau)] \mapsto [\mathrm{PB}_{E,\varphi^*E}(\zeta',\tau')]$. This may be considered as a transfer map on relative stratifold homology.

As above let $(W,M) \xrightarrow{(F,G)} (X,A)$ be a bordism from $(\zeta,\tau)$ to $(\zeta',\tau')$. Choose a triangulation of the stratifold $F^*E$ and the induced triangulations of $G^*(\varphi^*E)$. This yields a smooth singular chain $(a,b) \in C_{k-\dim F}(F^*E;\mathbb{Z}) \times$

$C_{k-\dim F-1}(G^*(\varphi^*E);\mathbb{Z})$. Restricting the bundle $G^*(\varphi^*E) \to M$ to the subspaces $S$ and $S'$ of $\partial W$, we obtain induced triangulations of $f^*E$ and $f'^*E$. Let $(x,y),(x',y') \in Z_k(\Phi;\mathbb{Z})$ be the induced fundamental cycles. Since $(F^*E, G^*(\varphi^*E)) \xrightarrow{(\mathbf{F},\mathbf{G})} (E,\varphi^*E)$ is a bordism from $\mathrm{PB}_{E,\varphi^*E}(\zeta,\tau)$ to $\mathrm{PB}_{E,\varphi^*E}(\zeta',\tau')$, we obtain from the bordism relation:

$$(x',y') - (x,y) = \partial_\Phi\big((\mathbf{F},\mathbf{G})_*(a,-b)\big)\,.$$

In particular, we obtain for the refined fundamental classes:

$$\begin{aligned}[\mathrm{PB}_{E,\varphi^*E}(\zeta',\tau')]_{\partial_\Phi S_{k+1}} &- [\mathrm{PB}_{E,\varphi^*E}(\zeta,\tau)]_{\partial_\Phi S_{k+1}} \\ &= \partial_\Phi\big((\mathbf{F},\mathbf{G})_*[(F^*E,\overline{G^*(\varphi^*E)})]_{S_{k+1}}\big)\,. \end{aligned} \tag{10}$$

We use this relation in the following section to construct transfer maps on the level of cycles.

Now let $\pi : (E,\partial E) \to X$ be a fiber bundle with compact oriented fibers with boundary. Then the pull-back operation $\mathrm{PB}_{E,\partial E}$ is compatible with the bordism relation: For geometric cycles $\zeta,\zeta' \in \mathcal{Z}_{k-\dim F-1}(X)$, represented by smooth maps $S \xrightarrow{f} X$ and $S' \xrightarrow{f'} X$, and a bordism from $\zeta$ to $\zeta'$, represented by a smooth map $F : W \to X$, we obtain a bordism $(F^*E, F^*\partial E) \xrightarrow{\mathbf{F}} (E,\partial E)$ from $\mathrm{PB}_{E,\partial E}(\zeta)$ to $\mathrm{PB}_{E,\partial E}(\zeta')$, with orientations on the second component chosen appropriately. Thus we obtain an induced map on stratifold homology $\mathcal{H}_{k-\dim F}(X) \to \mathcal{H}_k(i_{\partial E})$, $[\zeta] \mapsto [\mathrm{PB}_{E,\partial E}(\zeta)]$.

Similarly to the case above, we obtain for the refined fundamental classes:

$$\begin{aligned}[\mathrm{PB}_{E,\partial E}(\zeta')]_{\partial_{i_{\partial E}} S_{k+1}} &- [\mathrm{PB}_{E,\partial E}(\zeta)]_{\partial_{i_{\partial E}} S_{k+1}} \\ &= \partial_{i_{\partial E}}\big(\mathbf{F}_*[F^*E, F^*\partial E]_{S_{k+1}}\big)\,. \end{aligned} \tag{11}$$

### 2.6.3 Compatibility with Fiber Products

The pull-back operation for fiber bundles $\pi : E \to X$ and $\pi' : E' \to X'$ with closed oriented fibers is compatible with the cross product of geometric cycles and the pull-back operation for the fiber product $\pi \times \pi' : E \times E' \to X \times X'$. This means that the following diagram is graded commutative:

$$\begin{array}{ccc} \mathcal{Z}_{k+\dim F}(\Phi) \otimes \mathcal{Z}_{k'+\dim F'}(E') & \xrightarrow{\times} & \mathcal{Z}_{k+k'+\dim(F\times F')}(\Phi \times \mathrm{id}_{E'}) \\ \uparrow{\scriptstyle \mathrm{PB}_{E,\varphi^*E}} \quad \uparrow{\scriptstyle \mathrm{PB}_{E'}} & & \uparrow{\scriptstyle \mathrm{PB}_{(E,\varphi^*E)\times E'}} \\ \mathcal{Z}_k(\varphi) \otimes \mathcal{Z}_{k'}(X') & \xrightarrow[\times]{} & \mathcal{Z}_{k+k'}(\varphi \times \mathrm{id}_{X'})\,. \end{array} \tag{12}$$

The graded commutativity is caused by orientation conventions: cartesian products carry the ordinary product orientation while fiber bundles are oriented like products of first the base and then the fiber. The fiber product $\pi \times \pi' : E \times E' \to X \times X'$ carries the orientation of a fiber bundle over $X \times X'$ with fiber $F \times F'$. This orientation might differ from the product orientation of the total spaces (in case the total spaces carry an orientation). Explicitly, for cycles $(\zeta, \tau) \in \mathcal{Z}_k(\varphi)$ and $\zeta' \in \mathcal{Z}_{k'}(X')$, we have:

$$\mathrm{PB}_{(E,\varphi^*E)\times E'}((\zeta,\tau)\times\zeta') = (-1)^{k'\cdot\dim F}\cdot \mathrm{PB}_{E,\varphi^*E}(\zeta,\tau)\times \mathrm{PB}_{E'}(\zeta')\,. \quad (13)$$

We use this relation in the following section to construct transfer maps compatible with the cross product.

## 2.7 Transfer Maps

As above let $\pi : E \to X$ be a fiber bundle with closed oriented fibers and $\varphi : A \to X$ a smooth map. In [1, Chap. 4] we used geometric cycles to construct a transfer map $\lambda : C_k(X;\mathbb{Z}) \to C_{k+\dim F}(E;\mathbb{Z})$ that commutes with the boundary operator. Moreover, it is compatible with fiber integration of differential forms in the sense that for any smooth singular cycle $z \in Z_k(X;\mathbb{Z})$ and any closed differential form $\omega \in \Omega^{k+\dim F}(E)$, we have $\int_z \fint_F \omega = \int_{\lambda(z)} \omega$. Here we construct a transfer map $\lambda_\varphi : Z_k(\varphi;\mathbb{Z}) \to Z_{k+\dim F}(\Phi;\mathbb{Z})$ that commutes with the maps $i : Z_k(X;\mathbb{Z}) \to Z_k(\varphi;\mathbb{Z})$ and $p : Z_k(\varphi;\mathbb{Z}) \to Z_{k-1}(A;\mathbb{Z})$. These transfer maps are used in Sect. 5.1 to construct fiber integration on relative differential cohomology.

### 2.7.1 Representation by Geometric Cycles

The construction of the transfer map in [1, Chap. 4] is based on the pullback operation $\mathrm{PB}_\bullet$ on geometric cycles and a homomorphism that chooses geometric cycles to represent homomology classes. More precisely, for any singular cycle $z \in Z_k(X;\mathbb{Z})$ choose a geometric cycle $\zeta(z)$ and a singular chain $a(z)$ such that $[z - \partial a(z)]_{\partial S_{k+1}} = [\zeta(z)]_{\partial S_{k+1}}$. In other words, the refined fundamental class of $\zeta(z)$ represents the homology class of $z$. These choices can be made into homomorphisms $\zeta : Z_k(X;\mathbb{Z}) \to \mathcal{Z}_k(X)$ and $a : Z_k(X;\mathbb{Z}) \to C_{k+1}(X;\mathbb{Z})$ be first defining them on a basis and then extending linearly.

Now we do the same for cycles of the mapping cone complex of a smooth map $\varphi : A \to X$. By Theorem 3, the relative stratifold homology $\mathcal{H}_k(\varphi)$ is isomorphic to the mapping cone homology $H_k(\varphi;\mathbb{Z})$. Thus for any relative cycle $(s,t) \in Z_k(\varphi;\mathbb{Z})$ we may choose a geometric relative cycle $(\zeta,\tau) \in \mathcal{Z}_k(\varphi)$ such that its bordism class $[\zeta,\tau] \in \mathcal{H}_k(\varphi)$ maps to the mapping cone cohomology class $[s,t] \in H_k(\varphi;\mathbb{Z})$ under the isomorphism $\mathcal{H}_k(\varphi) \to H_k(\varphi;\mathbb{Z})$ from Theorem 3. In particular, we find a singular chain $(a,b) \in C_{k+1}(\varphi;\mathbb{Z})$

such that $[\zeta,\tau]_{\partial_\varphi S_{k+1}} = [(s,t) - \partial_\varphi(a,b)]_{\partial_\varphi S_{k+1}}$. We say that the geometric relative cycle $(\zeta,\tau) \in \mathcal{Z}_k(\varphi)$ *represents* the homology class $[s,t] \in H_k(\varphi;\mathbb{Z})$.

We may organize the choice of geometric relative cycles $(\zeta,\tau)$ and singular chains $(a,b)$ satisfying $[\zeta,\tau]_{\partial S_{k+1}} = [(s,t) - \partial_\varphi(a,b)]_{\partial S_{k+1}}$ into group homomorphisms

$$\begin{aligned} &(\zeta,\tau)_\varphi : Z_k(\varphi;\mathbb{Z}) \to \mathcal{Z}_k(\varphi)\,, \quad (s,t) \mapsto (\zeta,\tau)_\varphi(s,t)\,, \\ &(a,b)_\varphi : Z_k(\varphi;\mathbb{Z}) \to C_{k+1}(\varphi;\mathbb{Z})\,, \quad (s,t) \mapsto (a,b)_\varphi(s,t) = (a(s,t), b(s,t))\,, \end{aligned}$$

by first defining them on a basis of $\mathbb{Z}_k(\varphi;\mathbb{Z})$ and then extending linearly. These homomorphisms can be made compatible with the maps $i$ and $p$ as follows: The group $Z_k(\varphi;\mathbb{Z})$ of relative cycles sits in the split exact sequence

$$0 \longrightarrow Z_k(X;\mathbb{Z}) \xrightarrow[i]{} Z_k(\varphi;\mathbb{Z}) \underset{p}{\overset{\sigma}{\rightleftarrows}} Z_{k-1}(A,\mathbb{Z}) \longrightarrow 0 \tag{14}$$

where $i : s \mapsto (s,0)$ and $p : (s,t) \mapsto t$. Choose a splitting $\sigma : Z_{k-1}(A;\mathbb{Z}) \to Z_k(\varphi;\mathbb{Z})$. From bases of $Z_k(X;\mathbb{Z})$ and $Z_{k-1}(A;\mathbb{Z})$ and the splitting $\sigma$ we obtain a basis of $Z_k(\varphi;\mathbb{Z})$. We may thus choose the homomorphism $(\zeta,\tau)_\varphi : Z_k(\varphi;\mathbb{Z}) \to \mathcal{Z}_k(\varphi)$ compatible with the maps $p$ and $i$ and the homomorphism $\zeta$ defined on absolute cycles as follows: For basis elements $(s,t)$ in the image of $i : Z_k(X;\mathbb{Z}) \to Z_k(\varphi;\mathbb{Z})$, put $(\zeta,\tau)_\varphi(s,t) := (\zeta(s),0)$. For complementary basis elements $(s,t) = \sigma(t)$, obtained from a basis of $Z_{k-1}(A;\mathbb{Z})$, choose $(\zeta,\tau)_\varphi(s,t) \in \mathcal{Z}_k(\varphi)$ such that $p((\zeta,\tau)_\varphi(s,t)) = \tau(s,t) = \zeta(t) \in \mathcal{Z}_{k-1}(A)$. Then extend linearly. This yields a commutative diagram

$$\begin{array}{ccccc} Z_k(X;\mathbb{Z}) & \xrightarrow{i} & Z_k(\varphi;\mathbb{Z}) & \xrightarrow{p} & Z_{k-1}(A;\mathbb{Z}) \\ \downarrow{\scriptstyle \zeta} & & \downarrow{\scriptstyle (\zeta,\tau)_\varphi} & & \downarrow{\scriptstyle \zeta} \\ \mathcal{Z}_k(X) & \xrightarrow[i]{} & \mathcal{Z}_k(\varphi) & \xrightarrow[p]{} & \mathcal{Z}_{k-1}(A)\,. \end{array} \tag{15}$$

Similarly, we may choose the homomorphism $(a,b)_\varphi : Z_k(\varphi;\mathbb{Z}) \to C_{k+1}(\varphi;\mathbb{Z})$ compatible with the maps $i$ and $p$ and the homomorphism $a$ defined on absolute cycles. Using the splitting $\sigma$, we write a cycle $(s,t) \in Z_k(\varphi;\mathbb{Z})$ as $(s,t) = (z,0) + \sigma(p(s,t)) = i(z) + \sigma(t)$. For compatibility with the map $i$ we may simply put $(a,b)(i(z)) := (a(z),0)$. However, compatibility with the map $p$ involves a sign: Since

$$(s,t) - \partial_\varphi(a,b)_\varphi(s,t) = (s - \partial a(s,t) - \varphi_* a(s,t), t + \partial b(s,t))$$

represents the fundamental class of $(\zeta,\tau)_\varphi(s,t)$ and $t - \partial a(t)$ represents the fundamental class of $\zeta(t) = \zeta(p(s,t)) = p((\zeta,\tau)(s,t))$, we are forced to put $b(\sigma(t)) := -a(t) \in C_k(A;\mathbb{Z})$.

### 2.7.2 Compatibility of Transfer Maps

We define the mapping cone transfer map $\lambda_\varphi : Z_k(\varphi;\mathbb{Z}) \to Z_{k+\dim F}(\Phi;\mathbb{Z})$ as follows: for any cycle $(s,t)$ in a basis of $Z_k(\varphi;\mathbb{Z})$ choose a cycle $\lambda_\varphi(s,t) \in Z_{k+\dim F}(\Phi;\mathbb{Z})$ such that the equivalence classes modulo boundaries of thin chains satisfy $[\lambda_\varphi(s,t)]_{\partial_\Phi S_{k+\dim F+1}} = [\mathrm{PB}_{E,\varphi^*E}(\zeta,\tau)_\varphi(s,t)]_{\partial_\Phi S_{k+\dim F+1}}$. Then extend $\lambda_\varphi$ as a homomorphism.

By (3) and (5), the refined fundamental classes and the pull-back operations are compatible with the maps $i$ and $p$ that relate absolute and relative cycles. By the choice of homomorphisms $(\zeta,\tau)_\varphi$ and $(a,b)_\varphi$ above, we may also choose the transfer map $\lambda_\varphi$ compatible with $i$ and $p$. We thus obtain a commutative diagram of transfer maps:

$$\begin{array}{ccccc} Z_{k+\dim F}(E;\mathbb{Z}) & \xrightarrow{i} & Z_{k+\dim F}(\Phi;\mathbb{Z}) & \xrightarrow{p} & Z_{k+\dim F-1}(\varphi^*E;\mathbb{Z}) \\ \uparrow \lambda & & \uparrow \lambda_\varphi & & \uparrow \lambda \\ Z_k(X;\mathbb{Z}) & \xrightarrow[i]{} & Z_k(\varphi;\mathbb{Z}) & \xrightarrow[p]{} & Z_{k-1}(A;\mathbb{Z}) \,. \end{array} \tag{16}$$

Like for absolute cycles, the transfer map is compatible with fiber integration of forms in the mapping cone de Rham complex of the induced bundle map: Let $(\omega,\vartheta) \in \Omega^{k+\dim F}(\Phi)$ and $(s,t) \in Z_k(\varphi;\mathbb{Z})$. Then we have:

$$\begin{aligned} \int_{\lambda_\varphi(s,t)} (\omega,\vartheta) &= \int_{[\mathrm{PB}_{E,\varphi^*E}((\zeta,\tau)_\varphi(s,t))]_{\partial_\Phi S_{k+\dim F+1}}} (\omega,\vartheta) \\ &\overset{(8)}{=} \int_{[(\zeta,\tau)_\varphi(s,t)]_{\partial_\varphi S_{k+1}}} \fint_F (\omega,\vartheta) \\ &= \int_{(s,t)-\partial_\varphi((a,b)_\varphi(s,t))} \fint_F (\omega,\vartheta) \\ &= \int_{(s,t)} \fint_F (\omega,\vartheta) - \int_{(a,b)_\varphi(s,t)} d_\varphi \fint_F (\omega,\vartheta) \,. \end{aligned} \tag{17}$$

In particular, for a $d_\Phi$-closed pair $(\omega,\vartheta)$ we have:

$$\int_{\lambda_\varphi(s,t)} (\omega,\vartheta) = \int_{(s,t)} \fint_F (\omega,\vartheta) \,.$$

As explained in [1, Chap. 4], we can construct an extension of the transfer map $\lambda_\varphi : Z_k(\varphi;\mathbb{Z}) \to Z_{k+\dim F}(\Phi;\mathbb{Z})$ to a homomorphism $\lambda_\varphi : C_k(\varphi;\mathbb{Z}) \to C_{k+\dim F}(\Phi;\mathbb{Z})$ such that

$$\partial_\Phi \circ \lambda_\varphi = \lambda_\varphi \circ \partial_\varphi \,. \tag{18}$$

This is done by appropriate choices on a basis of $C_k(\varphi;\mathbb{Z})$. For basis elements in $Z_k(\varphi;\mathbb{Z})$ we choose $\lambda_\varphi$ as before. For complementary basis elements $(x,y) \in C_k(\varphi;\mathbb{Z})$ we choose $\lambda_\varphi(x,y) \in C_{k+\dim F}(\Phi;\mathbb{Z})$ such that (18) holds. By changing $\lambda_\varphi$ on the complementary basis elements if necessary[1], we may as well assume that for any $d_\Phi$-closed $(\omega,\vartheta) \in \Omega^{k+\dim F}(\Phi)$, we have:

$$\int_{\lambda_\varphi(x,y)} (\omega,\vartheta) = \int_{(x,y)-(a,b)(\partial_\varphi(x,y))} \fint_F (\omega,\vartheta) . \tag{19}$$

The transfer map $\lambda_\varphi$ will be used to define fiber integration for relative differential characters.

### 2.7.3 Multiplicativity of Transfer Maps

Let $\pi : E \to X$ and $\pi' : E' \to X'$ be fiber bundles with compact oriented fibers $F$ and $F'$ and let $\pi \times \pi' : E \times E' \to X \times X'$ be the fiber product. It carries the orientation of a fiber bundle over $X \times X'$ with fiber $F \times F'$. This orientation might differ from the product orientation of the total spaces.

Using multiplicativity of the pull-back operation (13) and a splitting of the Künneth sequence as constructed in Sect. 4.2 below, we may choose the transfer map for the product bundle in such a way that we obtain the following graded commutative diagram:

$$\begin{array}{ccc} Z_{k+\dim F}(\Phi;\mathbb{Z}) \otimes Z_{*+\dim F'}(E') & \xrightarrow{\times} & Z_{k+k'+\dim(F\times F')}(\Phi \times \mathrm{id}_{E'};\mathbb{Z}) \\ \lambda_\varphi \uparrow \quad\quad \lambda' \uparrow & & \uparrow \lambda_{\varphi\times\mathrm{id}_{X'}} \\ Z_k(\varphi;\mathbb{Z}) \otimes Z_{k'}(X';\mathbb{Z}) & \xrightarrow{\times} & Z_{k+k'}(\varphi \times \mathrm{id}_{X'};\mathbb{Z}) . \end{array} \tag{20}$$

More precisely, for cycles $(s,t) \in Z_k(\varphi;\mathbb{Z})$ and $z' \in Z_{k'}(X';\mathbb{Z})$, we have:

$$\lambda_\varphi(s,t) \times \lambda'(z') = (-1)^{k'\cdot\dim F} \cdot \lambda_{\varphi\times\mathrm{id}_{X'}}((s,t) \times z') . \tag{21}$$

This relation is used in the proof of the up-down formula in Sect. 5.3 below.

## 3 Differential Characters

In this chapter we discuss (absolute and relative) differential characters as models for (absolute and relative) differential cohomology classes. Differential characters were introduced in [16] as certain group homomorphisms $h : Z_{k-1}(X;\mathbb{Z}) \to \mathrm{U}(1)$ on the group of smooth singular cycles in a smooth

[1] This is explained in detail in [1, Chap. 4] for the case of absolute chains.

manifold $X$. The graded abelian group $\widehat{H}^*(X;\mathbb{Z})$ of differential characters was the first model for what is now called differential cohomology.[2]

The definition from [16] can be easily adapted to homomorphisms on relative cycles. As explained in the introduction, there are two ways to define relative singular homology $H_*(X, A;\mathbb{Z})$, either as the homology of the mapping cone complex of the inclusion $i_A : A \to X$ or as the homology of the quotient complex $C_*(X, A;\mathbb{Z}) := C_*(X,\mathbb{Z})/\mathrm{im}(i_{A*})$. Hence there arise two notions of relative cycles and thus two ways two adapt the notion of differential characters. The corresponding groups of differential characters are both refinements of the relative cohomology $H^*(X, A;\mathbb{Z})$ by differential forms. Both notions appear rather naturally.

In Sect. 3.1 we review the notion and elementary properties of *relative differential characters* as introduced in [7]. These are characters on the group of cycles $Z_*(\varphi;\mathbb{Z})$ in the mapping cone complex of a smooth map $\varphi : A \to X$. The graded abelian group of those characters is denoted by $\widehat{H}^*(\varphi;\mathbb{Z})$. We review the results from [1, Chap. 8], including a long exact secquence for $\widehat{H}^*(\varphi;\mathbb{Z})$ and the groups of absolute differential characters on $X$ and $A$.

In Sect. 3.3 we discuss differential characters on the group $Z_*(X, A;\mathbb{Z})$ of relative cycles. Here $A \subset X$ is an embedded smooth submanifold. The graded abelian group of differential characters on $Z_*(X, A;\mathbb{Z})$ is denoted by $\widehat{H}^*(X; A;\mathbb{Z})$. We prove a long exact sequence that relates the group $\widehat{H}^*(X; A;\mathbb{Z})$ to the groups of absolute differential characters on $X$ and $A$. Further, we show that $\widehat{H}^*(X, A;\mathbb{Z})$ coincides with the subgroup of *parallel* characters in $\widehat{H}^*(i_A;\mathbb{Z})$ . In Sect. 3.4 we clarify the relation of the groups $\widehat{H}^k(\varphi;\mathbb{Z})$ and $\widehat{H}^k(X, A;\mathbb{Z})$ to another notion of relative differential cohomology that has appeared in the literature: the relative Hopkins-Singer groups $\check{H}^k(\varphi;\mathbb{Z})$ for a smooth map $\varphi : A \to X$ and $\check{H}^k(i_A;\mathbb{Z})$ for the embedding $i_A : A \to X$ of a smooth manifold. These groups have been constructed in [7].

## 3.1 Relative Differential Characters

Differential characters on a smooth manifold $X$ were introduced by Cheeger and Simons in [16]. Differential characters relative to a smooth map $\varphi : A \to X$ were introduced in [7]. We briefly review the definition and elementary properties of relative differential characters, thereby treating the absolute differential characters of [16] as a special case.

[2] It is convenient to shift the degree of the differential characters by $+1$ as compared to the original definition from [16]. Thus a degree $k$ differential character has curvature and characteristic class of degree $k$.

### 3.1.1 Characters on Mapping Cone Cycles

Let $(C_*(X;\mathbb{Z}),\partial)$ be the complex of smooth singular chains in $X$. The *mapping cone complex* of a smooth map $\varphi : A \to X$ is the complex $C_k(\varphi;\mathbb{Z}) := C_k(X;\mathbb{Z}) \times C_{k-1}(A;\mathbb{Z})$ of pairs of smooth singular chains with the differential $\partial_\varphi(s,t) := (\partial s + \varphi_* t, -\partial t)$. The homology $H_k(\varphi;\mathbb{Z})$ of this complex coincides with the homology of the mapping cone of $\varphi$ in the topological sense. For the special case of an embedding $i_A : A \hookrightarrow X$ it coincides with the relative homology $H_k(X,A;\mathbb{Z})$.

As above, let $\Omega^*(\varphi)$ be the mapping cone de Rham complex with the differential $d_\varphi(\omega,\vartheta) := (d\omega, \varphi^*\omega - d\vartheta)$. The mapping cone de Rham cohomology $H^*_{\mathrm{dR}}(\varphi)$ is canonically identified with the real mapping cone cohomology $H^*(\varphi;\mathbb{R})$.

We denote by $Z_k(\varphi;\mathbb{Z})$ the group of $k$-cycles of the mapping cone complex and by $B_k(\varphi;\mathbb{Z})$ the group of $k$-boundaries. Let $k \geq 1$. The group of degree-$k$ relative differential characters is defined as follows:

$$\widehat{H}^k(\varphi;\mathbb{Z}) := \left\{ h \in \mathrm{Hom}(Z_{k-1}(\varphi;\mathbb{Z}),\mathrm{U}(1)) \,\middle|\, h \circ \partial_\varphi \in \Omega^k(\varphi) \right\}.$$

The notation $h \circ \partial_\varphi \in \Omega^k(\varphi)$ means that there exists $(\omega,\vartheta) \in \Omega^k(\varphi)$ such that for every smooth singular chain $(a,b) \in C_k(\varphi;\mathbb{Z})$ we have

$$h(\partial_\varphi(a,b)) = \exp\left(2\pi i \int_{(a,b)} (\omega,\vartheta)\right). \tag{22}$$

The form $\omega =: \mathrm{curv}(h) \in \Omega^k(X)$ is called the *curvature* of the relative differential character $h$. The form $\vartheta =: \mathrm{cov}(h) \in \Omega^{k-1}(A)$ is called its *covariant derivative*. The curvature is uniquely determined by the differential character. For $k \geq 2$, this is also true for the covariant derivative. For $k = 1$, the function $\vartheta$ is unique only up to addition of a locally constant integer valued function, see [1, Chap. 8].

We denote by $\Omega^k_0(\varphi)$ the set of all $d_\varphi$-closed forms $(\omega,\vartheta) \in \Omega^k(\varphi)$ with integral periods, i.e., such that $\int_{(s,t)}(\omega,\vartheta) \in \mathbb{Z}$ holds for all $(s,t) \in Z_k(\varphi;\mathbb{Z})$. Since $h \in \widehat{H}^k(\varphi;\mathbb{Z})$ is a homomorphism, condition (22) implies that

$$\int_{(s,t)} (\mathrm{curv},\mathrm{cov})(h) \in \mathbb{Z}$$

for any cycle $(s,t) \in Z_k(\varphi;\mathbb{Z})$. Moreover, since

$$\int_{\partial_\varphi(a,b)} (\mathrm{curv},\mathrm{cov})(h) = \int_{(a,b)} d_\varphi(\mathrm{curv},\mathrm{cov})(h) \in \mathbb{Z}$$

holds for all chains $(a,b) \in C_{k+1}(\varphi;\mathbb{Z})$, it follows that $(\mathrm{curv},\mathrm{cov})(h)$ is $d_\varphi$-closed. Thus $(\mathrm{curv},\mathrm{cov})(h) \in \Omega^k_0(\varphi)$.

Differential characters $h \in \widehat{H}^k(\varphi;\mathbb{Z})$ with $\mathrm{curv}(h) = 0$ are called *flat*, while characters with $\mathrm{cov}(h) = 0$ are called *parallel*. The condition $\varphi^*\mathrm{curv}(h) = d\mathrm{cov}(h)$ implies that parallel characters are in particular *flat along* $\varphi$.

It is shown in [7, p. 273f.] that relative differential characters $h \in \widehat{H}^k(\varphi;\mathbb{Z})$ have characteristic classes $c(h)$ in the mapping cone cohomology $H^k(\varphi;\mathbb{Z})$. The class $c(h)$ is defined as follows: Let $\tilde{h} \in C^{k-1}(\varphi;\mathbb{Z})$ be a real lift of $h$. Thus $h(s,t) = \exp(2\pi i \cdot \tilde{h}(s,t))$ holds for all cycles $(s,t) \in Z_{k-1}(\varphi;\mathbb{Z})$. By (22), the cocycle

$$\mu^{\tilde{h}} := (\mathrm{curv}, \mathrm{cov})(h) - \delta_\varphi \tilde{h} \tag{23}$$

satisfies $\exp(2\pi i \mu^{\tilde{h}}(a,b)) = 1$ for all $(a,b) \in C_k(\varphi;\mathbb{Z})$. Thus $\mu^{\tilde{h}} \in C^k(\varphi;\mathbb{Z})$. The *characteristic class* of $h$ is defined as $c(h) := [\mu^{\tilde{h}}] \in H^k(\varphi;\mathbb{Z})$. Characters $h \in \widehat{H}^k(\varphi;\mathbb{Z})$ with $c(h) = 0$ are called *topologically trivial*.

### 3.1.2 Exact Sequences

By [7, Thm. 2.4], the group $\widehat{H}^k(\varphi;\mathbb{Z})$ fits into the following short exact sequences:

$$0 \longrightarrow \frac{\Omega^{k-1}(\varphi)}{\Omega_0^{k-1}(\varphi)} \xrightarrow{\iota_\varphi} \widehat{H}^k(\varphi;\mathbb{Z}) \xrightarrow{c} H^k(\varphi;\mathbb{Z}) \longrightarrow 0$$

$$0 \longrightarrow H^{k-1}(\varphi;\mathrm{U}(1)) \xrightarrow{j} \widehat{H}^k(\varphi;\mathbb{Z}) \xrightarrow{(\mathrm{curv},\mathrm{cov})} \Omega_0^k(\varphi) \longrightarrow 0 \tag{24}$$

The map $j : H^{k-1}(\varphi;\mathrm{U}(1)) \to \widehat{H}^k_\varphi(X,A;\mathbb{Z})$ is defined by $j(\tilde{u})(s,t) := \langle \tilde{u}, [s,t] \rangle$. This is well-defined and injective, since divisibility of $\mathrm{U}(1)$ yields the identification $H^{k-1}(\varphi;\mathrm{U}(1)) \cong \mathrm{Hom}(H_{k-1}(\varphi;\mathbb{Z}),\mathrm{U}(1))$. The map $\iota_\varphi : \Omega^{k-1}(\varphi) \to \widehat{H}^k(\varphi;\mathbb{Z})$ is defined by $\iota_\varphi(\omega,\vartheta)(s,t) := \exp\left(2\pi i \int_{(s,t)} (\omega,\vartheta)\right)$. It descends to a map on the quotient $\frac{\Omega^{k-1}(\varphi)}{\Omega_0^{k-1}(\varphi)}$ which we also denote by $\iota_\varphi$. From the relative Stokes theorem (1) we conclude $(\mathrm{curv},\mathrm{cov}) \circ \iota_\varphi = d_\varphi$. A form $(\omega,\vartheta) \in \Omega^{k-1}(\varphi)$ such that $\iota_\varphi(\omega,\vartheta) = h$ is called a *topological trivialization* of the character $h$. Thus the map $\iota_\varphi$ provides topological trivializations, and the first sequence in (24) tells us that a character $h \in \widehat{H}^k(\varphi;\mathbb{Z})$ admits topological trivializations if and only if it is topologically trivial.

We denote by

$$R^k(\varphi;\mathbb{Z}) := \{(\omega,\vartheta,u) \in \Omega_0^k(\varphi) \times H^k(\varphi;\mathbb{Z}) \mid [\omega,\vartheta]_{\mathrm{dR}} = u_{\mathbb{R}}\} \tag{25}$$

the set of pairs of $d_\varphi$-closed differential forms with integral periods and integral mapping cone classes that match in the real mapping cone cohomology $H^k(\varphi;\mathbb{R})$. By definition of the characteristic class of a character $h \in \widehat{H}^k(\varphi;\mathbb{Z})$ we have $((\mathrm{curv},\mathrm{cov})(h), c(h)) \in R^k(\varphi;\mathbb{Z})$. Moreover, the exact sequences above may be joined to the exact sequence

$$0 \longrightarrow \frac{H^{k-1}(\varphi;\mathbb{R})}{H^{k-1}(\varphi;\mathbb{Z})_{\mathbb{R}}} \longrightarrow \widehat{H}^k_\varphi(X,A;\mathbb{Z}) \xrightarrow{(\mathrm{curv},\mathrm{cov},c)} R^k(\varphi;\mathbb{Z}) \longrightarrow 0. \quad (26)$$

Here $H^{k-1}(\varphi;\mathbb{Z})_{\mathbb{R}}$ denotes the image of $H^{k-1}(\varphi;\mathbb{Z})$ in $H^{k-1}(\varphi;\mathbb{R})$ under the change of coefficients homomorphism induced by $\mathbb{Z} \hookrightarrow \mathbb{R}$.

### 3.1.3 Naturality, Thin invariance, Torsion cycles

The following properties are used in several constructions throughout this work.

*Remark 4. Pull-back of relative differential characters.* Let $\varphi : A \to X$ and $\psi : B \to Y$ be smooth maps. A smooth map of pairs $(Y,B) \xrightarrow{(f,g)} (X,A)$ is a pair of smooth maps such that $\varphi \circ g = f \circ \psi$. Thus we have the commutative diagram:

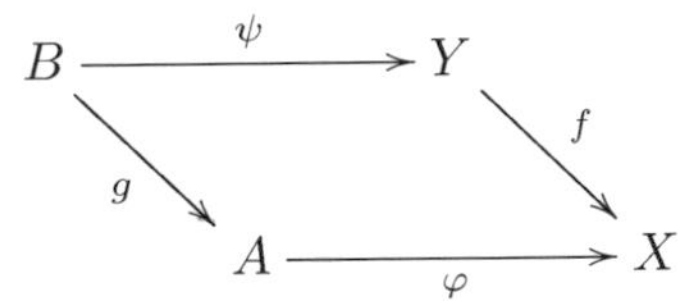


We define the pull-back of relative characters along a smooth map $(Y,B) \xrightarrow{(f,g)} (X,A)$ by:

$$(f,g)^* : \widehat{H}^k(\varphi;\mathbb{Z}) \to \widehat{H}^k(\psi;\mathbb{Z}), \quad h \mapsto h \circ (f,g)_* \, .$$

Here $(f,g)_*$ denotes the induced map on relative cycles: for $(s,t) \in Z_{k-1}(\psi;\mathbb{Z})$, we have $(f,g)_*(s,t) := (f_*s, g_*t)$ and hence $\big((f,g)^*h\big)(s,t) = h(f_*s, g_*t)$.

*Remark 5. Thin invariance.* By definition of thin chains, relative differential characters vanish on boundaries of thin chains of the mapping cone complex. We term this property the *thin invariance* of differential characters. In particular, we have a well-defined evaluation of characters $h \in \widehat{H}^k(\varphi;\mathbb{Z})$ on the refined fundamental class $(f,g)_*[M,N]_{\partial S_k} \in Z_{k-1}(\varphi;\mathbb{Z})/\partial_\varphi S_k(\varphi;\mathbb{Z})$ of a geometric relative cycle $(M,N) \xrightarrow{(f,g)} (X,A)$.

*Remark 6. Evaluation on torsion cycles.* Let $z \in Z_{k-1}(X;\mathbb{Z})$ be a *torsion cycle*, i.e. a cycle that represents a torsion class in $H_{k-1}(X;\mathbb{Z})$. If $z$ is a boundary then by definition, the evaluation of a differential character $h \in \widehat{H}^k(X;\mathbb{Z})$ on $z$ only depends upon $\mathrm{curv}(h)$. In [1, Chap. 5] we show that the evaluation of $h \in \widehat{H}^k(X;\mathbb{Z})$ on a torsion cycle $z$ only depends upon $\mathrm{curv}(h)$ and $c(h)$.

An analogous statement holds for relative characters and mapping cone cycles: Let $h \in \widehat{H}^k(\varphi;\mathbb{Z})$ and let $\tilde{h} \in C^{k-1}(\varphi;\mathbb{R})$ be a real lift as in the definition of the characteristic class. Suppose that $(s,t) \in Z_{k-1}(\varphi;\mathbb{Z})$ is a torsion cycle. Thus we find an integer $N \in \mathbb{N}$ and a chain $(a,b) \in C_k(\varphi;\mathbb{Z})$ such that $N \cdot (s,t) = \partial_\varphi(a,b)$. Then we have:

$$\begin{aligned} h(s,t) &= \exp\Big(2\pi i \cdot \tilde{h}\Big(\frac{1}{N}\partial_\varphi(a,b)\Big)\Big) \\ &= \exp\Big(\frac{2\pi i}{N} \cdot \big(\delta_\varphi \tilde{h}\big)(a,b)\Big) \\ &= \exp\Big(\frac{2\pi i}{N} \cdot \big((\mathrm{curv},\mathrm{cov})(h) - \mu^{\tilde{h}}\big)(a,b)\Big) \\ &= \exp\Big(\frac{2\pi i}{N} \cdot \Big(\int_{(a,b)} (\mathrm{curv},\mathrm{cov})(h) - \langle c(h),(a,b)\rangle\Big)\Big). \end{aligned} \tag{27}$$

Note that the evaluation of the characteristic class $c(h)$ on the chain $(a,b)$ is not well-defined. But the term in (27) is well-defined: any two cocycles that represent $c(h)$ differ by an integral coboundary $\delta_\varphi \ell$ for $\ell \in C^{k-1}(\varphi;\mathbb{Z})$ and $\frac{1}{N}\langle \delta_\varphi \ell,(a,b)\rangle = \langle \ell,(s,t)\rangle \in \mathbb{Z}$.

### 3.1.4 Absolute Differential Characters

Let $x \in X$ be any point. We may consider $x$ as a smooth map $\varphi = x : \{*\} \to X$, $* \mapsto x$. We have the canonical identification $C_k(\{*\};\mathbb{Z}) \cong \mathbb{Z}$ for $k \geq 0$. The boundary map $\partial : C_k(\{*\};\mathbb{Z}) \to C_{k-1}(\{*\};\mathbb{Z})$ is the identity for positive even $k$ and identically 0 else. For $k \geq 2$, we obtain canonical identifications

$$Z_{k-1}(x;\mathbb{Z}) \cong Z_{k-1}(X;\mathbb{Z}) \oplus Z_{k-2}(\{*\};\mathbb{Z}) \cong \begin{cases} Z_{k-1}(X;\mathbb{Z}) & \text{if } k \text{ even} \\ Z_{k-1}(X;\mathbb{Z}) \oplus \mathbb{Z} & \text{if } k \text{ odd.} \end{cases}$$

For relative differential forms, we have $\Omega^k_0(x) = \Omega^k_0(X) \times \{0\}$ for any $k \geq 2$.

*Remark 7. Absolute differential characters.* Let $k \geq 2$. Let $h : Z_{k-1}(x;\mathbb{Z}) \to \mathrm{U}(1)$ be a relative differential character. Then we have $\mathrm{cov}(h) = 0$ for dimensional reasons. In particular, $\mathrm{curv}(h) \in \Omega^k_0(X)$. For even $k$, the relative character $h$ is a homomorphism $h : Z_{k-1}(X;\mathbb{Z}) \to \mathrm{U}(1)$. For odd $k$, condition (22) implies that the homomorphism $h : Z_{k-1}(X;\mathbb{Z}) \oplus \mathbb{Z} \to \mathrm{U}(1)$ vanishes on the second factor, since any $(0,t) \in Z_{k-1}(x;\mathbb{Z})$ is a boundary. Thus $h$ induces a homomorphism $h : Z_{k-1}(X;\mathbb{Z}) \to \mathrm{U}(1)$ that satisfies $h(\partial a) = \exp(2\pi i \int_a \mathrm{curv}(h))$.

We thus obtain a canonical identification of $\widehat{H}^k(x;\mathbb{Z})$ with the group

$$\widehat{H}^k(X;\mathbb{Z}) := \big\{\, h \in \mathrm{Hom}(Z_{k-1}(X;\mathbb{Z}),\mathrm{U}(1)) \,\big|\, h \circ \partial \in \Omega^k(X) \,\big\} \tag{28}$$

of absolute differential characters on $X$, as defined in [16]. The group $\widehat{H}^k(X;\mathbb{Z})$ fits into short exact sequences analogous to the sequences in (24) with mapping cone cohomology groups replaced by the corresponding absolute cohomology groups, and similarly for the spaces of differential forms.

### 3.1.5 Long Exact Sequence

Pre-composition with the maps $i$ and $p$ in the exact sequence (14) induces homomorphisms $\breve{p}_\varphi$ and $\breve{\imath}_\varphi$ on differential characters groups

$$\widehat{H}^{k-1}(A;\mathbb{Z}) \xrightarrow{\breve{\imath}_\varphi} \widehat{H}^k_\varphi(X,A;\mathbb{Z}) \xrightarrow{\breve{p}_\varphi} \widehat{H}^k(X;\mathbb{Z}) .$$

Thus for a character $h \in \widehat{H}^{k-1}(A;\mathbb{Z})$ and a relative cycle $(s,t) \in Z_{k-1}(\varphi;\mathbb{Z})$, we have $\breve{\imath}_\varphi(h)(s,t) := h(t)$. Likewise for a relative character $h \in \widehat{H}^k(\varphi;\mathbb{Z})$ and a cycle $z \in Z_{k-1}(X;\mathbb{Z})$, we have $\breve{p}_\varphi(h)(z) := h(z,0)$. One easily checks that

$$\operatorname{curv} \circ \breve{\imath}_\varphi \equiv 0 \tag{29}$$

$$\operatorname{cov} \circ \breve{\imath}_\varphi \equiv -\operatorname{curv} \tag{30}$$

$$\text{and} \qquad \operatorname{curv} \circ \breve{p}_\varphi \equiv \operatorname{curv} . \tag{31}$$

Let $\psi : B \to Y$ be another smooth map. The homomorphisms $\breve{\imath}$ and $\breve{p}$ are natural with respect pull-back along smooth maps $(Y,B) \xrightarrow{(f,g)} (X,A)$: For a character $h \in \widehat{H}^{k-1}(A;\mathbb{Z})$ and a relative cycle $(s,t) \in Z_{k-1}(\psi;\mathbb{Z})$ we have $\big((f,g)^*\breve{\imath}_\varphi(h)\big)(s,t) = \breve{\imath}_\varphi(h)((f,g)_*(s,t)) = h(g_*t) = g^*h(t)$ and hence

$$(f,g)^*\breve{\imath}_\varphi(h) = \breve{\imath}_\psi(g^*h) . \tag{32}$$

Similarly, for a relative character $h \in \widehat{H}^k(\varphi;\mathbb{Z})$ and a cycle $z \in Z_{k-1}(Y;\mathbb{Z})$ we have $\breve{p}_\psi((f,g)^*h)(z) = (f,g)^*h(z,0) = h(f_*z,0) = \breve{p}_\varphi(h)(f_*z) = f^*(\breve{p}_\varphi(h))(z)$ and hence

$$\breve{p}_\psi((f,g)^*h) = f^*\breve{p}_\varphi(h) . \tag{33}$$

In [1, Chap. 8] we show that for $k \geq 2$ the absolute and relative differential characters groups fit into the following long exact sequence:

$$\begin{array}{l}
\cdots \longrightarrow H^{k-3}(A;\mathrm{U}(1)) \longrightarrow H^{k-2}(\varphi;\mathrm{U}(1)) \longrightarrow H^{k-2}(X;\mathrm{U}(1)) \longrightarrow \\
\xrightarrow{j\circ\varphi^*} \widehat{H}^{k-1}(A;\mathbb{Z}) \xrightarrow{\breve{\imath}} \widehat{H}^k(\varphi;\mathbb{Z}) \xrightarrow{\breve{p}} \widehat{H}^k(X;\mathbb{Z}) \longrightarrow \\
\xrightarrow{\varphi^*\circ c} H^k(A;\mathbb{Z}) \longrightarrow H^{k+1}(\varphi;\mathbb{Z}) \longrightarrow H^{k+1}(X;\mathbb{Z}) \longrightarrow \cdots
\end{array} \tag{34}$$

The sequence proceeds as the long exact sequence for singular cohomology with U(1)-coefficients on the left and with integer coefficients on the right.

## 3.2 Sections and Topological Trivializations

Let $h \in \widehat{H}^k(X;\mathbb{Z})$ be a differential character and $\varphi : A \to X$ a smooth map. As in [1, Chap. 8] we say that $h$ *admits sections* along $\varphi$ if $h$ lies in the image of the map $\breve{p} : H^k(\varphi;\mathbb{Z}) \to H^k(X;\mathbb{Z})$. Any preimage $\breve{p}^{-1}(h)$ of $h$ is called a *section* of $h$ along $\varphi$. From the exact sequence (34) we conclude that $h$ admits setions along $\varphi$ if and only if $\varphi^*c(h) = 0$, i.e. if and only if $h$ is topologically trivial along $\varphi$.

### 3.2.1 Sections and Covariant Derivative

We discuss the role that sections and their covariant derivatives play for topological trivializations. We briefly recall the following basic example from [1].

*Example 8.* It is well-known that the group $\widehat{H}^2(X;\mathbb{Z})$ is canonically isomorphic to the group of isomorphism classes of hermitian line bundles with connection (under connection preserving isomorphisms). A differential character $h \in \widehat{H}^k(X;\mathbb{Z})$ corresponds to the holonomy map of a bundle $(L,\nabla)$ under this isomorphism. Holonomy is invariant under connection preserving isomorphisms. Moreover, the characteristic class $c(h) \in \widehat{H}^2(X;\mathbb{Z})$ coincides with the first Chern class of the bundle. For the curvature we have $\operatorname{curv}(h) = \frac{i}{2\pi} \cdot R^\nabla$, where $R^\nabla \in \Omega^2(X; i\mathbb{R})$ is the curvature 2-form of the connection $\nabla$. The image of a differential form $\omega \in \Omega^1(X)$ under the map $\iota : \Omega^1(X) \to \widehat{H}^2(X;\mathbb{Z})$ corresponds to a topologically trivial line bundle with connection 1-form $\omega$. Hence the name *topological trivialization* for the map $\iota$.

In [1, Chap. 8] we show that the group $\widehat{H}^2(\varphi;\mathbb{Z})$ is isomorphic to the group of isomorphism classes of hermitian line bundles with connection $(L,\nabla)$ and section $\sigma : A \to \varphi^*L$ along $\varphi$. The isomorphisms are bundle isomorphisms of $L$ that preserve both the connection and the section. The map $\breve{p}$ corresponds to the forgetful map that ignores the section. The covariant derivative of the character is related to the covariant derivative of the section by $\nabla\sigma = \operatorname{cov}(h) \cdot \sigma \in \Gamma(T^*A \otimes \varphi^*L)$.

Thus for a differential character $h \in \widehat{H}^2(X;\mathbb{Z})$, any preimage $\breve{p}^{-1}(h) \in \widehat{H}^2(\varphi;\mathbb{Z})$ corresponds to an isomorphism class of sections along the map $\varphi$. Hence the name.

By the exact sequence (34), a differential character admits sections along a smooth map $\varphi$ if and only if it is topologically trivial along $\varphi$. Consequently, the character $\varphi^* h$ is topologically trivial. We show that it is trivialized by the covariant derivative of any section of $h$. The special case $\varphi = \mathrm{id}_X$ was discussed in [1, Chap. 8].

**Proposition 9 (Topological Trivialization by Cov. Derivative I).** *Let $\varphi : A \to X$ be a smooth map. Then we have the following commutative diagram:*

$$\begin{array}{ccc} \widehat{H}^k(\varphi;\mathbb{Z}) & \xrightarrow{\breve{p}_\varphi} & \widehat{H}^k(X;\mathbb{Z}) \\ {\scriptstyle \mathrm{cov}}\downarrow & & \downarrow{\scriptstyle \varphi^*} \\ \Omega^{k-1}(A) & \xrightarrow[\iota]{} & \widehat{H}^k(A;\mathbb{Z}) . \end{array} \tag{35}$$

*Thus covariant derivatives of sections along a smooth map yield topological trivializations of the pulled back characters.*

*Proof.* Let $h \in \widehat{H}^k(\varphi;\mathbb{Z})$ be a relative character and $z \in Z_{k-1}(A;\mathbb{Z})$ a cycle. Then we have:

$$\begin{aligned} (\varphi^* \breve{p}_\varphi(h))(z) &= (\breve{p}_\varphi(h))(\varphi_* z) \\ &= h(\varphi_* z, 0) \\ &= h(\partial_\varphi(0, z)) \\ &= \exp\Big(2\pi i \int_{(0,z)} (\mathrm{curv}, \mathrm{cov})(h)\Big) \\ &= \exp\Big(2\pi i \int_z \mathrm{cov}(h)\Big) \\ &= \iota(\mathrm{cov}(h))(z). \end{aligned}$$

□

In [1, Chap. 8] we show that for $\varphi = \mathrm{id}_X$ the covariant derivative is an isomorphism $\mathrm{cov} : \widehat{H}^k(\mathrm{id}_X;\mathbb{Z}) \to \Omega^{k-1}(X)$. We have the followinmg generalization in case $\varphi$ only has a right inverse:

**Proposition 10 (Topological Trivialization by Cov. Derivative II).** *Let $\varphi : A \to X$ and $\psi : X \to A$ be smooth maps such that $\varphi \circ \psi = \mathrm{id}_X$. Then any character $h \in \widehat{H}^k(\varphi;\mathbb{Z})$ is uniquely determined by its covariant derivative. More explicitly, we have:*

$$h = \iota_\varphi(\psi^* \mathrm{cov}(h), 0). \tag{36}$$

*In particular, the evaluation of $h$ on a cycle $(s,t) \in Z_{k-1}(\varphi;\mathbb{Z})$ does not depend on $t$.*

*Proof.* We have $(\varphi, \mathrm{id}_A) \circ (\psi, \mathrm{id}_A) = \mathrm{id}_{(X,A)}$. This yields

$$h = (\psi, \mathrm{id}_A)^* \big((\varphi, \mathrm{id}_A)^* h\big) = (\psi, \mathrm{id}_A)^* \big(\iota_{\mathrm{id}_A}(\mathrm{cov}(h), 0)\big) = \iota_\varphi(\psi^* \mathrm{cov}(h), 0).$$

Since the left hand side is independent of the right inverse $\psi$, so is the right hand side.

In particular, for any cycle $(s,t) \in Z_{k-1}(\varphi;\mathbb{Z})$ we have

$$h(s,t) = \exp\Big(2\pi i \int_s \psi^* \mathrm{cov}(h)\Big). \qquad \square$$

### 3.2.2 Bundle Gerbes

In this section we briefly discuss how bundle gerbes with connection represent relative differential cohomology classes. Recall that a bundle gerbe $\mathcal{G}$ on a smooth manifold $X$ consists of a surjective submersion $\pi : Y \to X$ together with a hermitian line bundle $L \to Y^{[2]}$ over the two-fold fiber product of the submersion, subject to compatibility conditions on the tensor products of the various pull-backs of $L$ to $Y^{[3]}$ and $Y^{[4]}$ (see [43] for a nice overview of bundle gerbes). A bundle gerbe represents a cohomology class in $H^3(X;\mathbb{Z})$, usually called the Dixmier-Douady class of $\mathcal{G}$. Stable isomorphism classes of bundle gerbes on $X$ are in 1-1 correspondence with $H^3(X;\mathbb{Z})$.

A connection on a bundle gerbe $\mathcal{G}$ over $X$ consists of a hermitian connection $\nabla$ on $L$ and a 2-form $B \in \Omega^2(Y)$ subject to the condition $F_\nabla = \pi_2^*B - \pi_1^*B$. Here $F_\nabla \in \Omega^2(Y^{[2]})$ denotes the curvature of the connection $\nabla$ and $\pi_1, \pi_2 : Y^{[2]} \to Y$ denote the projections to the first and second fiber, respectively. The 2-form $B \in \Omega^2(Y)$ in the notion of a bundle gerbe connection is called the *curving*. A connection on a bundle gerbe determines a 3-form $H \in \Omega^3_0(X)$, called the *curvature*. The curvature and curving of a connection satisfy $\pi^*H = dB$. Since the curvature has integral periods, by [1, Cor. 76] the pair $(H,B) \in \Omega^3(\pi)$ is closed with integral periods. Stable isomorphism classes of bundle gerbes on $X$ with connection are classified by the differential cohomology $\widehat{H}^3(X;\mathbb{Z})$.

The trivial bundle gerbe $\mathcal{I}_\varrho$ associated with a 2-form $\varrho \in \Omega^2(X)$ is the bundle gerbe with connection, defined by the submersion $\mathrm{id}_X : X \to X$, the trivial line bundle with trivial connection, and the 2-form $\varrho$. A trivialization of a bundle gerbe is an isomorphism $\mathcal{G} \to \mathcal{I}_\varrho$ to a trivial bundle gerbe. More in detail, a trivialization consists of a hermitian line bundle with connection $(\mathcal{L}, A) \to Y$ together with a 2-form $\varrho \in \Omega^2(X)$ subject to the condition $F_A = \pi^*\varrho - B$ and further compatibility conditions on tensor products of the pull-backs of $L$ and $\mathcal{L}$ to $Y^{[2]}$ and $Y^{[3]}$. A bundle gerbe admits a trivialization if and only if its Dixmier-Douady class vanishes.

A bundle gerbe $\mathcal{G}$ with submersion $\pi : Y \to X$ comes together with a *canonical* trivialization of the pull-back bundle gerbe $\pi^*\mathcal{G}$. In particular, a cohomology class $c \in H^3(X;\mathbb{Z})$ is the Dixmier-Douady class of a bundle $\mathcal{G}$ with submersion $\pi : Y \to X$ if and only if $\pi^*c = 0$.

Since stable isomorphisms of bundle gerbes do not preserve the submersion $\pi : Y \to X$, the associated differential cohomology class in $\widehat{H}^3(X;\mathbb{Z})$, represented by the bundle gerbe, does not inherit any information about the submersion. The additional data provided by the actual submersion can be captured into a *relative* differential cohomology class: Any bundle gerbe $\mathcal{G}$ with connection, represented by a submersion $\pi : Y \to X$, determines a relative differential character $h_{\mathcal{G}} \in \widehat{H}^3(\pi;\mathbb{Z})$. The character $h_{\mathcal{G}}$ maps to the stable ismomorphism class of $\mathcal{G}$ under the map $\breve{p} : \widehat{H}^3(\pi;\mathbb{Z}) \to \widehat{H}^3(X;\mathbb{Z})$. We define the character $\mathcal{G}$ by its evaluations on relative cycles:

**Definition 11.** Let $\mathcal{G}$ be a bundle gerbe with connection, represented by a submersion $\pi : Y \to X$. Let $(s,t) \in Z_2(\pi;\mathbb{Z})$ be a cycle. Choose a geometric relative cycle $(\zeta,\tau) \in \mathcal{Z}_2(\pi)$, represented by a smooth map $(S,T) \xrightarrow{(f,g)} (X,Y)$, such that $(\zeta,\tau)$ represents the homology class of $(s,t)$. Choose a relative chain $(a,b) \in C_3(\pi;\mathbb{Z})$ such that $[\zeta,\tau]_{\partial_\pi S_3} = [(s,t) - \partial_\pi(a,b)]_{\partial_\pi S_3}$. Since $\dim(S) = 2$, the pull-back bundle gerbe $f^*\mathcal{G}$ is trivial. Choose a trivialization $\mathcal{G} \to \mathcal{I}_\varrho$ with line bundle $(\mathcal{L},A) \to f^*Y$. The map $g : T \to Y$ induces a map $g : T \to f^*Y$. Since $\dim(T) = 1$, the pull-back bundle $g^*\mathcal{L} \to T$ is topologically trivial. Any topological trivialization provides a 1-form $\vartheta \in \Omega^1(T)$ such that $d\vartheta = g^*F_A$.[3] Now put

$$h_{\mathcal{G}}(s,t) := \exp\Big(2\pi i\Big(\int_{(S,T)} (\varrho,\vartheta) + \int_{(a,b)} (H,B)\Big)\Big) \tag{37}$$

Since the choices of $(S,T) \in \mathcal{Z}_2(\pi)$ and $(a,b) \in C_3(\pi;\mathbb{Z})$ can be organized to be homomorphisms, we have defined a homomorphism $h_{\mathcal{G}} : Z_2(\pi;\mathbb{Z}) \to \mathrm{U}(1)$. It remains to check that $h_{\mathcal{G}}$ is indeed a differential character. Thus let $(s,t) = \partial_\pi(v,u)$ for some relative chain $(v,u) \in C_3(\pi;\mathbb{Z})$. Since $(\zeta,\tau)$ represents the trivial homology class, we find a bordism $(W,M) \xrightarrow{(F,G)} (X,Y)$ from the empty cycle to $(\zeta,\tau)$. By definition, the map $F : W \to X$ extends the map $f : S \to X$. Since $\partial W = \overline{M} \cup S$ is 2-dimensional, the pull-back bundle gerbe $F|_{\partial W}^*\mathcal{G}$ is still trivial, and we find a trivialization $F^*\mathcal{G} \to \mathcal{I}_{\varrho'}$ compatible with the one on $S \subset \partial W$. Then we have

$$\int_{\overline{M}\cup S} \varrho' = \int_{\partial W} \varrho' = \int_W d\varrho' = \int_W H.$$

Since $\overline{T} = \partial S = \partial M$, and $G : M \to f^*Y$ extends the map $g : T \to f^*Y$, we obtain a trivialization of $G^*\mathcal{L}$ along $\partial M$ by a 1-form $\vartheta' \in \Omega^1(\partial M)$, compatible with the trivialization of $g^*\mathcal{L}$ over $T$. This yields:

$$\int_T \vartheta = -\int_{\partial M} \vartheta' = -\int_M d\vartheta' = -\int_M G^*F_A.$$

Thus we obtain

[3] In other words, the isomorphism class $[g^*(\tilde{L},A)] \in \widehat{H}^2(S;\mathbb{Z})$ is given by $\iota(\vartheta)$, as explained in detail in [1, Chap. 5].

$$\exp\Big(2\pi i\int_{(S,T)}(\varrho,\vartheta)\Big)=\exp\Big(2\pi i\int_{\partial W}\varrho'+\int_M\varrho'-G^*F_A\Big)$$
$$=\exp\Big(2\pi i\int_{(W,M)}(H,B)\Big).$$

By definition of the bordism relation, we have

$$\begin{aligned}\partial_\pi(F,G)_*[W,M]_{S_3}&=(f,g)_*[S,T]_{\partial_\pi S_3}\\&=[\zeta,\tau]_{\partial_\pi S_3}\\&=[\partial_\pi(v,u)-\partial_\pi(a,b)]_{\partial_\pi S_3}.\end{aligned}$$

Thus there exists a cycle $(x,y)\in Z_3(\pi;\mathbb{Z})$ such that

$$(F,G)_*[W,M]_{S_3}=[(v,u)-(a,b)-(x,y)]_{S_3}.$$

Putting everything together, we obtain:

$$\begin{aligned}h_{\mathcal{G}}(\partial_\pi(v,u))&=\exp\Big(2\pi i\Big(\int_{(S,T)}(\varrho,\vartheta)+\int_{(a,b)}(H,B)\Big)\Big)\\&=\exp\Big(2\pi i\Big(\int_{(W,M)}(H,B)+\int_{(a,b)}(H,B)\Big)\Big)\\&=\exp\Big(2\pi i\int_{(v,u)}(H,B)\Big).\end{aligned}$$

In the last step we used the fact that the form $(H,B)\in\Omega^3(\pi)$ is closed with integral periods. This follows from $\pi^*H-dB=0$ and the fact that $H$ has integral periods. Thus $h_{\mathcal{G}}$ is indeed a relative differential character. It follows immediately from the definition, that the character $h_{\mathcal{G}}$, evaluated on relative cycles of the form $(z,0)$ for $z\in Z_2(X;\mathbb{Z})$, coincides with the stable isomorphism class of the bundle gerbe. For in this case, the formula (37) reduces to the well-known description of surface holonomy of the bundle gerbe.

Thus we have proved:

**Proposition 12.** *Let $\mathcal{G}$ be a bundle gerbe with connection, represented by a submersion $\pi: Y\to X$. Let $(H,B)\in\Omega^3(\pi)$ denote the curvature and curving of $\mathcal{G}$. Then $h_{\mathcal{G}}$ as defined in (37) is a relative differential character in $\widehat{H}^3(\pi;\mathbb{Z})$ with*

$$(\mathrm{curv},\mathrm{cov})(h_{\mathcal{G}})=(H,B). \tag{38}$$

*Moreover, $\breve{p}(h_{\mathcal{G}})\in\widehat{H}^3(X;\mathbb{Z})$ coincides with the stable isomorphism class of the bundle gerbe with connection $\mathcal{G}$.*

Analogously, a bundle 2-gerbe with connection $\mathcal{G}$, represented by a submersion $\pi: Y\to X$, defines a relative differential character $h_{\mathcal{G}}\in\widehat{H}^4(\pi;\mathbb{Z})$, see [2].

### 3.2.3 The Cheeger-Chern-Simons Construction

A particular example of relative differential characters as sections of absolute characters along a smooth map arises by the differential character valued refinement of the Chern-Weil construction, due to Cheeger and Simons:

*Example 13.* Let $G$ be a compact Lie group with Lie algebra $\mathfrak{g}$. An invariant polynomial, homogeneous of degree $k$, is a symmetric $\mathrm{Ad}_G$-invariant multilinear map $q : \mathfrak{g}^{\otimes k} \to \mathbb{R}$. The Chern-Weil construction associates to any principal $G$-bundle with connection $(P, \nabla) \to X$ a closed differential form $CW(q) = q(R^\nabla) \in \Omega^{2k}(X)$ by applying the polynomial $q$ to the curvature 2-form $R^\nabla$ of the connection $\nabla$. Consider those polynomials $q$ for which the Chern-Weil form $CW(q)$ has integral periods. Let $u \in H^{2k}(X;\mathbb{Z})$ be a universal characteristic class for principal $G$-bundles that coincides in $H^{2k}(X;\mathbb{R})$ with the de Rham class of $CW(q)$. The *Cheeger-Simons construction* [16, Thm 2.2] associates to this setting a differential character $\widehat{CW}(q,u) \in \widehat{H}^{2k}(X;\mathbb{Z})$ with curvature $\mathrm{curv}(\widehat{CW}(q,u)) = CW(q)$, the Chern-Weil form, and characteristic class $c(\widehat{CW}(q,u)) = u$, the fixed universal characteristic class. The construction is natural with respect to bundle maps.

Since the total space $EG$ of the universal principal $G$-bundle is contractible, universal characteristic classes vanish upon pull-back to the total space. By the long exact sequence (34) the Cheeger-Simons character $\widehat{CW}(q,u)$ thus admits sections along the bundle projection $\pi : P \to X$. The so-called *Cheeger-Chern-Simons construction* of [2] yields a canonical section $\widehat{CCS}(q,u) \in \widehat{H}^{2k}(\pi;\mathbb{Z})$ with covariant derivative $\mathrm{cov}(\widehat{CCS}(q,u)) = CS(q) \in \Omega^{2k-1}(P)$, the Chern-Simons form. The construction is natural with respect to bundle maps.

Thus the Cheeger-Chern-Simons construction is a relative differential character valued refinement of the Chern-Weil and Chern-Simons constructions in the same way as the Cheeger-Simons construction is a differential character valued refinement of the Chern-Weil construction alone.

### 3.2.4 Parallel Sections

In general, the property for a given character to admit sections with prescribed covariant derivatives depends on the character. For example, a hermitian line bundle with connection $(L, \nabla)$ and with sections along a smooth map $\varphi : A \to X$ admits *parallel* sections if and only if the pull-back $\varphi^*(L, \nabla)$ is isomorphic to the trivial bundle with trivial connection.

The analogous statement holds for any differential characters, as we shall prove next:

**Theorem 14 (Parallel Sections).** *A differential character $h \in \widehat{H}^k(X;\mathbb{Z})$ admits parallel sections along a smooth map $\varphi : A \to X$ if and only if $\varphi^*h = 0$.*

*Proof.* Let $h \in \widehat{H}^k(X;\mathbb{Z})$ with $\varphi^* h = 0$. Then in particular $h$ is topologically trivial along $\varphi$ and hence admits sections along $\varphi$. By the commutative diagram (35) and the exact sequence (24), the covariant derivative of any such section $h' \in \widehat{H}^k(\varphi;\mathbb{Z})$ satisfies $\mathrm{cov}(h') \in \Omega_0^{k-1}(A)$, i.e. it is closed with integral periods. Choose a character $h'' \in \widehat{H}^{k-1}(A;\mathbb{Z})$ with $\mathrm{curv}(h'') = \mathrm{cov}(h')$. Now put $h''' := h' + \breve{\imath}(h'')$. Then we have $\breve{p}(h''') = \breve{p}(h') = h$ and $\mathrm{cov}(h''') = \mathrm{cov}(h') - \mathrm{curv}(h'') = 0$. Thus $h'''$ is a parallel section of $h$.

Conversely let $h' \in \widehat{H}^k(\varphi;\mathbb{Z})$ be a parallel section of $h \in \widehat{H}^k(X;\mathbb{Z})$. By the commutative diagram (35) we find $\varphi^* h = \iota(\mathrm{cov}(h')) = 0$. □

## 3.3 Parallel Characters

Throughout this section let $i_A : A \to X$ be the embedding of a smooth submanifold. As explained in the introduction, there is another notion of relative differential cohomology, based on homomorphisms on the group of relative cycles $Z_{*-1}(X,A;\mathbb{Z})$. This notion has appeared in [21] for the special case where $A = \partial X$ is the boundary of $X$.

Let $X$ be a smooth manifold and $A \subset X$ an embedded smooth submanifold. In this section, we introduce differential characters on the group $Z_{*-1}(X,A;\mathbb{Z})$ of relative cycles. We denote the corresponding group of differential characters by $\widehat{H}^*(X,A;\mathbb{Z})$. We prove a long exact sequence that relates this group to differential characters groups on $X$ and $A$. An analogous sequence has appeared in [42] for generalized differential cohomology. Further we show that $\widehat{H}^*(X,A;\mathbb{Z})$ is in 1-1 correspondence with the subgroup of parallel characters in $\widehat{H}^*(i_A;\mathbb{Z})$.

### 3.3.1 Characters on Relative Cycles

Let $X$ be a smooth manifold and $A \subset X$ an embedded smooth submanifold. Denote the embedding by $i_A : A \to X$. Let $Z_*(X,A;\mathbb{Z})$ be the group of relative cycles, i.e. cycles in the quotient complex $C_*(X,A;\mathbb{Z}) := C_*(X;\mathbb{Z})/C_*(A;\mathbb{Z})$.

We put

$$\widehat{H}^k(X,A;\mathbb{Z}) := \left\{ h \in \mathrm{Hom}(Z_{k-1}(X,A;\mathbb{Z}), \mathrm{U}(1)) \,\middle|\, f \circ \partial \in \Omega^k(X) \right\}.$$

The notation $f \circ \partial_\varphi \in \Omega^k(X)$ means that there exists a differential form $\omega \in \Omega^k(X)$ such that for every relative chain $x \in C_k(X,A;\mathbb{Z})$ we have

$$f(\partial c) = \exp\left(2\pi i \int_c \omega\right). \tag{39}$$

The condition in particular implies that the integral of $\omega$ over chains $c \in C_k(X,A;\mathbb{Z})$ is well-defined. Hence $i_A^*\omega \equiv 0$.

Since condition (39) holds for all chains $c \in C_k(X,A;\mathbb{Z})$, the differential form $\omega$ is uniquely determined. We call it the *curvature* of $h$ and denote it by $\mathrm{curv}(h)$. A character $h \in \widehat{H}^k(X,A;\mathbb{Z})$ with $\mathrm{curv}(h) = 0$ is called *flat*.

Let $\Omega^k(X,A) := \{\omega \in \Omega^k(X) \,|\, i_A^*\omega \equiv 0\}$ be the space of $k$-forms on $X$ relative $A$. As we have seen, a character $h \in \widehat{H}^k(X,A;\mathbb{Z})$ has curvature $\mathrm{curv}(h) \in \Omega^k(X,A)$. We define the relative de Rham cohomology $H^*_{\mathrm{dR}}(X,A)$ as the cohomology of the de Rham subcomplex $(\Omega^*(X,A),d) \subset (\Omega^*(X),d)$. The short exact sequence of de Rham complexes

$$0 \longrightarrow \Omega^k(X,A) \longrightarrow \Omega^k(X) \xrightarrow{i_A^*} \Omega^k(A) \longrightarrow 0$$

gives rise to a long exact sequence relating absolute and relative de Rham cohomology groups.

Integration of differential forms over smooth singular chains in $X$ yields a well-defined homomorphism $\Omega^k(X,A) \to C^k(X,A;\mathbb{R})$. By the de Rham theorem and the five lemma, applied to the long exact sequences, this induces a canonical isomorphism $H^*_{\mathrm{dR}}(X,A) \xrightarrow{\cong} H^*(X,A;\mathbb{R})$. Denote by $\Omega_0^k(X,A)$ the subgroup of differential forms $\omega \in \Omega^k(X,A)$ with integral periods, i.e. such that $\int_y \omega \in \mathbb{Z}$ holds for any $y \in Z_k(X,A;\mathbb{Z})$. Then we have $\Omega_0^k(X,A) = \{\omega \in \Omega_0^k(X) \,|\, i_A^*\omega \equiv 0\}$.

Since $h$ is a homomorphism, condition (39) implies that $\int_z \mathrm{curv}(h) \in \mathbb{Z}$ holds for any cycle $z \in Z_k(X;\mathbb{Z})$. Stokes theorem implies that $\int_c d\mathrm{curv}(h) = \int_{\partial c} \mathrm{curv}(h) \in \mathbb{Z}$ holds for any chain $c \in C_{k+1}(X;\mathbb{Z})$, hence $\mathrm{curv}(h)$ is closed. Thus $\mathrm{curv}(h) \in \Omega_0^k(X,A)$.

The *characteristic class* $c(h) \in H^k(X,A;\mathbb{Z})$ of a character $h \in \widehat{H}^k(X,A;\mathbb{Z})$ is defined as follows: The curvature defines a cocycle $\mathrm{curv}(h) : C_k(X,A;\mathbb{R}) \to \mathbb{R}$, $c \mapsto \int_c \mathrm{curv}(h)$. Choose a real lift $\tilde{h} \in C^{k-1}(X,A;\mathbb{R})$ of $h$, i.e. $h(z) = \exp\left(2\pi i \tilde{h}(z)\right)$. Put $\mu^{\tilde{h}} := \mathrm{curv}(h) - \delta\tilde{h} \in C^k(X,A;\mathbb{Z})$. In fact, $\mu^{\tilde{h}}$ is an integral cochain because of (39). Since the curvature is a closed form, $\mu^{\tilde{h}}$ is a cocycle. Now define $c(h) := [\mu^{\tilde{h}}] \in H^k(X,A;\mathbb{Z})$. It is easy to see that $c(h)$ does not depend upon the choice of real lift $\tilde{h}$: The difference between two choices of real lifts is an integral cochain. Thus the cocycles for two choices of real lifts differ by an integral coboundary. A character $h \in \widehat{H}^k(X,A;\mathbb{Z})$ with $c(h) = 0$ is called *topologically trivial*.

### 3.3.2 Exact Sequences

We have a natural map $\iota : \Omega^{k-1}(X,A) \to \widehat{H}^k(X,A;\mathbb{Z})$, defined by $\iota(\vartheta)(z) := \exp\left(2\pi i \int_z \vartheta\right)$. By the Stokes theorem, the induced character $\iota(\vartheta)$ satisfies

$\mathrm{curv}(\iota(\vartheta)) = d\vartheta$. The map $\iota$ descends to an injective map $\iota : \frac{\Omega^{k-1}(X,A)}{\Omega_0^{k-1}(X,A)} \to \widehat{H}^k(X,A;\mathbb{Z})$. A form $\omega \in \Omega^{k-1}(X,A)$ such that $\iota(\omega) = h$ is called a *topological trivialization* of $h$.

Finally, we have an obvious injection $j : H^{k-1}(X,A;\mathrm{U}(1)) \to \widehat{H}^k(X,A;\mathbb{Z})$, defined by $j(u)(z) := \langle u, [z] \rangle$.

The above maps fit into the following exact sequences:

$$0 \longrightarrow H^{k-1}(X,A;\mathrm{U}(1)) \xrightarrow{j} \widehat{H}^k(X,A;\mathbb{Z}) \xrightarrow{\mathrm{curv}} \Omega_0^k(X,A) \longrightarrow 0$$

$$0 \longrightarrow \frac{\Omega^{k-1}(X,A)}{\Omega_0^{k-1}(X,A)} \xrightarrow{\iota} \widehat{H}^k(X,A;\mathbb{Z}) \xrightarrow{c} H^k(X,A;\mathbb{Z}) \longrightarrow 0\,. \tag{40}$$

Exactness of the curvature sequence at $\widehat{H}^k(X,A;\mathbb{Z})$ is clear, since by (39) flat characters are precisely those that vanish on boundaries and hence descend to homomorphisms on $H_{k-1}(X;\mathbb{Z})$. Surjectivity of the curvature follows e.g. from surjectivity of the curvature map $\mathrm{curv} : \widehat{H}^k(X;\mathbb{Z}) \to \Omega_0^k(X)$ and the exact sequence (42) below.

Exactness of the characteristic class sequence at $\widehat{H}^k(X,A;\mathbb{Z})$ follows from (39) and the definition of the characteristic class. Surjectivity of the curvature follows e.g. from surjectivity of the characteristic class map $c : \widehat{H}^k(X;\mathbb{Z}) \to H^k(X;\mathbb{Z})$ and the exact sequence (42) below.

The second sequence in (40) tells us that a character $h \in \widehat{H}^k(X,A;\mathbb{Z})$ admits a topological trivilization if and only if it is topologically trivial.

Put $R^k(X,A;\mathbb{Z}) := \{(\omega,u) \in \Omega_0^k(X,A) \times H^k(X,A;\mathbb{Z}) \,|\, u_{\mathbb{R}} = [\omega]_{dR} \in H^k(X,A;\mathbb{R})\}$. Then the two sequences above may be joined to give the following exact sequence:

$$0 \longrightarrow \frac{H^{k-1}(X,A;\mathbb{R})}{H^{k-1}(X,A;\mathbb{R})_{\mathbb{Z}}} \xrightarrow{j} \widehat{H}^k(X,A;\mathbb{Z}) \xrightarrow{(\mathrm{curv},c)} R^k(X,A;\mathbb{Z}) \longrightarrow 0\,. \tag{41}$$

### 3.3.3 Absolute Differential Characters

Let $x \in X$ be an arbitrary point. We write $x$ instead of $\{x\} \subset X$. For positive even $k$, we have $Z_{k-1}(x;\mathbb{Z}) = \mathbb{Z} = B_{k-1}(x;\mathbb{Z})$, while for odd $k$, we have $Z_{k-1}(x;\mathbb{Z}) = \{0\}$. Since differential forms of degree $k \geq 1$ vanish upon pull-back to $x$, we obtain a canonical identification

$$\widehat{H}^k(X,x;\mathbb{Z}) \xrightarrow{\cong} \widehat{H}^k(X,\mathbb{Z})\,.$$

### 3.3.4 Long Exact Sequence

Pre-composition of a differential character $h : Z_{k-1}(X,A;\mathbb{Z}) \to \mathrm{U}(1)$ with the quotient map $Z_{k-1}(X;\mathbb{Z}) \to Z_{k-1}(X,A;\mathbb{Z})$, $z \mapsto z + \mathrm{im}(i_{A*})$, yields a homomorphis $h' : Z_{k-1}(X;\mathbb{Z}) \to \mathrm{U}(1)$. This homomorphism is in fact a differential character in $\widehat{H}^k(X;\mathbb{Z})$, since for any chain $c \in C_k(X;\mathbb{Z})$, we have

$$h'(\partial c) := h(\partial c + \mathrm{im}(i_{A*})) \overset{(39)}{=} \exp\Big(2\pi i \int_c \mathrm{curv}(h)\Big) .$$

Hence $\mathrm{curv}(h') = \mathrm{curv}(h) \in \Omega^k_0(X,A) \subset \Omega^k_0(X)$. We thus obtain a homomorphism $\widehat{H}^k(X,A;\mathbb{Z}) \to \widehat{H}^k(X;\mathbb{Z})$ that preserves the curvature. Moreover, a real lift for $h \in \widehat{H}^k(X,A;\mathbb{Z})$ also defines a real lift of its image in $\widehat{H}^k(X;\mathbb{Z})$. Thus the homomorphism is also compatible with the characteristic class and we obtain the following commutative diagram:

$$\begin{array}{ccc} \widehat{H}^k(X,A;\mathbb{Z}) & \longrightarrow & \widehat{H}^k(X;\mathbb{Z}) \\ \Big\downarrow c & & \Big\downarrow c \\ H^k(X,A;\mathbb{Z}) & \longrightarrow & H^k(X;\mathbb{Z}) , \end{array}$$

Here the lower horizontal map is the usual map in the long exact sequence for absolute and relative cohomology.

We denote the connecting homomorphism in the long exact sequence for relative and absolute cohomology by $\beta : H^*(A;\mathbb{Z}) \to H^{*+1}(X,A;\mathbb{Z})$ (and likewise for $\mathrm{U}(1)$ coefficients). Concatenation with $j$ yields a map : $H^{k-2}(A;\mathrm{U}(1)) \xrightarrow{\beta} H^{k-1}(X,A;\mathrm{U}(1)) \xrightarrow{j} \widehat{H}^k(X,A;\mathbb{Z})$. Likewise, we obtain a map $\widehat{H}^k(A;\mathbb{Z}) \xrightarrow{c} H^k(A;\mathbb{Z}) \xrightarrow{\beta} H^{k+1}(A;\mathbb{Z})$. These maps fit into the following long exact sequence:

**Theorem 15 (Long Exact Sequence).** *Let $i_A : A \hookrightarrow X$ be the embedding of a smooth submanifold. Let $k \geq 1$. Then we have the following long exact sequence for the groups of differential characters:*

$$\begin{aligned} \ldots \longrightarrow H^{k-2}(X,A;\mathrm{U}(1)) \longrightarrow H^{k-2}(X;\mathrm{U}(1)) \longrightarrow H^{k-2}(A;\mathrm{U}(1)) \\ \xrightarrow{j\circ\beta} \widehat{H}^k(X,A;\mathbb{Z}) \longrightarrow \widehat{H}^k(X;\mathbb{Z}) \xrightarrow{i_A^*} \widehat{H}^k(A;\mathbb{Z}) \\ \xrightarrow{\beta\circ c} H^{k+1}(X,A;\mathbb{Z}) \longrightarrow H^{k+1}(X;\mathbb{Z}) \longrightarrow H^{k+1}(A;\mathbb{Z}) \longrightarrow \ldots \end{aligned} \tag{42}$$

*The sequence proceeds as the long exact sequence for singular cohomology with* U(1)*-coefficients on the left and with integer coefficients on the right.*

*Remark 16.* Exactness at $\widehat{H}^k(X;\mathbb{Z})$ also follows from Theorem 14 above and Theorem 17 below.

*Proof.* Exactness at the first two and the last two groups is of course well-known. We give a direct proof of the exactness at the remaining groups:

a) Exactness of the sequence (42) at $H^{k-2}(A;\mathrm{U}(1))$ follows from exactness of the sequence $H^{k-2}(X;\mathrm{U}(1)) \xrightarrow{i_A^*} H^{k-2}(A,\mathrm{U}(1)) \xrightarrow{\beta} H^{k-1}(X,A;\mathrm{U}(1))$ and injectivity of the map $H^{k-1}(X,A;\mathrm{U}(1)) \xrightarrow{j} \widehat{H}^k(X,A;\mathbb{Z})$.

b) We prove exactness at $\widehat{H}^k(X,A;\mathbb{Z})$: Let $u \in H^{k-2}(A;\mathrm{U}(1))$ and $z \in Z_{k-1}(X;\mathbb{Z})$. From divisibility of U(1) and the universal coefficient theorem, we obtain the identification $H^{k-2}(A;\mathrm{U}(1)) = \mathrm{Hom}(H_{k-2}(A;\mathbb{Z}),\mathrm{U}(1))$. The connecting homomorphism $\beta : H^{k-2}(A;\mathrm{U}(1)) \to H^{k-1}(X,A;\mathrm{U}(1))$ is dual to the connecting homomorphism $\beta : H_{k-1}(X,A;\mathbb{Z}) \to H_{k-2}(A;\mathbb{Z})$ in homology. This yields:

$$\begin{aligned}
(j\circ\beta(u))(z) &= (j\circ\beta(u))(z+\mathrm{im}(i_{A*})) \\
&= \langle \beta(u), [z+\mathrm{im}(i_{A*})]\rangle \\
&= \langle u, \underbrace{\beta([z+\mathrm{im}(i_{A*})])}_{=0}\rangle \\
&= 1\,.
\end{aligned}$$

Here we use the fact that $H_{k-1}(X,\mathbb{Z}) \to H_{k-1}(X,A;\mathbb{Z}) \xrightarrow{\beta} H_{k-2}(A;\mathbb{Z})$, $[z] \mapsto \beta([z+\mathrm{im}(i_{A*})])$, is the trivial map.

Conversely, let $h \in \widehat{H}^k(X,A;\mathbb{Z})$ such that the induced character in $\widehat{H}^k(X;\mathbb{Z})$ vanishes. In particular, we have $\mathrm{curv}(h) = 0$. By the exact sequence (40) we find $\tilde{u} \in H^{k-1}(X,A;\mathrm{U}(1))$ such that $h = j(\tilde{u})$. By assumption, $h$ vanishes on cycles in $X$, hence $\tilde{u}$ lies in the kernel of the map $H^{k-1}(X,A;\mathrm{U}(1)) \to H^k(X;\mathrm{U}(1))$. Thus we find $u \in H^{k-2}(A;\mathrm{U}(1))$ such that $\tilde{u} = \beta(u)$ and hence $h = j\circ\beta(u)$.

c) We show exactness at $\widehat{H}^k(X;\mathbb{Z})$: The map $\widehat{H}^k(X,A;\mathbb{Z}) \to \widehat{H}^k(X;\mathbb{Z}) \xrightarrow{i_A^*} \widehat{H}^k(A;\mathbb{Z})$ is trivial, since cycles in $Z_{k-1}(A;\mathbb{Z})$ represent 0 in $Z_{k-1}(X,A;\mathbb{Z})$.

Conversely, let $h \in \widehat{H}^k(X;\mathbb{Z})$ with $i_A^*h = 0$. Let $\sigma : B_{k-2}(X;\mathbb{Z}) \to C_{k-1}(X;\mathbb{Z})$ be a splitting of the exact sequence

$$0 \to Z_{k-1}(X;\mathbb{Z}) \to C_{k-1}(X;\mathbb{Z}) \to B_{k-2}(X;\mathbb{Z}) \to 0\,.$$

Let $V := \sigma(B_{k-2}(X;\mathbb{Z})) \cap \{c \in C_{k-1}(X;\mathbb{Z}) \,|\, \partial c \in \mathrm{im}(i_{A*})\} \subset C_{k-1}(X;\mathbb{Z})$. Since $V$ is a submodule of $C_{k-1}(X;\mathbb{Z})$, it is a free $\mathbb{Z}$-module, and we have the splittings

$$\{c \in C_{k-1}(X;\mathbb{Z}) \,|\, \partial c \in \mathrm{im}(i_{A*})\} = Z_{k-1}(X;\mathbb{Z}) \oplus V$$

and

$$Z_{k-1}(X,A;\mathbb{Z}) = \frac{Z_{k-1}(X;\mathbb{Z})}{\operatorname{im}(i_{A*})} \oplus \frac{V}{\operatorname{im}(i_{A*})}\,.$$

By assumption the character $h : Z_{k-1}(X;\mathbb{Z}) \to \mathrm{U}(1)$ vanishes upon pull-back to $A$. Hence it descends to a homomorphism $\overline{h}$ on the first factor. The above splitting allows us to extend $\overline{h}$ to a homomorphism $\overline{h} : Z_{k-1}(X,A;\mathbb{Z}) \to \mathrm{U}(1)$. By construction, any such extension $\overline{h}$ is a differential character in $\widehat{H}^k(X,A;\mathbb{Z})$ which maps to $h$ under the map $\widehat{H}^k(X,A;\mathbb{Z}) \to \widehat{H}^k(X;\mathbb{Z})$.

d) We show exactness at $\widehat{H}^k(A;\mathbb{Z})$: Since the characteristic class is natural with respect to pull-back, we have $\beta \circ c \circ i_A^* = (\beta \circ i_A^*) \circ c = 0$.

Conversely, let $h \in \widehat{H}^k(A;\mathbb{Z})$ such that $\beta \circ c(h) = 0$. By exactness of the sequence $H^k(X;\mathbb{Z}) \xrightarrow{i_A^*} H^k(A;\mathbb{Z}) \xrightarrow{\beta} H^{k+1}(X,A;\mathbb{Z})$ and surjectivity of the characteristic class we find a character $h' \in \widehat{H}^k(X;\mathbb{Z})$ such that $i_A^*h' - h = \iota(\vartheta)$ for some differential form $\vartheta \in \Omega^{k-1}(A)$. Choose a differential form $\vartheta' \in \Omega^{k-1}(X)$ such that $\vartheta = i_A^*\vartheta'$. Now put $h'' := h' + \iota(\varrho') \in \widehat{H}^k(X;\mathbb{Z})$. Then we have $i_A^*h'' = i_A^*h' + \iota(\varrho) = h$.

e) Finally, exactness at $H^{k+1}(X,A;\mathbb{Z})$ follows from exactness of the sequence $H^k(X;\mathbb{Z}) \xrightarrow{i_A^*} H^k(A;\mathbb{Z}) \xrightarrow{\beta} H^{k+1}(X,A;\mathbb{Z})$ and surjectivity of the characteristic class. □

### 3.3.5 Comparison of Two Notions of Relative Differential Cohomology

Now we compare the two notions of relative differential cohomology, based on differential characters on $Z_{k-1}(i_A;\mathbb{Z})$ and $Z_{k-1}(X,A;\mathbb{Z})$, respectively. Pre-composition of relative differential characters $h : Z_{k-1}(i_A;\mathbb{Z}) \to \mathrm{U}(1)$ with the projection map $q : Z_{k-1}(i_A;\mathbb{Z}) \twoheadrightarrow Z_{k-1}(X,A;\mathbb{Z})$, $(s,t) \mapsto s + \operatorname{im}(i_{A*})$, yields a homomorphism

$$\widehat{H}^k(X,A;\mathbb{Z}) \to \widehat{H}^k(i_A;\mathbb{Z})\,, \quad h \mapsto h \circ q\,.$$

**Theorem 17 (Comparison of Relative Differential Cohomologies).** *Let $i_A : A \to X$ be the embedding of a smooth submanifold. Let $k \geq 2$. Then the homomorphism*

$$\widehat{H}^k(X,A;\mathbb{Z}) \to \widehat{H}^k(i_A;\mathbb{Z})\,, \quad h \mapsto h \circ q\,,$$

*commutes with curvature, characteristic class and the inclusion of cohomology classes in $H^{k-1}(X,A;\mathrm{U}(1))$. It provides a 1-1 correspondence of $\widehat{H}^k(X,A;\mathbb{Z})$ with the subgroup of parallel characters in $\widehat{H}^k(i_A;\mathbb{Z})$.*

*Proof.* Let $h \in \widehat{H}^k(X,A;\mathbb{Z})$ and $(a,b) \in C_k(i_A;\mathbb{Z})$. Then we have

$$(h \circ q)(\partial_{i_A}(a,b)) = h(\partial a + \mathrm{im}(i_{A*})) \overset{(39)}{=} \exp\Big(2\pi i \int_a \mathrm{curv}(h)\Big).$$

Thus the composition $h \circ q$ is indeed a relative differential character in $\widehat{H}^k(i_A;\mathbb{Z})$ and we have $(\mathrm{curv},\mathrm{cov})(h \circ q) = (\mathrm{curv}(h),0)$.

The projection $q : Z_{k-1}(i_A;\mathbb{Z}) \to Z_{k-1}(X,A;\mathbb{Z})$ induces isomorphisms on homology and cohomology. Moreover, since $\mathrm{U}(1)$ is divisible, we have the identification $H^{k-1}(X,A;\mathrm{U}(1)) \cong \mathrm{Hom}(H_{k-1}(X,A;\mathbb{Z}),\mathrm{U}(1))$. Hence the above homomorphism commutes with the inclusion of cohomology classes $u \in H^{k-1}(X,A;\mathrm{U}(1))$.

Now let $\tilde{h} \in C^k(X,A;\mathbb{Z})$ be a real lift for $h \in \widehat{H}^k(X,A;\mathbb{Z})$. Then $\tilde{h} \circ q \in C^k(i_A;\mathbb{Z})$ is a real lift for $h \circ q$. Since $\mathrm{curv}(h \circ q) = \mathrm{curv}(h) \in \Omega^k_0(X,A)$, we conclude that $\mu^{\tilde{h}} \circ q$ represents the characteristic class of $h \circ q$. Hence $c(h \circ q) = c(h)$.

Since the projection $q : Z_{k-1}(i_A;\mathbb{Z}) \to Z_{k-1}(X,A;\mathbb{Z})$, $(s,t) \mapsto s + \mathrm{im}(i_{A*})$, is surjective, the homomorphism $\widehat{H}^k(X,A;\mathbb{Z}) \to \widehat{H}^k(i_A;\mathbb{Z})$ is injective. As we have seen, its image is contained in the subgroup of parallel characters in $\widehat{H}^k(i_A;\mathbb{Z})$.

It remains to show that any parallel character in $\widehat{H}^k(i_A;\mathbb{Z})$ lies in the image. Thus let $h' \in \widehat{H}^k(i_A;\mathbb{Z})$ with $\mathrm{cov}(h') = 0$. In particular, $i_A^*\mathrm{curv}(h') = 0$, thus $\mathrm{curv}(h') \in \Omega^k_0(X,A)$. We construct a character $h \in \widehat{H}^k(X,A;\mathbb{Z})$ such that $h' = h \circ q$. From the exact sequences (26) and (41) we obtain the following commutative diagram with exact rows and injective vertical maps:

$$\begin{array}{ccccccccc}
0 & \longrightarrow & \frac{H^{k-1}(X,A;\mathbb{R})}{H^{k-1}(X,A;\mathbb{R})_{\mathbb{Z}}} & \xrightarrow{j} & \widehat{H}^k(X,A;\mathbb{Z}) & \xrightarrow{(\mathrm{curv},c)} & R^k(X,A;\mathbb{Z}) & \longrightarrow & 0 \\
 & & \downarrow & & \downarrow & & \downarrow & & \\
0 & \longrightarrow & \frac{H^{k-1}(X,A;\mathbb{R})}{H^{k-1}(X,A;\mathbb{R})_{\mathbb{Z}}} & \xrightarrow{j} & \widehat{H}^k(i_A;\mathbb{Z}) & \xrightarrow{((\mathrm{curv},\mathrm{cov}),c)} & R^k(i_A;\mathbb{Z}) & \longrightarrow & 0
\end{array}$$

Since $\mathrm{curv}(h') \in \Omega^k_0(X,A)$, we have $(\mathrm{curv}(h'),c(h')) \in R^k(X,A;\mathbb{Z})$. By exactness of the upper row we may choose a character $h'' \in \widehat{H}^k(X,A;\mathbb{Z})$ with $\mathrm{curv}(h'') = \mathrm{curv}(h') \in \Omega^k_0(X,A)$ and $c(h'') = c(h') \in H^k(X,A;\mathbb{Z})$. By exactness of the lower row we find $u \in \frac{H^{k-1}(X,A;\mathbb{R})}{H^{k-1}(X,A;\mathbb{R})_{\mathbb{Z}}}$ such that $h' - h'' \circ q = j(u)$. Now put $h := h'' + j(u) \in \widehat{H}^k(X,A;\mathbb{Z})$. This yields $h \circ q = h'' \circ q + j(u) = h'$. □

*Example 18.* Let $\varphi : A \to X$ be a smooth map. As in Example 8 we have the identification of the group $\widehat{H}^2(\varphi;\mathbb{Z})$ with the group of isomorphism classes of hermitian line bundles with connection $(L,\nabla) \to X$ and a section $\sigma : A \to \varphi^*L$ along the smooth map $\varphi : A \to X$.

Now let $i_A : A \to X$ be the inclusion of a smooth submanifold. Theorem 17 yields an identification of the group $\widehat{H}^2(X,A;\mathbb{Z})$ with the group of isomorphism classes of hermitian line bundles with connection $(L,\nabla) \to X$

and *parallel* sections $\sigma : A \to L|_A$. In both cases, the isomorphisms are bundle isomorphisms of $L$ that preserve both the connection $\nabla$ and the section $\sigma$. In particular, they preserve the property of the section $\sigma$ to be parallel with respect to the connection $\nabla$.

## *3.4 Relative Differential Cocycles*

In this section, we discuss the relation of the group $\widehat{H}^k(X, A; \mathbb{Z})$ to another notion of relative differential cohomology that has appeared in the literature. As above let $\varphi : A \to X$ be a smooth map. The Hopkins-Singer complex of differential cocycles is a cochain complex, the $k$-th homology of which is isomorphic to the differential cohomology group $\widehat{H}^k(X; \mathbb{Z})$. In the original definition of differential cocycles in [26], the complex that computes the $k$-th differential cohomology group $\widehat{H}^k(X; \mathbb{Z})$ depends on the degree $k$: for each degree of differential cohomology, one has to consider a different complex.[4] However, the Hopkins-Singer complex can be modified such that differential cohomology groups of all degrees arise as homology groups of a single complex, see [7].

The *relative Hopkins-Singer differential cohomology group* $\check{H}^k(\varphi; \mathbb{Z})$ is defined in [7] as the $k$-th homology group of the mapping cone complex of the modified Hopkins-Singer complex. The cocycles of this mapping cone complex are referred to as *relative differential cocycles*.

The main feature of the relative Hopkins-Singer groups $\check{H}^*(\varphi; \mathbb{Z})$ is the long exact sequence they fit into: The complex of relative differential cochains sits in the usual short exact sequence of cochain complexes which relates the modified Hopkins-Singer complexes on $X$ and $A$ to the corresponding mapping cone complex. Thus the relative Hopkins-Singer groups fit into the following long exact sequence [7]:

$$\ldots \longrightarrow \widehat{H}^{k-1}(A; \mathbb{Z}) \longrightarrow \check{H}^k(\varphi; \mathbb{Z}) \longrightarrow \widehat{H}^k(X; \mathbb{Z}) \xrightarrow{\varphi^*} \widehat{H}^k(A; \mathbb{Z}) \longrightarrow \ldots \tag{43}$$

Comparison of (43) with the long exact sequences (34) and (42) for the relative groups $\widehat{H}^k(\varphi; \mathbb{Z})$ and $\widehat{H}^k(X, A; \mathbb{Z})$ (for $\varphi = i_A$) shows that the relative Hopkins-Singer group $\check{H}^k(i_A; \mathbb{Z})$ differs from both.

In [7] it is shown that the relative Hopkins-Singer group $\check{H}^k(\varphi; \mathbb{Z})$ is a subquotient of the group of relative differential characters $\widehat{H}^k(\varphi; \mathbb{Z})$. More precisely, it is a quotient of the subgroup $\widehat{H}^k_0(\varphi; \mathbb{Z}) := \{h \in \widehat{H}^k(\varphi; \mathbb{Z}) \,|\, \varphi^* \breve{p}_\varphi(h) = 0\}$. From the results of Sect. 3.3, we easily obtain the following identification:

[4] The same holds for smooth Deligne cohomology: the smooth Deligne complex, the $k$-th homology of which is isomorphic to $\widehat{H}^k(X; \mathbb{Z})$, is the total complex of a Čech-de Rham double complex, truncated at the de Rham order $(k - 1)$.

**Proposition 19.** *Let $\varphi : A \to X$ be a smooth map. Then the subgroup $\widehat{H}^k_0(\varphi;\mathbb{Z}) \subset \widehat{H}^k(\varphi;\mathbb{Z})$ coincides with the subgroup of characters with covariant derivative in $\Omega^{k-1}_0(A)$, i.e. a closed form with integral periods. In particular, we have the exact sequence:*

$$0 \longrightarrow \widehat{H}^k_0(\varphi;\mathbb{Z}) \longrightarrow \widehat{H}^k(\varphi;\mathbb{Z}) \xrightarrow{\text{cov}} \frac{\text{pr}_2(\Omega^{k-1}_0(\varphi))}{\Omega^{k-1}_0(A)} \longrightarrow 0 \tag{44}$$

*Here $\text{pr}_2(\Omega^{k-1}_0(\varphi)) \subset \Omega^{k-1}(A)$ denotes the image of the projection $\text{pr}_2 : \Omega^{k-1}_0(\varphi) \to \Omega^{k-1}(A)$, $(\omega,\vartheta) \mapsto \vartheta$.*

*Proof.* The identification of $\widehat{H}^k_0(\varphi;\mathbb{Z}) = \ker(\varphi^* \circ \breve{p}_\varphi)$ follows from the commutative diagram (35) and the exact sequence (24) for topological trivializations. The exact sequence follows from this identification. □

The sequence (44) appeared in [7, Prop. 4.1]. As mentioned above, it is shown in [7] that the relative Hopkins-Singer group $\check{H}^k(\varphi;\mathbb{Z})$ is a quotient of $\widehat{H}^k_0(\varphi;\mathbb{Z})$. More precisely, we have the exact sequence [7, Thm. 4.2]:

$$0 \longrightarrow \frac{\Omega^{k-1}_0(X)}{\widetilde{\Omega}^{k-1}(X)} \longrightarrow \widehat{H}^k_0(\varphi;\mathbb{Z}) \longrightarrow \check{H}^k(\varphi;\mathbb{Z}) \longrightarrow 0\,. \tag{45}$$

Here we have $\widetilde{\Omega}^{k-1}(X) := \{\omega \in \Omega^{k-1}_0(X) \,|\, (\omega,0) \in \Omega^{k-1}_0(\varphi)\}$ with the homomorphism $\frac{\Omega^{k-1}_0(X)}{\widetilde{\Omega}^{k-1}(X)} \to \widehat{H}^k_0(\varphi;\mathbb{Z})$, $\omega \mapsto \iota_\varphi(\omega,0)$.

It remains to determine the relation between the relative Hopkins-Singer group $\check{H}^k(i_A;\mathbb{Z})$ and the group $\widehat{H}^k(X,A;\mathbb{Z})$. It turns out that the latter is a subgroup of the former:

**Proposition 20.** *Let $i_A : A \to X$ be the embedding of a smooth submanifold. Then the following sequences are exact:*

$$0 \longrightarrow \widehat{H}^k(X,A;\mathbb{Z}) \longrightarrow \widehat{H}^k_0(i_A;\mathbb{Z}) \xrightarrow{\text{cov}} \Omega^{k-1}_0(A) \longrightarrow 0$$

$$0 \longrightarrow \widehat{H}^k(X,A;\mathbb{Z}) \longrightarrow \check{H}^k(i_A;\mathbb{Z}) \xrightarrow{\text{cov}} \frac{\Omega^{k-1}_0(A)}{i_A^*\Omega^{k-1}_0(X)} \longrightarrow 0\,.$$

*Proof.* Exactness of the first sequence follows from the results of Sect. 3.3: Clearly, $\widehat{H}^k(X,A;\mathbb{Z})$ is the kernel of $\text{cov} : \widehat{H}^k_0(i_A;\mathbb{Z}) \to \Omega^{k-1}_0(A)$ by Theorem 17 and Proposition 19. The latter is surjective since for any $\vartheta \in \Omega^{k-1}_0(A)$, we may choose a differential character $h \in \widehat{H}^{k-1}(A;\mathbb{Z})$ with $\text{curv}(h) = \vartheta$. Then $\breve{\imath}_{i_A}(-h) \in \widehat{H}^k_0(i_A;\mathbb{Z})$ and $\text{cov}(\breve{\imath}_{i_A}(-h)) = \vartheta$.

The second sequence is obtained from the first by dividing out the action of $\Omega^{k-1}_0(X)$ on $\widehat{H}^k_0(i_A;\mathbb{Z})$. Vanishing of the composition $\widehat{H}^k(X,A;\mathbb{Z}) \to \check{H}^k(i_A;\mathbb{Z}) \to \frac{\Omega^{k-1}_0(A)}{i_A^*\Omega^{k-1}_0(X)}$ follows from the first sequence.

We show exactness at $\widehat{H}^k(X,A;\mathbb{Z})$: For $\varphi = i_A$, we have $\widetilde{\Omega}^{k-1}(X) = \Omega_0^{k-1}(X,A)$. Let $h \in \widehat{H}^k(X,A;\mathbb{Z})$ which maps to 0 in $\check{H}^k(i_A;\mathbb{Z})$. By Theorem 17 we may consider $h$ as a parallel character in $\widehat{H}_0^k(i_A;\mathbb{Z})$. From the exact sequence (45) we conclude $h = \iota_{i_A}(\omega,0)$ for some $\omega \in \Omega_0^{k-1}(X)$. Now we have $0 = \mathrm{cov}(h) = \mathrm{cov}(\iota_{i_A}(\omega,0)) = \varphi^*\omega$. Thus $(\omega,0) \in \Omega_0^{k-1}(X,A)$, hence $h = \iota_{i_A}(\omega,0) = 0 \in \widehat{H}_0^k(i_A;\mathbb{Z})$. From the first sequence we conclude $h = 0$.

Next we show exactness at $\check{H}^k(i_A;\mathbb{Z})$: Let $[h] \in \check{H}^k(i_A;\mathbb{Z})$ be an equivalence class of characters in the kernel of the map $\check{H}^k(i_A;\mathbb{Z}) \to \frac{\Omega_0^{k-1}(A)}{i_A^*\Omega_0^{k-1}(X)}$. Choose a representant $h \in \widehat{H}_0^k(i_A;\mathbb{Z})$ of the equivalence class $[h]$. Then there exists a differential form $\omega \in \Omega_0^{k-1}(X)$ such that $\mathrm{cov}(h) = \varphi^*\omega$. Now put $h' := h - \iota_{i_A}(\omega,0)$. Then we have $\mathrm{cov}(h') = \mathrm{cov}(h) - \varphi^*\omega = 0$, thus $h' \in \widehat{H}^k(X,A;\mathbb{Z})$. From the exact sequence (45) we conclude that $[h]$ is the image of $h'$ under the map $\widehat{H}^k(X,A;\mathbb{Z}) \to \check{H}^k(i_A;\mathbb{Z})$.

Finally, exactness at $\frac{\Omega_0^{k-1}(A)}{i_A^*\Omega_0^{k-1}(X)}$ is clear from exactness of the first sequence and the sequence (45). □

# 4 Internal and External Products

In this chapter we discuss internal and external products in differential cohomology. The internal product of differential characters and the induced ring structure on differential cohomology $\widehat{H}^*(X;\mathbb{Z})$ has first been constructed in [16]. Uniqueness of the ring structure is proved in [36] and [1]. The proof in [1, Chap. 6] starts from an axiomatic definition of internal and external products, similar to the one in [36], and ends up with an explicit formula. In this sense the proof is constructive. Simple formulas for the ring structure are obtained in models of differential cohomology based on differential forms with singularities [15], de Rham-Federer currents as in [24, Sec. 3], and differential cocycles [26], [11].

Uniqueness of (the external product and) the ring structure has been shown in [36] and [1, Chap. 6]. Our proof there is constructive in the sense that it yields a formula for the external product, starting from an abstract definition. In Sect. 4.1, we derive that formula from the original construction of the ring structure in [16]. In Sect. 4.2 we use the methods of [1, Chap. 6] to construct a cross product between relative and absolute differential characters. This in turn provides the graded abelian group $\widehat{H}^*(\varphi;\mathbb{Z})$ with the structure of a right module over the ring $\widehat{H}^*(X;\mathbb{Z})$. The module structure is constructed from the cross product by pull-back along a version of the diagonal map.

In [1, Chap. 6] we focussed only on *uniqueness* of the cross product and ring structure of absolute differential characters since existence of the products is well-known. Here we only prove *existence* of the cross product between relative and absolute characters and the module structure on $\widehat{H}^*(\varphi;\mathbb{Z})$. We do not prove uniqueness of the cross product. However, the uniqueness proof from [1, Chap. 6] for absolute differential cohomology carries over directly to the notion of relative differential cohomology considered here.

## 4.1 The Ring Structure on Differential Cohomology

In this section we briefly recall the original construction of the internal product

$$* : \widehat{H}^k(X;\mathbb{Z}) \times \widehat{H}^{k'}(X;\mathbb{Z}) \to \widehat{H}^{k+k'}(X;\mathbb{Z})$$

from [16]. We derive another formula for the induced external product based on representation of smooth singular homology by geometric cycles. This new formula is proved in [1, Chap. 6] by means of an abstract definition of internal and external products of differential characters. Here we derive the new formula for the external product from the original Cheeger-Simons formula for the internal product.

### 4.1.1 The Cheeger-Simons Internal Product

Let $h \in \widehat{H}^k(X;\mathbb{Z})$ and $h' \in \widehat{H}^{k'}(X;\mathbb{Z})$ be differential characters. Choose real lifts $\tilde{h} \in C^{k-1}(X;\mathbb{R})$ and $\tilde{h}' \in C^{k'-1}(X;\mathbb{R})$. Denote by $B : C_*(X;\mathbb{Z}) \to C_*(X;\mathbb{Z})$ the barycentric subdivision and by $H : C_*(X;\mathbb{Z}) \to C_{*+1}(X;\mathbb{Z})$ a chain homotopy from $B$ to the identity, hence

$$\mathrm{id} - B = \partial \circ H + H \circ \partial. \tag{46}$$

By construction, the image of the characteristic class $c(h)$ in $H^k(X;\mathbb{R})$ coincides with the de Rham cohomology class of $\mathrm{curv}(h)$ under the de Rham isomorphism. The wedge product of closed differential forms, regarded as smooth singular cocycles, descends to the cup product on $H^*_{\mathrm{dR}}(X) \cong H^*(X;\mathbb{R})$. Thus the differential form $\mathrm{curv}(h)\wedge\mathrm{curv}(h')$, regarded as a smooth singular cocycle, differs from the cocycle $\mathrm{curv}(h)\cup\mathrm{curv}(h')$ by a real coboundary. An explicit construction of a cochain $E(h,h') \in C^{k+k'-1}(X;\mathbb{R})$ such that $\delta E(h,h') = \mathrm{curv}(h)\wedge\mathrm{curv}(h') - \mathrm{curv}(h)\cup\mathrm{curv}(h')$ is given in [16, p. 55f.] by:

$$E(h,h')(x) := -\sum_{j=0}^{\infty}(\mathrm{curv}(h)\cup\mathrm{curv}(h'))(H(B^j x))\,.$$

Here $x \in C_{k+k'-1}(X;\mathbb{R})$.

Now put $\nu(\tilde{h},\tilde{h}') := (\tilde{h}\cup\mu^{\tilde{h}'} + (-1)^k\mathrm{curv}(h)\cup\tilde{h}') \in C^{k+k'-1}(X;\mathbb{R})$. Then the differential character $h * h' \in \widehat{H}^{k+k'}(X;\mathbb{Z})$ is defined by

$$(h * h')(z) := \exp\left(2\pi i\Big(\nu(\tilde{h}, \tilde{h}') + E(h, h')\Big)(z)\right), \tag{47}$$

where $z \in Z_{k+k'-1}(X; \mathbb{Z})$. As observed in [16], the internal product $*$ is well-defined, i.e. it does not depend upon the choice of real lifts $\tilde{h}, \tilde{h}'$ and chain homotopy $H$. Moreover, the product $*$ is associative and graded commutative, and it is natural with respect to smooth maps. It is compatible with the exact sequences in (24) in the sense that curvature and characteristic class are multiplicative and

$$\iota(\varrho) * h' = \iota(\varrho \wedge \mathrm{curv}(h')). \tag{48}$$

By definition, the internal product is $\mathbb{Z}$-bilinear. In particular, if $h = 0$ or $h' = 0$, then $h * h' = 0$.

Using these properties, we derive an expression for the internal product that no longer involves the cochain $E(h, h')$. For a similar formula, see [16, p. 57].

**Proposition 21.** *Let $h \in \widehat{H}^k(X; \mathbb{Z})$ and $h' \in \widehat{H}^{k'}(X; \mathbb{Z})$ be differential characters on $X$. Let $z \in Z_{k+k'-1}(X; \mathbb{Z})$ be cycle. Choose a geometric cycle $M \xrightarrow{f} X$ that represents the homology class $[z]$. Let $y \in Z_{k+k'-1}(M; \mathbb{Z})$ be a fundamental cycle of $M$. Choose a chain $a(z) \in C_{k+k'}(X; \mathbb{Z})$ such that $z = f_* y + \partial a(z)$. Then we have*

$$(h*h')(z) = \lim_{j\to\infty} \exp\left(2\pi i\Big(f^*\nu(\tilde{h}, \tilde{h}')(B^j y) + \int_{a(z)} \mathrm{curv}(h) \wedge \mathrm{curv}(h')\Big)\right). \tag{49}$$

*Proof.* Since $z = f_* y + \partial a(z)$, we have:

$$\begin{aligned}(h * h')(z) &= f^*(h * h')(y) \cdot \exp\left(2\pi i \int_{a(z)} \mathrm{curv}(h * h')\right) \\ &= (f^* h * f^* h')(y) \cdot \exp\left(2\pi i \int_{a(z)} \mathrm{curv}(h) \wedge \mathrm{curv}(h')\right).\end{aligned}$$

We compute $f^*(h*h')([M])$ using (47) on the stratifold $M$. The characteristic class $c(h)$ is represented by the cocycle $\mu^{\tilde{h}} := \mathrm{curv}(h) - \delta\tilde{h} \in C^k(M; \mathbb{Z})$, and similarly for $h'$. Now we have:

$$\begin{aligned}\mu^{\tilde{h}} \cup \mu^{\tilde{h}'} &= (\mathrm{curv}(h) - \delta\tilde{h}) \cup (\mathrm{curv}(h') - \delta\tilde{h}') \\ &= \mathrm{curv}(h) \cup \mathrm{curv}(h') - \delta(\tilde{h} \cup (\mathrm{curv}(h') - \delta\tilde{h}') + (-1)^k \mathrm{curv}(h) \cup \tilde{h}') \\ &= \mathrm{curv}(h) \cup \mathrm{curv}(h') - \delta\nu(\tilde{h}, \tilde{h}').\end{aligned}$$

Since $M$ is $(k + k' - 1)$-dimensional, the cocycle $f^*(\mu^{\tilde{h}} \cup \mu^{\tilde{h}'})$ is an integral coboundary for dimensional reasons. Thus we have $f^*(\mu^{\tilde{h}} \cup \mu^{\tilde{h}'}) = \delta t$ for some $t \in C^{k+k'-1}(M; \mathbb{Z})$ and hence $f^*(\mathrm{curv}(h) \cup \mathrm{curv}(h')) = \delta t + \delta\nu(f^*\tilde{h}, f^*\tilde{h}')$.

Evaluating the cochain $E(f^*h, f^*h')$ on the fundamental cycle $y$ of $M$, we obtain:

$$
\begin{aligned}
E(f^*h,&f^*h')(y) \\
&= -\sum_{j=0}^{\infty}(f^*\mathrm{curv}(h) \cup f^*\mathrm{curv}(h'))(H(B^j y)) \\
&= -\sum_{j=0}^{\infty} \delta(t + \nu(f^*\tilde{h}, f^*\tilde{h}'))(H(B^j y)) \\
&= -\sum_{j=0}^{\infty}(t + \nu(f^*\tilde{h}, f^*\tilde{h}'))(\partial H(B^j y)) \\
&\overset{(46)}{=} -\sum_{j=0}^{\infty}(t + \nu(f^*\tilde{h}, f^*\tilde{h}'))((\mathrm{id} - B)B^j y - H(\partial B^j y)) \\
&= -(t + \nu(f^*\tilde{h}, f^*\tilde{h}'))(y) + \lim_{j\to\infty}(t + \nu(f^*\tilde{h}, f^*\tilde{h}'))(B^j y) \\
&= \underbrace{-t(y) + \lim_{j\to\infty} t(B^j y)}_{\in\mathbb{Z}} - \nu(f^*\tilde{h}, f^*\tilde{h}')(y) + \lim_{j\to\infty} \nu(f^*\tilde{h}, f^*\tilde{h}')(B^j y).
\end{aligned}
\tag{50}
$$

We thus have:

$$
\begin{aligned}
f^*&(h * h')(y) \\
&\overset{(47)}{=} \exp\Big(2\pi i\Big(f^*\nu(\tilde{h}, \tilde{h}') + E(f^*h, f^*h')\Big)(y)\Big) \\
&\overset{(50)}{=} \exp\Big(2\pi i\Big((f^*\nu(\tilde{h}, \tilde{h}') - \nu(f^*\tilde{h}, f^*\tilde{h}'))(y) + \lim_{j\to\infty} \nu(f^*\tilde{h}, f^*\tilde{h}')(B^j y)\Big)\Big) \\
&= \exp\Big(2\pi i\Big(\lim_{j\to\infty} \nu(f^*\tilde{h}, f^*\tilde{h}')(B^j y)\Big)\Big).
\end{aligned}
$$

□

### 4.1.2 The External or Cross Product

Similar to singular cohomology, the internal product of differential characters on a manifold $X$ gives rise to an external or cross product

$$
\times : \widehat{H}^k(X;\mathbb{Z}) \times \widehat{H}^{k'}(X';\mathbb{Z}) \to \widehat{H}^{k+k'}(X \times X';\mathbb{Z}), \quad (h, h') \mapsto (\mathrm{pr}_1^* h) * (\mathrm{pr}_2^* h').
$$

Here $\mathrm{pr}_1, \mathrm{pr}_2$ denote the projection on the first and second factor of $X \times X'$, respectively. Conversely, the internal product can be recovered from the external product by pull-back along the diagonal $\Delta_X : X \to X \times X$, $x \mapsto (x, x)$: for characters $h, h' \in \widehat{H}^k(X;\mathbb{Z})$, we have:

$$\Delta_X^*(h \times h') = \Delta_X^*(\mathrm{pr}_1^* h * \mathrm{pr}_2^* h') = (\mathrm{pr}_1 \circ \Delta_X)^* h * (\mathrm{pr}_2 \circ \Delta_X)^* h' = h * h' .$$

The external product is $\mathbb{Z}$-bilinear. Moreover, since curvature and characteristic class are multiplicative for the internal product, the same holds for the external product:

$$\mathrm{curv}(h \times h') = \mathrm{curv}(h) \times \mathrm{curv}(h') \in \Omega^{k+k'}(X \times X'), \tag{51}$$

$$c(h \times h') = c(h) \times c(h') \in H^{k+k'}(X \times X'; \mathbb{Z}). \tag{52}$$

### 4.1.3 A Formula for the Cross Product

To understand the external product $h \times h'$ of differential characters $h \in \widehat{H}^k(X;\mathbb{Z})$ and $h' \in \widehat{H}^{k'}(X';\mathbb{Z})$, the following special case is crucial. It is the key step in the uniqueness proof in [1, Chap. 6]. We give another proof here, based on Proposition 21.

**Lemma 22.** *Let $M$ and $M'$ be closed oriented regular stratifolds satisfying $\dim(M) + \dim(M') = k + k' - 1$. Let $h \in \widehat{H}^k(M;\mathbb{Z})$ and $h' \in \widehat{H}^{k'}(M';\mathbb{Z})$. Then we have:*

$$(h \times h')([M \times M']) = \begin{cases} h([M])^{\langle c(h'),[M']\rangle} & \textit{if } \dim(M) = k-1, \\ h'([M'])^{(-1)^k \langle c(h),[M]\rangle} & \textit{if } \dim(M) = k, \\ 1 & \textit{otherwise}. \end{cases} \tag{53}$$

*Proof.* If $\dim(M) < k-1$, we have $\widehat{H}^k(M;\mathbb{Z}) = \{0\}$. If $\dim(M') < k'-1$, we have $\widehat{H}^{k'}(M';\mathbb{Z}) = \{0\}$. Thus if $(\dim(M), \dim(M')) \notin \{(k-1,k'), (k,k'-1)\}$ then either $h = 0$ or $h' = 0$. Hence $h \times h' = 0$.

Now let $y \in Z_{k+k'-1}(M \times M';\mathbb{Z})$ be a fundamental cycle of $M \times M'$. By (49), we have

$$\begin{aligned} (h \times h')([M \times M']) &= (\mathrm{pr}_1^* h * \mathrm{pr}_2^* h')(y) \\ &\overset{(49)}{=} \lim_{j\to\infty} \exp\big(2\pi i \cdot \nu(\mathrm{pr}_1^* \widetilde{h}, \mathrm{pr}_2^* \widetilde{h'})(B^j y)\big) \\ &= \lim_{j\to\infty} \exp\Big(2\pi i \cdot \big(\widetilde{h} \times \mu^{\widetilde{h'}} + (-1)^k \mathrm{curv}(h) \times \widetilde{h'}\big)(B^j y)\Big) \end{aligned}$$

Since $(h \times h')([M \times M'])$ does not depend upon the choice of fundamental cycle, we may choose $y = x \times x'$, where $x$ and $x'$ are fundamental cycles of $M$ and $M'$, respectively. Moreover, we have $B^j y = y + \partial b_j$ for some $b_j \in C_{k+k'}(M \times M';\mathbb{Z}) = S_{k+k'}(M \times M';\mathbb{Z})$.

If $\dim(M) = k-1$, we may choose $\widetilde{h}$ to be a cocycle. Then $\widetilde{h} \times \mu^{\widetilde{h'}}$ is a cocycle, too. Moreover, $\mathrm{curv}(h) = 0$ in this case, and we obtain:

$$
\begin{aligned}
(h \times h')([M \times M']) \\
&= \lim_{j\to\infty} \exp\Big(2\pi i \cdot \big(\widetilde{h} \times \mu^{\widetilde{h}'} + (-1)^k \mathrm{curv}(h) \times \widetilde{h}'\big)(x \times x' + \partial b)\Big) \\
&= \exp\Big(2\pi i \cdot \widetilde{h}(x) \cdot \mu^{\widetilde{h}'}(x')\Big) \\
&= \big(h([M])\big)^{\langle c(h'),[M']\rangle}
\end{aligned}
$$

Similarly, for $\dim(M) = k$, we have $\dim(M') = k' - 1$, hence $\mathrm{curv}(h') = 0$. We may choose $\widetilde{h}'$ to be a cocycle. Then $\mathrm{curv}(h) \times \widetilde{h}'$ is a cocycle and $\mu^{\widetilde{h}'} = 0$. This yields:

$$
\begin{aligned}
(h \times h')([M \times M']) \\
&= \lim_{j\to\infty} \exp\Big(2\pi i \cdot \big(\widetilde{h} \times \mu^{\widetilde{h}'} + (-1)^k \mathrm{curv}(h) \times \widetilde{h}'\big)(x \times x' + \partial b)\Big) \\
&= \exp\Big(2\pi i \cdot (-1)^k \cdot \widetilde{h}'(x') \cdot \int_M \mathrm{curv}(h)\Big) \\
&= \big(h'([M'])\big)^{(-1)^k \cdot \langle c(h),[M]\rangle}.
\end{aligned}
$$

□

From Lemma 22 and Remark 6 we easily obtain a formula for the external product as in [1, Chap. 6]. The Künneth sequence allows us to decompose cycles in $X \times X'$ into a sum of cross products of cycles in $X$ and $X'$ respectively, and a torsion cycle. This is done by carefully constructing splittings of the Künneth sequence at the level of cycles as explained in detail in [1, Chap. 6]. The construction is briefly reviewed in the appendix. The construction for the relative Künneth sequence is given in detail there. We use it in the following section to construct the module structure on $\widehat{H}^*(\varphi;\mathbb{Z})$ in the following section.

Let $h \in \widehat{H}^k(X;\mathbb{Z})$ and $h' \in \widehat{H}^{k'}(X';\mathbb{Z})$ be differential characters. To evaluate the character $h \times h' \in \widehat{H}^{k+k'}(X \times X';\mathbb{Z})$ on a cycle $z \in Z_{k+k'-1}(X \times X';\mathbb{Z})$ we use the Künneth decomposition of $z$ into a sum of cross products of cycles on $X$ and $X'$ and a torsion cycle. Then the two types of summands are treated separately. This yields:

**Corollary 23 (Formula for the Cross Product).** *Let $h \in \widehat{H}^k(X;\mathbb{Z})$ and $h' \in \widehat{H}^{k'}(X';\mathbb{Z})$, and let $z \in Z_{k+k'-1}(X \times X';\mathbb{Z})$. Decompose it according to the Künneth sequence into a sum of cross products of cycles in $X$ and $X'$, respectively, and torsion cycles. Then $h \times h'$ is evaluated on the two types of summands separately:*

*If $z$ represents an $N$-torsion class, choose a chain $b \in C_{k+k'}(X \times X';\mathbb{Z})$ such that $\partial b = N \cdot z$. Then we have:*

$$
(h \times h')(z) = \exp\Big(\frac{2\pi i}{N}\Big(\int_b \mathrm{curv}(h) \times \mathrm{curv}(h') - \langle c(h) \times c(h'), b\rangle\Big)\Big). \tag{54}
$$

*If $z = y_i \times y'_j$ with $y_i \in Z_i(X;\mathbb{Z})$ and $y'_j \in Z_j(X';\mathbb{Z})$ we have:*

$$(h \times h')(y_i \times y'_j) = \begin{cases} h(y_i)^{\langle c(h'), y'_j \rangle} & \textit{if } (i,j) = (k-1, k') \\ h'(y'_j)^{(-1)^k \langle c(h), y_i \rangle} & \textit{if } (i,j) = (k, k'-1) \\ 1 & \textit{otherwise}. \end{cases} \quad (55)$$

*Proof.* The representation (54) on torsion cycles follows from Remark 6 and (51), (52).

Now let $z = y_i \times y'_j$ where $y_i \in Z_i(X;\mathbb{Z})$ and $y'_j \in Z'_j(X';\mathbb{Z})$. Choose geometric cycles $\zeta(y_i) \in \mathcal{Z}_i(X)$ and $\zeta'(y'_j) \in \mathcal{Z}_j(X')$ and chains $a(y_i) \in C_{i+1}(X;\mathbb{Z})$ and $a(y'_j) \in C_{j+1}(X';\mathbb{Z})$ such that $[y_i - \partial a(y_i)]_{\partial S_{i+1}} = [\zeta(y_i)]_{\partial S_{i+1}}$ and $[y'_j - \partial a(y'_j)]_{\partial S_{j+1}} = [\zeta(y'_j)]_{\partial S_{j+1}}$.

Now apply Lemma 22 to the fundamental cycles of $\zeta(y_i)$ and $\zeta'(y'_j)$: For degrees $(i,j)$ different from $(k-1,k')$ and $(k,k'-1)$ we have $(h \times h')(y_i \times y'_j) = 1$. For $(i,j) = (k-1,k')$ we obtain:

$$\begin{aligned} &(h \times h')(y_{k-1} \times y'_{k'}) \\ &= h([\zeta(y_{k-1})]_{\partial S_k})^{\langle c(h'), y'_{k'} \rangle} \cdot \exp\Big(2\pi i \int_{(a(y_i) \times y'_{k'})} \mathrm{curv}(h \times h')\Big) \\ &= h([\zeta(y_{k-1})]_{\partial S_k})^{\langle c(h'), y'_{k'} \rangle} \cdot \exp\Big(2\pi i \int_{a(y_{k-1})} \mathrm{curv}(h) \cdot \int_{y'_{k'}} \mathrm{curv}(h')\Big) \\ &= h([\zeta(y_{k-1})]_{\partial S_k})^{\langle c(h'), y'_{k'} \rangle} \cdot h(\partial a(y_{k-1}))^{\langle c(h'), y'_{k'} \rangle} \\ &= h(y_{k-1})^{\langle c(h'), y'_{k'} \rangle}. \end{aligned}$$

Similarly for $(i,j) = (k-1,k')$ we obtain:

$$(h \times h')(y_k \times y'_{k'-1}) = h(y'_{k'-1})^{(-1)^k \cdot \langle c(h), y_{k'} \rangle}. \qquad \square$$

## 4.2 The Module Structure on Relative Differential Cohomology

In this section, we use the method developed in [1, Chap. 6] to construct an external and internal product between relative and absolute differential characters. This provides the graded abelian group $\widehat{H}^*(\varphi;\mathbb{Z})$ of relative differential characters with the structure of a right module over the ring $\widehat{H}^*(X;\mathbb{Z})$. The module structure is natural with respect to smooth maps. It is compatible with the module structures on relative cohomology and the relative de Rham complex in the sense that the structure maps (i.e. curvature, covariant derivative, characteristic class and topological trivializations) are multiplicative. Moreover, the module structure is compatible with the maps $\breve{p}$ and $\breve{\imath}$ between absolute and relative differential characters groups.

### 4.2.1 The Cross Product

As above let $\varphi : A \to X$ be a smooth map. We consider the induced map $\varphi \times \mathrm{id}_{X'} : A \times X' \to X \times X'$. The cup product on smooth singular cochains induces an internal product between relative and absolute cochains

$$\cup : C_*(\varphi;\mathbb{Z}) \otimes C_*(X;\mathbb{Z}) \to C_*(\varphi;\mathbb{Z}) , \quad (\mu,\nu) \cup \sigma := (\mu \cup \sigma, \nu \cup \sigma) .$$

Likewise, the cup product induces an external or cross product

$$\times : C_*(\varphi;\mathbb{Z}) \otimes C_*(X';\mathbb{Z}) \to C_*(\varphi \times \mathrm{id}_{X'};\mathbb{Z}) , \quad (\mu,\nu) \times \sigma := (\mu \times \sigma, \nu \times \sigma) .$$

Since cross and cup products are natural chain maps, so are the induced internal and external products between absolute and relative cochains. Clearly, the products are invariant under the boundary operator of the mapping cone complex and thus descend to the cup and cross products on cohomology. The cup product in particular provides the mapping cone cohomology $H^*(\varphi;\mathbb{Z})$ with the structure of a right module over the cohomology ring $H^*(X;\mathbb{Z})$.

Likewise, the wedge product[5] of differential forms induces an internal product between relative differential forms $(\omega,\vartheta) \in \Omega^k(\varphi)$ and differential forms $\omega' \in \Omega^{k'}(X)$:

$$(\omega,\vartheta) \wedge \omega' := (\omega \wedge \omega', \vartheta \wedge \varphi^*\omega') \in \Omega^{k+k'}(\varphi)$$

This provides the mapping cone de Rham complex $\Omega^*(\varphi)$ with the structure of a right module over the ring $\Omega^*(X)$ of differential forms on $X$. Similarly, we have the external product of $(\omega,\vartheta) \in \Omega^k(\varphi)$ with $\omega' \in \Omega^{k'}(X')$:

$$(\omega,\vartheta) \times \omega' := (\omega \times \omega', \vartheta \times \varphi^*\omega') \in \Omega^{k+k'}(\varphi \times \mathrm{id}_{X'}) .$$

The internal and external products on the de Rham complex $\Omega^*(\varphi)$ and the cochain complex $C^*(\varphi;\mathbb{R})$ coincide in cohomology under the de Rham isomorphism. Hence they induce the same module structure on $H^*_{\mathrm{dR}}(\varphi) \cong H^*(\varphi;\mathbb{R})$.

Now we construct the external product between a relative character $h \in \widehat{H}^*(\varphi;\mathbb{Z})$ and an absolute character $h' \in \widehat{H}^*(X';\mathbb{Z})$. The construction is completely analogous to the one for absolute characters reviewed in the previous section.

We have the relative Künneth sequence

$$0 \to \big[H_*(\varphi;\mathbb{Z}) \otimes H_*(X';\mathbb{Z})\big]_n \xrightarrow{\times} H_n(\varphi \times \mathrm{id}_{X'};\mathbb{Z}) \to \mathrm{Tor}(H_*(\varphi;\mathbb{Z}), H_*(X';\mathbb{Z}))_{n-1} \to 0 .$$

---

[5] We avoid the familiar term "exterier product" to avoid confusion with the external product. The wedge product clearly defines an *internal* rather than an *external* product on the de Rham complex.

As is well-known, the sequence splits on the level of cycles. A construction of a splitting $S : Z(C_*(\varphi \times \mathrm{id}_{X'};\mathbb{Z}) \to Z_*(\varphi;\mathbb{Z}) \otimes Z_*(X';\mathbb{Z})$ is given in the appendix. We denote the complement of the image of the cross product $Z_*(\varphi;\mathbb{Z}) \otimes Z_*(X';\mathbb{Z}) \xrightarrow{\times} Z_{k+k'-1}(\varphi \times \mathrm{id}_{X'};\mathbb{Z})$, obtained from the Künneth splitting, by $T_{k+k'-1}(\varphi \times \mathrm{id}_{X'};\mathbb{Z})$. It will be referred to as the *Künneth complement.*

Now let $(s,t) \in Z_{k+k'-1}(\varphi \times \mathrm{id}_{X'};\mathbb{Z})$ be a cycle. The Künneth splitting allows us to decompose $(s,t)$ into a sum of cross products of cycles $(x,y) \in Z_*(\varphi;\mathbb{Z})$ and $y' \in Z_*(X';\mathbb{Z})$ and torsion cycles in $T_{k+k'-1}(\varphi \times \mathrm{id}_{X'};\mathbb{Z})$. Analogously to Corollary 23 we define the external product as follows:

**Definition 24 (Cross Product).** The *cross product* of differential characters $h \in \widehat{H}^k(\varphi;\mathbb{Z})$ and $h' \in H^{k'}(X';\mathbb{Z})$ is the homomorphism $h \times h' : Z_{k+k'-1}(\varphi \times \mathrm{id}_{X'};\mathbb{Z}) \to \mathrm{U}(1)$ defined as follows:

For cycles $(x,y) \in Z_i(\varphi;\mathbb{Z})$ and $y' \in Z_j(X';\mathbb{Z})$, put

$$(h \times h')((x,y) \times y') := \begin{cases} h(x,y)^{\langle c(h'),y'\rangle} & \text{if } (i,j) = (k-1,k') \\ h'(y')^{(-1)^k \cdot \langle c(h),(x,y)\rangle} & \text{if } (i,j) = (k,k'-1) \\ 1 & \text{otherwise.} \end{cases} \tag{56}$$

For an $N$-torsion cycle $(s,t) \in T_{k+k'-1}(\varphi \times \mathrm{id}_{X'};\mathbb{Z})$ in the Künneth complement choose a chain $(a,b) \in C_{k+k'}(\varphi \times \mathrm{id}_{X'};\mathbb{Z})$ such that $N \cdot (s,t) = \partial_{\varphi \times \mathrm{id}_{X'}}(a,b)$. Then put

$$\begin{aligned}&(h \times h')(s,t) \\ &:= \exp\Big(\frac{2\pi i}{N}\Big(\int_{(a,b)} (\mathrm{curv},\mathrm{cov})(h) \times \mathrm{curv}(h') - \langle c(h) \times c(h'),(a,b)\rangle\Big)\Big). \end{aligned} \tag{57}$$

The homomorphism $h \times h' : Z_{k+k'-1}(\varphi \times \mathrm{id}_{X'};\mathbb{Z}) \to \mathrm{U}(1)$ is uniquely determined by these two cases.

Some comments on the notations in Definition 24 are in order. First of all, we write $\langle c(h),(x,y)\rangle = \langle c(h),[x,y]\rangle = \int_{(x,y)} (\mathrm{curv},\mathrm{cov})(h)$ and $\langle c(h'),y'\rangle = \langle c(h'),[y']\rangle = \int_{y'} \mathrm{curv}(h')$ for the Kronecker pairing between cohomology and homology in (56).

Secondly, the term $\langle c(h) \times c(h'),(a,b)\rangle$ in (57) is not well-defined. Replacing the cohomology class $c(h) \times c(h')$ by a cocycle $\mu \in C^{k+k'}(\varphi \times \mathrm{id}_{X'};\mathbb{Z})$ representing it, the term $\exp \frac{2\pi i}{N}\langle \mu,(a,b)\rangle$ is independent of the choice of cocycle. This is because $\frac{1}{N}\langle \delta_{\varphi \times \mathrm{id}_{X'}}\ell,(a,b)\rangle = \langle \ell,(s,t)\rangle \in \mathbb{Z}$ holds for any cochain $\ell \in C^{k+k'-1}(\varphi \times \mathrm{id}_{X'};\mathbb{Z})$.

Thirdly, the value of $h \times h'$ on a torsion cycle $(s,t)$ obtained from the Künneth splitting is independent of the choice of chain $(a,b)$ satisfying $N \cdot (s,t) = \partial_{\varphi \times \mathrm{id}_{X'}}(a,b)$. For if we change $(a,b)$ by adding a cycle $(v',w') \in Z_{k+k'}(\varphi \times \mathrm{id}_{X'};\mathbb{Z})$, the result in (57) changes by multiplication with

$$\exp\Big(\frac{2\pi i}{N}\Big(\underbrace{\int_{(v',w')}(\mathrm{curv},\mathrm{cov})(h)\times\mathrm{curv}(h')-\langle c(h)\times c(h'),(v',w')\rangle}_{=0}\Big)\Big)=1\,.$$

The Künneth complement $T_{k+k'-1}(\varphi\times\mathrm{id}_{X'};\mathbb{Z})\subset Z_{k+k'-1}(\varphi\times\mathrm{id}_{X'};\mathbb{Z})$ is the sum over $N\in\mathbb{N}$ of its subgroups of $N$-torsion cycles. This sum is of course not direct. However, it is easy to see that the homomorphism $h\times h'$ in (57) is well-defined: for a cycle $(s,t)$ in the complement choose $N'$ minimal such that $N'\cdot(s,t)=\partial_{\varphi\times\mathrm{id}_{X'}}(a,b)$. Then the homology class $[s,t]$ has order $N'$ in $H_{k+k'-1}(\varphi\times\mathrm{id}_{X'};\mathbb{Z})$ and all other possible choices of $N$ divide $N'$. Thus the values in (57) for all such choices coincide.

### 4.2.2 Well-Definedness

Clearly, the map $h\times h':Z_{k+k'-1}(\varphi\times\mathrm{id}_{X'};\mathbb{Z})\to\mathrm{U}(1)$ defined by (56) and (57) is a homomorphism. We show that it satisfies condition (22) and thus defines a differential character in $\widehat{H}^{k+k'}(\varphi\times\mathrm{id}_{X'};\mathbb{Z})$.

**Proposition 25.** *Let $h\in\widehat{H}^k(\varphi;\mathbb{Z})$ and $h'\in\widehat{H}^{k'}(X';\mathbb{Z})$ be differential characters. Then the homomorphism $h\times h':Z_{k+k'-1}(\varphi\times\mathrm{id}_{X'};\mathbb{Z})\to\mathrm{U}(1)$ in Definition 24 is a differential character in $\widehat{H}^{k+k'}(\varphi\times\mathrm{id}_{X'};\mathbb{Z})$ with $(\mathrm{curv},\mathrm{cov})(h\times h')=(\mathrm{curv},\mathrm{cov})(h)\times\mathrm{curv}(h')$.*

*Proof.* We check condition (22) for the two cases separately. Since the cross product is injective on cohomology, a cross product of cycles is a boundary if and only if one of the factors is a boundary. For $(x,y)=\partial_\varphi(a,b)\in Z_{k-1}(\varphi;\mathbb{Z})$ and $y'\in Z_{k'}(X';\mathbb{Z})$, we have:

$$\begin{aligned}
&(h\times h')(\partial_{\varphi\times\mathrm{id}_{X'}}((a,b)\times y')))\\
&\quad=(h\times h')(\partial_\varphi(a,b)\times y')\\
&\quad\overset{(56)}{=}(h\times h')(\partial_\varphi(a,b))^{\langle c(h'),y'\rangle}\\
&\quad=\exp\Big(2\pi i\Big(\int_{(a,b)}(\mathrm{curv}(h),\mathrm{cov}(h))\cdot\int_{y'}\mathrm{curv}(h')\Big)\Big)\\
&\quad=\exp\Big(2\pi i\int_{(a,b)\times y'}(\mathrm{curv},\mathrm{cov})(h)\times\mathrm{curv}(h')\Big)\,.
\end{aligned}$$

For $(x,y)\in Z_{k-1}(\varphi;\mathbb{Z})$ and $y'=\partial b'\in Z_{k'}(X';\mathbb{Z})$, we have:

$$\begin{aligned}
&(h\times h')(\partial_{\varphi\times\mathrm{id}_{X'}}((x,y)\times b')))\\
&\quad=(h\times h')((-1)^{k'-1}(x,y)\times\partial b')\\
&\quad\overset{(56)}{=}h((-1)^{k'-1}(x,y))^{\langle c(h'),\partial b'\rangle}\\
&\quad=1
\end{aligned}$$

$$= \exp\left(2\pi i \int_{(x,y)\times b'} (\mathrm{curv},\mathrm{cov})(h) \times \mathrm{curv}(h')\right).$$

The last equality follows from the fact that the differential form $\mathrm{curv}(h') \in \Omega_0^{k'}(X')$ and the chain $b' \in C_{k'+1}$ have different degrees (and similarly for the other factor).

Similarly, for $(x,y) \in Z_k(\varphi;\mathbb{Z})$ and $y' = \partial b' \in Z_{k'-1}(X';\mathbb{Z})$, we have:

$$\begin{aligned}
(h \times h')(\partial_{\varphi\times\mathrm{id}_{X'}}&(x,y) \times b') \\
&= (h \times h')((-1)^k (x,y) \times \partial b') \\
&\overset{(56)}{=} h'(\partial b')^{\langle c(h),(x,y)\rangle} \\
&= \exp\left(2\pi i \int_{b'} \mathrm{curv}(h') \cdot \int_{(x,y)} (\mathrm{curv},\mathrm{cov})(h)\right) \\
&= \exp\left(2\pi i \int_{(x,y)\times b'} (\mathrm{curv},\mathrm{cov})(h) \times \mathrm{curv}(h')\right).
\end{aligned}$$

Finally, for $(x,y) = \partial_\varphi(a,b) \in Z_k(\varphi;\mathbb{Z})$ and $y' \in Z_{k'-1}(X';\mathbb{Z})$, we have:

$$\begin{aligned}
(h \times h')(\partial_{\varphi\times\mathrm{id}_{X'}}(a,b) \times y') &= (h \times h')(\partial_\varphi(a,b) \times \partial y') \\
&\overset{(56)}{=} h'(\partial b')^{\langle c(h),\partial_\varphi(a,b)\rangle} \\
&= 1 \\
&= \exp\left(2\pi i \int_{(a,b)\times y'} (\mathrm{curv},\mathrm{cov})(h) \times \mathrm{curv}(h')\right).
\end{aligned}$$

If $(s,t) = \partial_{\varphi\times\mathrm{id}_{X'}}(v,w) \in T_{k+k'-1}(\varphi\times\mathrm{id}_{X'};\mathbb{Z})$ is a boundary, we may choose $N = 1$ in (57). This yields

$$\begin{aligned}
&(h \times h')(\partial_{\varphi\times\mathrm{id}_{X'}}(v,w)) \\
&\overset{(57)}{=} \exp\Big(2\pi i \Big(\int_{(v,w)} (\mathrm{curv},\mathrm{cov})(h) \times \mathrm{curv}(h') - \underbrace{\langle c(h) \times c(h'),(v,w)\rangle}_{\in\mathbb{Z}}\Big)\Big) \\
&= \exp\left(2\pi i \int_{(v,w)} (\mathrm{curv},\mathrm{cov})(h) \times \mathrm{curv}(h')\right).
\end{aligned}$$

Thus the homomorphism $h \times h' : Z_{k+k'-1}(\varphi \times \mathrm{id}_{X'};\mathbb{Z}) \to \mathrm{U}(1)$ is a relative differential character in $\widehat{H}^{k+k'}(\varphi \times \mathrm{id}_{X'};\mathbb{Z})$ with $(\mathrm{curv},\mathrm{cov})(h \times h') = (\mathrm{curv},\mathrm{cov})(h) \times \mathrm{curv}(h')$. $\square$

### 4.2.3 Naturality and Compatibilities

We show that the cross product of relative and absolute differential characters is natural with respect to smooth maps. Moreover, it is compatible with the structure maps (curvature, covariant derivative, characteristic class and topological trivializations) and with the homomorphisms $\breve{p}$ and $\breve{\imath}$ between absolute and relative characters groups.

**Theorem 26 (Cross Product: Naturality and Compatibilities).** *The cross product between relative and absolute differential characters*

$$\times : \widehat{H}^k(\varphi;\mathbb{Z}) \times \widehat{H}^{k'}(X';\mathbb{Z}) \to \widehat{H}^{k+k'}(\varphi \times \mathrm{id}_{X'};\mathbb{Z}), \quad (h,h') \mapsto h \times h',$$

*is $\mathbb{Z}$-bilinear and associative with respect to absolute characters: for a relative character $h \in \widehat{H}^k(\varphi;\mathbb{Z})$ and absolute characters $h' \in \widehat{H}^{k'}(X';\mathbb{Z})$ and $h'' \in \widehat{H}^{k''}(X'';\mathbb{Z})$, we have*

$$(h \times h') \times h'' = h \times (h' \times h'') \in \widehat{H}^{k+k'+k''}(\varphi \times \mathrm{id}_{X' \times X''}). \tag{58}$$

*The cross product is natural: for smooth maps $(Y,B) \xrightarrow{(f,g)} (X,A)$ and $Y' \xrightarrow{f'} X'$, we have:*

$$((f,g) \times f')^*(h \times h') = (f,g)^*h \times f'^*h'. \tag{59}$$

*Curvature, covariant derivative, characteristic class and topological trivializations are multiplicative:*

$$(\mathrm{curv},\mathrm{cov})(h \times h') = (\mathrm{curv},\mathrm{cov})(h) \times \mathrm{curv}(h'). \tag{60}$$

$$c(h \times h') = c(h) \times c(h'). \tag{61}$$

$$\iota_\varphi(\omega,\vartheta) \times h' = \iota_{\varphi \times \mathrm{id}_{X'}}((\omega,\vartheta) \times \mathrm{curv}(h')), \tag{62}$$

*where $(\omega,\vartheta) \in \Omega^{k-1}(\varphi)$, $h' \in \widehat{H}^{k'}(X';\mathbb{Z})$ and hence $(\omega,\vartheta) \times \mathrm{curv}(h') \in \Omega^{k+k'-1}(\varphi \times \mathrm{id}_{X'})$.*

*The homomorphism $\breve{\imath}$ is multiplicative: for $h \in \widehat{H}^{k-1}(A;\mathbb{Z})$ and $h' \in \widehat{H}^{k'}(X';\mathbb{Z})$, we have:*

$$\breve{\imath}_\varphi(h) \times h' = \breve{\imath}_{\varphi \times \mathrm{id}_{X'}}(h \times h'). \tag{63}$$

*Likewise, the map $\breve{p}$ is multiplicative: for $h \in \widehat{H}^k(\varphi;\mathbb{Z})$ and $h' \in \widehat{H}^{k'}(X';\mathbb{Z})$, we have:*

$$\breve{p}_{\varphi \times \mathrm{id}_{X'}}(h \times h') = \breve{p}_\varphi(h) \times h'. \tag{64}$$

*Proof.* The cross product is obviously $\mathbb{Z}$-bilinear. Moreover, (60) was observed in the proof of Proposition 25. The other properties have to be checked now.

a) We compute the characteristic class $c(h \times h')$. Let $\tilde{h}$ and $\tilde{h}'$ be real lifts of $h$ and $h'$ and $\mu^{\tilde{h}} := (\mathrm{curv},\mathrm{cov})(h) - \delta_\varphi \tilde{h} \in C^k(\varphi;\mathbb{Z})$ and $\mu^{\tilde{h}'} := \mathrm{curv}(h') - \delta\tilde{h}' \in C^{k'}(X';\mathbb{Z})$ be the corresponding cocyles for the characteristic classes. We first compute a real lift $\widetilde{h \times h'}$ for the character $h \times h'$.

We have the Künneth splitting:[6]

$$\begin{aligned} &Z_{k+k'-1}(\varphi \times \mathrm{id}_{X'};\mathbb{Z}) \\ &\qquad = \big(Z_*(\varphi;\mathbb{Z}) \otimes Z_*(X';\mathbb{Z})\big)_{k+k'-1} \oplus T_{k+k'-1}(\varphi \times \mathrm{id}_{X'};\mathbb{Z})\,. \end{aligned} \tag{65}$$

On the first factor we obtain from (56) the real lift

$$\tilde{h} \times \mu^{\tilde{h}'} + (-1)^k (\mathrm{curv},\mathrm{cov})(h) \times \tilde{h}'\,.$$

On $N$-torsion cycles in the second factor we obtain from (57) the real lift

$$\frac{1}{N}\big((\mathrm{curv},\mathrm{cov})(h) \times \mathrm{curv}(h') - \mu^{\tilde{h}} \times \mu^{\tilde{h}'}\big) \circ \partial^{-1}_{\varphi \times \mathrm{id}_{X'}} \circ \big(N \cdot (\cdot)\big)\,.$$

In particular, the coboundary of the lift on the second factor is given by

$$\delta_{\varphi \times \mathrm{id}_{X'}}\big(\widetilde{h \times h'}|_{T_{k+k'-1}(\varphi \times \mathrm{id}_{X'};\mathbb{Z})}\big) = (\mathrm{curv},\mathrm{cov})(h) \times \mathrm{curv}(h') - \mu^{\tilde{h}} \times \mu^{\tilde{h}'}\,. \tag{66}$$

Now we compute the cocycle $\mu^{\widetilde{h \times h'}} \in C^{k+k'}(\varphi \times \mathrm{id}_{X'};\mathbb{Z})$ that represents the characteristic class $c(h \times h') \in H^{k+k'}(\varphi \times \mathrm{id}_{X'};\mathbb{Z})$. We use the decomposition

$$\begin{aligned} &C_{k+k'}(\varphi \times \mathrm{id}_{X'};\mathbb{Z}) \\ &= Z_{k+k'}(\varphi \times \mathrm{id}_{X'};\mathbb{Z}) \oplus \mathrm{im}\Big(C_{k+k'}(\varphi \times \mathrm{id}_{X'};\mathbb{Z}) \xleftarrow{s_{\varphi \times \mathrm{id}_{X'}}} B_{k+k'-1}(\varphi \times \mathrm{id}_{X'};\mathbb{Z})\Big) \end{aligned} \tag{67}$$

obtained from a splitting of the boundary map $\partial_{\varphi \times \mathrm{id}_{X'}}$. On the first factor in (67), we have

$$\begin{aligned} \mu^{\widetilde{h \times h'}}\big|_{Z_{k+k'}(\varphi \times \mathrm{id}_{X'};\mathbb{Z})} &:= (\mathrm{curv},\mathrm{cov})(h \times h') - \delta_{\varphi \times \mathrm{id}_{X'}} \widetilde{h \times h'} \\ &= (\mathrm{curv},\mathrm{cov})(h) \times \mathrm{curv}(h') \\ &= \big(\mu^{\tilde{h}} \times \mu^{\tilde{h}'}\big)\big|_{Z_{k+k'}(\varphi \times \mathrm{id}_{X'};\mathbb{Z})} \end{aligned}$$

The second factor in (67) inherits a further splitting from (65). With respect to this splitting, we obtain:

$$\begin{aligned} &\mu^{\widetilde{h \times h'}} \\ &:= (\mathrm{curv},\mathrm{cov})(h \times h') - \delta_{\varphi \times \mathrm{id}_{X'}} \widetilde{h \times h'} \\ &\overset{(66)}{=} (\mathrm{curv},\mathrm{cov})(h) \times \mathrm{curv}(h') \\ &\quad - \delta_{\varphi \times \mathrm{id}_{X'}}\Big(\tilde{h} \times \mu^{\tilde{h}'} + (-1)^k (\mathrm{curv},\mathrm{cov})(h) \times \tilde{h}'\Big) \end{aligned}$$

[6] More precisely, the first factor of the right hand side of (65) is the image of the cross product $\big(Z_*(\varphi;\mathbb{Z}) \otimes Z_*(X';\mathbb{Z})\big)_{k+k'-1} \xrightarrow{\times} Z_{k+k'-1}(\varphi \times \mathrm{id}_{X'};\mathbb{Z})$. Therefor, we write cross products instead of tensor products for the real lifts on this factor.

$$
\begin{aligned}
&\qquad\qquad\qquad\qquad \oplus \Big( (\mathrm{curv}, \mathrm{cov})(h \times h') - \mu^{\tilde h} \times \mu^{\tilde h'} \Big) \\
&= \Big( (\mathrm{curv}, \mathrm{cov})(h) \times \mathrm{curv}(h') - \delta_\varphi \tilde h \times \mu^{\tilde h'} - (\mathrm{curv}, \mathrm{cov})(h) \times \delta \tilde h' \Big) \\
&\qquad\qquad\qquad\qquad \oplus \Big( \mu^{\tilde h} \times \mu^{\tilde h'} \Big) \\
&= \Big( (\mathrm{curv}, \mathrm{cov})(h) \times \mu^{\tilde h'} - \delta_\varphi \tilde h \times \mu^{\tilde h'} \Big) \oplus \Big( \mu^{\tilde h} \times \mu^{\tilde h'} \Big) \\
&= \mu^{\tilde h} \times \mu^{\tilde h'} .
\end{aligned}
$$

In conclusion, we have $\mu^{\widetilde{h \times h'}} = \mu^{\tilde h} \times \mu^{\tilde h'}$ and thus (61) holds.

b) Next we prove associativity. To apply Definition 24 we need to first derive an appropriate Künneth splitting of $Z_{k+k'+k''-1}(\varphi \times \mathrm{id}_{X' \times X''}; \mathbb{Z})$. The cross product of cycles and the classical Alexander-Whitney and Eilenberg-Zilber maps are associative. This implies that the induced Alexander-Whitney and Eilenberg-Zilber maps for mapping cone complexes are also associative. More explicitly, we have the following commutative diagram:

$$
\begin{array}{ccc}
Z_*(\varphi;\mathbb{Z}) \otimes Z_*(X';\mathbb{Z}) \otimes Z_*(X'';\mathbb{Z}) & \underset{S_\varphi \otimes \mathrm{id}}{\overset{\times \otimes \mathrm{id}}{\rightleftarrows}} & Z_*(\varphi \times \mathrm{id}_{X'};\mathbb{Z}) \otimes Z_*(X'';\mathbb{Z}) \\
\mathrm{id} \otimes S \uparrow \downarrow \mathrm{id} \otimes \times & \nwarrow \mathbf{S} & S_{\varphi \times \mathrm{id}_{X'} \times \mathrm{id}_{X''}} \uparrow \downarrow \times \\
Z_*(\varphi;\mathbb{Z}) \otimes Z_*(X' \times X'';\mathbb{Z}) & \underset{S_{\varphi \times \mathrm{id}_{X \times X'}}}{\overset{\times}{\rightleftarrows}} & Z_*(\varphi \times \mathrm{id}_{X' \times X''};\mathbb{Z}) .
\end{array}
$$

The induced splitting $\mathbf{S} : Z_*(\varphi \times \mathrm{id}_{X' \times X''}; \mathbb{Z}) \to Z_*(\varphi;\mathbb{Z}) \otimes Z_*(X';\mathbb{Z}) \otimes Z_*(X'';\mathbb{Z})$ of the concatenation $\times \circ (\mathrm{id} \otimes \times) = \times \circ (\times \otimes \mathrm{id})$ yields a direct sum decomposition

$$
Z_*(\varphi \times \mathrm{id}_{X' \times X''};\mathbb{Z}) = Z_*(\varphi;\mathbb{Z}) \otimes Z_*(X';\mathbb{Z}) \otimes Z_*(X'';\mathbb{Z}) \oplus \ker(\mathbf{S}) . \tag{68}
$$

By the relative Künneth theorem, the cycles in $\ker(\mathbf{S})$ represent torsion classes in $H_{k+k'+k''}(\varphi \times \mathrm{id}_{X' \times X''};\mathbb{Z})$.

The cross products of relative and absolute differential forms and cohomology classes are associative. Thus for a relative differential character $h \in \widehat H^k(\varphi;\mathbb{Z})$ and absolute characters $h' \in \widehat H^{k'}(X';\mathbb{Z})$ and $h'' \in \widehat H^{k''}(X'';\mathbb{Z})$, the cross products $(h \times h') \times h''$ and $h \times (h' \times h'')$ have the same curvature, covariant derivative and characteristic class. Hence they coincide on torsion cycles, and in particular on cycles in $\ker(\mathbf{S})$.

Now we compare the two characters $(h \times h') \times h''$ and $h \times (h' \times h'')$ on cross products $(x, y) \times y' \times y''$, where $(x,y) \in Z_i(\varphi;\mathbb{Z})$ and $y' \in Z_j(X';\mathbb{Z})$ and $y'' \in Z_l(X'';\mathbb{Z})$. By (56) both characters vanish on cross products with

$(i,j,l)$ different from $(k-1,k',k'')$, $(k,k'-1,k'')$ and $(k,k',k''-1)$. Now we compute the remaining cases. For $(i,j,l) = (k-1,k',k'')$, we have:

$$\begin{aligned}((h \times h') \times h'')((x,y) \times y' \times y'') &= (h \times h')((x,y) \times y')^{\langle c(h''),y''\rangle} \\ &= h(x,y)^{\langle c(h'),y'\rangle \cdot \langle c(h''),y''\rangle} \\ &\overset{(52)}{=} h(x,y)^{\langle c(h' \times h''),y' \times y''\rangle} \\ &= (h \times (h' \times h''))((x,y) \times y' \times y'') \,.\end{aligned}$$

Similarly, for $(i,j,l) = (k-1,k',k'')$, we have:

$$\begin{aligned}((h \times h') \times h'')((x,y) \times y' \times y'') &= (h \times h')((x,y) \times y')^{\langle c(h''),y''\rangle} \\ &= h'(y')^{(-1)^k \cdot \langle c(h),(x,y)\rangle \cdot \langle c(h''),y''\rangle} \\ &= (h' \times h'')(y' \times y'')^{(-1)^k \cdot \langle c(h),(x,y)\rangle} \\ &= (h \times (h' \times h''))((x,y) \times y' \times y'') \,.\end{aligned}$$

Finally, for $(i,j,l) = (k,k',k''-1)$, we have:

$$\begin{aligned}((h \times h') \times h'')((x,y) \times y' \times y'') &= h''(y'')^{(-1)^{k+k'} \cdot \langle c(h \times h'),(x,y) \times y'\rangle} \\ &\overset{(61)}{=} h''(y'')^{(-1)^k \cdot \langle c(h),(x,y)\rangle \cdot (-1)^{k'} \cdot \langle c(h'),y'\rangle} \\ &= (h \times (h' \times h''))((x,y) \times y' \times y'') \,.\end{aligned}$$

Thus $(h \times h') \times h'' = h \times (h' \times h'')$.

c) Now we consider topological trivializations. Let $(\omega,\vartheta) \in \Omega^{k-1}(\varphi)$. We compare the characters $\iota_\varphi(\omega,\vartheta) \times h'$ and $\iota_{\varphi \times \mathrm{id}_{X'}}((\omega,\vartheta) \times \mathrm{curv}(h'))$. For curvature and covariant derivative, we have:

$$\begin{aligned}(\mathrm{curv},\mathrm{cov})(\iota_\varphi(\omega,\vartheta) \times h') &= (\mathrm{curv},\mathrm{cov})(\iota_\varphi(\omega,\vartheta)) \times \mathrm{curv}(h') \\ &= d_\varphi(\omega,\vartheta) \times \mathrm{curv}(h') \\ &= d_{\varphi \times \mathrm{id}_{X'}}\big((\omega,\vartheta) \times \mathrm{curv}(h')\big) \\ &= (\mathrm{curv},\mathrm{cov})\big(\iota_{\varphi \times \mathrm{id}_{X'}}((\omega,\vartheta) \times \mathrm{curv}(h'))\big) \,.\end{aligned}$$

For the characteristic class, we have:

$$c(\iota_\varphi(\omega,\vartheta) \times h') = \underbrace{c(\iota_\varphi(\omega,\vartheta))}_{=0} \times c(h') = 0 = c\big(\iota_{\varphi \times \mathrm{id}_{X'}}((\omega,\vartheta) \times \mathrm{curv}(h'))\big) \,.$$

By (57) the characters $\iota_\varphi(\omega,\vartheta) \times h'$ and $\iota_{\varphi \times \mathrm{id}_{X'}}((\omega,\vartheta) \times \mathrm{curv}(h'))$ thus coincide on the factor $T_{k+k'-1}(\varphi \times \mathrm{id}_{X'};\mathbb{Z})$ in $Z_{k+k'-1}(\varphi \times \mathrm{id}_{X'};\mathbb{Z})$.

Let $(s,t) = (x,y) \times y'$ be a cross product of cycles $(x,y) \in Z_i(\varphi;\mathbb{Z})$ and $y' \in Z_j(X';\mathbb{Z})$. By (57), we have $(\iota_\varphi(\omega,\vartheta) \times h')((x,y) \times y') = 1$ if $(i,j) \neq (k-1,k')$. The same holds for the character $\iota_{\varphi \times \mathrm{id}_{X'}}((\omega,\vartheta) \times \mathrm{curv}(h'))$, since the differential form $(\omega, \vartheta) \times \mathrm{curv}(h')$ vanishes upon integration over cross products of cycles of degrees different from $(k-1, k')$. For $(i,j) = (k-1, k')$, we have:

$$\begin{aligned}(\iota_\varphi(\omega,\vartheta) \times h')((x,y) \times y') &= (\iota_\varphi(\omega,\vartheta)(x,y))^{\langle c(h'),y'\rangle} \\ &= \exp\Big(2\pi i \int_{(x,y)} (\omega,\vartheta) \int_{y'} \mathrm{curv}(h')\Big) \\ &= \iota_{\varphi\times \mathrm{id}_{X'}}((\omega,\vartheta) \times \mathrm{curv}(h'))((x,y) \times y') \,.\end{aligned}$$

This proves (62).

d) Now we prove naturality. Let $\psi : Y \to B$ and $f' : Y' \to X'$ be smooth maps. Let $(Y,B) \xrightarrow{(f,g)} (X,A)$ be a smooth map. Let $h \in \widehat{H}^k(\varphi;\mathbb{Z})$ and $h' \in \widehat{H}^{k'}(X';\mathbb{Z})$. The relative classical Alexander-Whitney and Eilenberg-Zilber maps are natural with respect to smooth maps. Thus so are the Künneth splittings (65) constructed in the appendix. More explicity, the map induced by $(Y,B) \times Y' \xrightarrow{(f,g)\times f'} (X,A) \times X'$ maps the splitting

$$Z_{k+k'-1}(\psi \times \mathrm{id}_{Y'};\mathbb{Z}) = \big(Z_*(\psi;\mathbb{Z}) \otimes Z_*(Y';\mathbb{Z})\big)_{k+k'-1} \oplus T_{k+k'-1}(\psi \times \mathrm{id}_{Y'};\mathbb{Z})$$

to the splitting

$$Z_{k+k'-1}(\varphi \times \mathrm{id}_{X'};\mathbb{Z}) = \big(Z_*(\varphi;\mathbb{Z}) \otimes Z_*(X';\mathbb{Z})\big)_{k+k'-1} \oplus T_{k+k'-1}(\varphi \times \mathrm{id}_{X'};\mathbb{Z}) \,.$$

Since curvature, covariant derivative and characteristic class are natural, by (57) we have for any cycle $(s,t) \in T_{k+k'-1}(\psi \times \mathrm{id}_{Y'};\mathbb{Z})$:

$$\begin{aligned}(((f,g) \times f')^*(h \times h'))(s,t) &= (h \times h')(((f,g) \times f')_*(s,t)) \\ &= ((f,g)^*h \times f'^*h')(s,t) \,.\end{aligned}$$

Similarly, for cross product cycles $(s,t) = (x,y) \times y'$ with $(x,y) \in Z_i(\psi;\mathbb{Z})$ and $y' \in Z_j(Y';\mathbb{Z})$, we obtain from (57):

$$\begin{aligned}(((f,g) \times f')^*(h \times h'))((x,y) \times y') &= (h \times h')(((f,g) \times f')_*((x,y) \times y')) \\ &= (h \times h')((f,g)_*(x,y) \times f'_*y')) \\ &= ((f,g)^*h \times f'^*h')(s,t) \,.\end{aligned}$$

e) Finally we consider compatibility with the maps $\breve{\imath}$ and $\breve{p}$ that relate absolute and relative differential characters groups. For $h \in \widehat{H}^{k-1}(A;\mathbb{Z})$ and $h' \in \widehat{H}^{k'}(X';\mathbb{Z})$, we have:

$$\begin{aligned}(\mathrm{curv},\mathrm{cov})(\breve{\imath}_\varphi(h) \times h') &= (0, -\mathrm{curv}(h) \times \mathrm{curv}(h')) \\ &= (0, -\mathrm{curv}(h \times h')) \\ &= (\mathrm{curv},\mathrm{cov})(\breve{\imath}_{\varphi\times \mathrm{id}_{X'}}(h \times h')) \,.\end{aligned}$$

Moreover, the characteristic classes of both $\breve{\imath}_\varphi(h) \times h'$ and $\breve{\imath}_{\varphi\times\mathrm{id}_{X'}}(h \times h')$ equal the image of the class $c(h \times h')$ under the map $H^{k+k'-1}(A \times X';\mathbb{Z}) \to H^{k+k'}(\varphi\times\mathrm{id}_{X'};\mathbb{Z})$. Thus the characters $\breve{\imath}_\varphi(h)\times h'$ and $\breve{\imath}_{\varphi\times\mathrm{id}_{X'}}(h\times h')$ coincide on the second factor in (65).

Let $(s,t) = (x,y) \times y'$ be a cross product of cycles $(x,y) \in Z_i(\varphi;\mathbb{Z})$ and $y' \in Z_j(X';\mathbb{Z})$. By (55) and (56) both $\breve{\imath}_\varphi(h) \times h'$ and $\breve{\imath}_{\varphi\times\mathrm{id}_{X'}}(h \times h')$ vanish on cross products with $(k-1,k') \neq (i,j) \neq (k,k'-1)$. For $(i,j) = (k-1,k')$ or $(i,j) = (k,k'-1)$, we have:

$$(\breve{\imath}(h) \times h')((x,y) \times y') = (h \times h')(y \times y') = \breve{\imath}(h \times h')((x,y) \times y')\,.$$

Thus (63) holds. The proof of (64) is completely analogous. □

### 4.2.4 The Module Structure

As is well-known, the cup product provides relative cohomology with the structure of a right module over the absolute cohomology ring. In the same way, the mapping cone cohomology $H^*(\varphi;\mathbb{Z})$ of a (smooth) map $\varphi : A \to X$ is a right module over the cohomology ring $H^*(X;\mathbb{Z})$. Similarly, we have an internal product on the mapping cone de Rham complex $\Omega^*(\varphi)$ defined by

$$(\omega,\vartheta) \wedge \omega' := (\omega \wedge \omega', \vartheta \wedge \varphi^*\omega')\,, \tag{69}$$

where $(\omega,\vartheta) \in \Omega^*(\varphi)$ and $\omega' \in \Omega^*(X)$. Thus the abelian group $\Omega^*(\varphi)$ is a right module over the ring $\Omega^*(X)$ of differential forms on $X$.

From the external product between relative and absolute differential characters we derive an internal product by pull-back along a version of the diagonal map. By the analogue of Theorem 26, the internal product provides the graded abelian group $\widehat{H}^*(\varphi;\mathbb{Z})$ with a natural structure of a right module over the ring $\widehat{H}^*(X;\mathbb{Z})$ such that the structure maps (curvature, covariant derivative and characteristic class) become ring homomorphisms.

Denote by $\Delta_X : X \to X \times X$, $x \mapsto (x,x)$, the diagonal map, and similarly for $A$. Let $\varphi : A \to X$ be a smooth map and $\varphi \times \mathrm{id}_X : A \times X \to X \times X$. Similar to the diagonal map, let

$$\begin{aligned}\Delta_{(X,A)} := (\Delta_X, (\mathrm{id}_A \times\varphi) \circ \Delta_A) : (X,A) &\to (X,A) \times X = (X \times X, A \times X)\,,\\ (x,a) &\mapsto ((x,x),(a,\varphi(a)))\,.\end{aligned}$$

Since the external product of relative and absolute characters is natural, we may use $\Delta_{(X,A)}$ to pull-back cross products $h \times h' \in \widehat{H}^{k+k'}(\varphi \times \mathrm{id}_{X'};\mathbb{Z})$ to $(X,A)$.

**Definition 27 (Internal product).** Let $h \in \widehat{H}^k(\varphi;\mathbb{Z})$ and $h' \in \widehat{H}^{k'}(X;\mathbb{Z})$ be differential characters. Their internal product is the character $h * h' \in \widehat{H}^{k+k'}(\varphi;\mathbb{Z})$, defined by

$$h * h' := \Delta^*_{(X,A)}(h \times h') .$$

The properties of the external product proved in Theorem 26 directly translate into properties of the internal product. This establishes the module structure on $\widehat{H}^*(\varphi;\mathbb{Z})$ and turns the curvature, covariant derivative and characteristic into module homomorphisms:

**Theorem 28 (Module Structure: Naturality and Compatibilities).** *The internal product between relative and absolute differential characters*

$$* : \widehat{H}^k(\varphi;\mathbb{Z}) \times \widehat{H}^{k'}(X';\mathbb{Z}) \to \widehat{H}^{k+k'}(\varphi \times \mathrm{id}_{X'};\mathbb{Z}) , \quad (h,h') \mapsto h \times h' ,$$

*is $\mathbb{Z}$-bilinear and associative with respect to absolute characters: for a relative character $h \in \widehat{H}^k(\varphi;\mathbb{Z})$ and absolute characters $h' \in \widehat{H}^{k'}(X;\mathbb{Z})$ and $h'' \in \widehat{H}^{k''}(X;\mathbb{Z})$, we have*

$$(h * h') * h'' = h * (h' * h'') .$$

*The internal product is natural: for a smooth map $(Y,B) \xrightarrow{(f,g)} (X,A)$ we have:*

$$(f,g)^*(h * h') = (f,g)^*h * f^*h' .$$

*Curvature, covariant derivative, characteristic class and topological trivializations are multiplicative:*

$$(\mathrm{curv},\mathrm{cov})(h * h') = (\mathrm{curv},\mathrm{cov})(h) \wedge \mathrm{curv}(h') .$$
$$c(h * h') = c(h) \cup c(h') .$$
$$\iota_\varphi(\omega,\vartheta) * h' = \iota_\varphi((\omega,\vartheta) \wedge \mathrm{curv}(h')) ,$$

*where $(\omega,\vartheta) \in \Omega^{k-1}(\varphi)$.*
*The group homomorphism $\breve{\imath}_\varphi : \widehat{H}^{*-1}(A;\mathbb{Z}) \to \widehat{H}^*(\varphi;\mathbb{Z})$ is multiplicative: for characters $h \in \widehat{H}^{k-1}(A;\mathbb{Z})$ and $h' \in \widehat{H}^{k'}(X';\mathbb{Z})$, we have:*

$$\breve{\imath}_\varphi(h) * h' = \breve{\imath}_\varphi(h * \varphi^*h') . \tag{70}$$

*Likewise, the homomorphism $\breve{p}_\varphi : \widehat{H}^*(\varphi;\mathbb{Z}) \to \widehat{H}^*(X;\mathbb{Z})$ is multiplicative: for characters $h \in \widehat{H}^k(\varphi;\mathbb{Z})$ and $h' \in \widehat{H}^{k'}(X';\mathbb{Z})$, we have:*

$$\breve{p}_\varphi(h * h') = \breve{p}_\varphi(h) * h' . \tag{71}$$

*Proof.* To prove associativity, we need to keep track of the various pull-backs:

$$\begin{aligned}(h * h') * h'' &= \Delta^*_{(X,A)}\Big(\big(\Delta^*_{(X,A)}h \times h'\big) \times h''\Big) \\ &\overset{(59)}{=} \Delta^*_{(X,A)}\Big(\big(\Delta_{(X,A)} \times \mathrm{id}_X\big)^*(h \times h') \times h''\Big)\end{aligned}$$

$$
\begin{aligned}
&= \Big( (\Delta_{(X,A)} \times \mathrm{id}_X) \circ \Delta_{(X,A)} \Big)^* \big( (h \times h') \times h'' \big) \\
&\overset{(58)}{=} \Big( (\mathrm{id}_{(X,A)} \times \Delta_X) \circ \Delta_{(X,A)} \Big)^* \big( h \times (h' \times h'') \big) \\
&\overset{(59)}{=} \Delta^*_{(X,A)} \big( h \times \Delta^*_X (h' \times h'') \big) \\
&= h * (h' * h'') \,.
\end{aligned}
$$

In the third last equation we used equality of the maps

$$
\begin{aligned}
(\Delta_{(X,A)} \times \mathrm{id}_X) &\circ \Delta_{(X,A)} \\
&= (\mathrm{id}_{(X,A)} \times \Delta_X) \circ \Delta_{(X,A)} : (X,A) \to (X,A) \times (X \times X) \,, \\
(x,a) &\mapsto ((x,x,x),(a,\varphi(a),\varphi(a))) \,.
\end{aligned}
$$

Naturality of the internal product follows from naturality of the cross product together with the equality of maps

$$
\Delta_{(X,A)} \circ (f,g) = ((f,g) \times f) \circ \Delta_{(Y,B)} : (Y,B) \to (X,A) \times X \,.
$$

Thus for characters $h \in \widehat{H}^k(\varphi;\mathbb{Z})$ and $h' \in \widehat{H}^{k'}(X';\mathbb{Z})$, we have:

$$
\begin{aligned}
(f,g)^* h * h' &= (f,g)^* \Delta^*_X (h \times h') \\
&= (\Delta_X \circ (f,g))^* (h \times h') \\
&= ((f,g) \times f) \circ \Delta_Y)^* (h \times h') \\
&\overset{(59)}{=} \Delta^*_Y ((f,g)^* h \times f^* h') \\
&= (f,g)^* h * f^* h' \,.
\end{aligned}
$$

For curvature and covariant derivative, we have:

$$
\begin{aligned}
(\mathrm{curv},\mathrm{cov})(h * h') &\overset{(60)}{=} (\mathrm{curv},\mathrm{cov})(\Delta^*_{(X,A)}(h \times h')) \\
&= \Delta^*_{X,A}((\mathrm{curv},\mathrm{cov})(h) \times \mathrm{curv}(h')) \\
&= (\Delta^*_X \mathrm{curv}(h) \times \mathrm{curv}(h'), \Delta^*_A (\mathrm{id}_A \times \varphi)^* \mathrm{cov}(h) \times \mathrm{curv}(h')) \\
&\overset{(59)}{=} (\mathrm{curv}(h) \wedge \mathrm{curv}(h'), \mathrm{cov}(h) \wedge \varphi^* \mathrm{curv}(h')) \\
&\overset{(69)}{=} (\mathrm{curv},\mathrm{cov})(h) \wedge \mathrm{curv}(h') \,.
\end{aligned}
$$

Likewise, for topological trivializations we have:

$$
\begin{aligned}
\iota_\varphi(\omega,\vartheta) * h' &= \Delta^*_{(X,A)} (\iota_\varphi(\omega,\vartheta) \times h') \\
&\overset{(62)}{=} \Delta^*_{(X,A)} \iota_\varphi((\omega,\vartheta) \times \mathrm{curv}(h')) \\
&= \iota_\varphi((\omega,\vartheta) \wedge \mathrm{curv}(h')) \,.
\end{aligned}
$$

Multiplicativity of the characteristic class follows from (61) and the fact that the cup product is the pull-back along $\Delta_{(X,A)}$ of the cross product.

It remains to prove multiplicativity of the homomorphisms $\breve{\imath}_\varphi$ and $\breve{p}_\varphi$. For characters $h \in \widehat{H}^{k-1}(A;\mathbb{Z})$ and $h' \in \widehat{H}^{k'}(X';\mathbb{Z})$, and a cycle $(s,t) \in Z_{k+k'-1}(\varphi;\mathbb{Z})$ we have:

$$\begin{aligned}
(\breve{\imath}_\varphi(h) * h')(s,t) &= \big(\Delta^*_{(X,A)}(\breve{\imath}_\varphi(h) \times h')\big)(s,t) \\
&\overset{(63)}{=} \big(\Delta^*_{(X,A)}\breve{\imath}_\varphi(h \times h')\big)(s,t) \\
&= (h \times h')\big((\mathrm{id}_A \times \varphi) \circ \Delta_A\big)_*(t) \\
&\overset{(59)}{=} \Delta^*_A(h \times \varphi^* h')(t) \\
&= \breve{\imath}_\varphi(h * \varphi^* h')(s,t)\,.
\end{aligned}$$

Likewise, for characters $h \in \widehat{H}^k(\varphi;\mathbb{Z})$ and $h' \in \widehat{H}^{k'}(X';\mathbb{Z})$ and a cycle $z \in Z_{k+k'-1}(X;\mathbb{Z})$ we have:

$$\begin{aligned}
(\breve{p}_\varphi(h) * h')(z) &= \Delta^*_X(\breve{p}_\varphi(h) \times h')(z) \\
&\overset{(64)}{=} (h \times h')(\Delta_{X*}z, 0) \\
&= \big(\Delta^*_{(X,A)}(h \times h')\big)(z,0) \\
&= \breve{p}_\varphi(h * h')(z)\,.
\end{aligned}$$

□

### 4.2.5 Uniqueness of the Cross Product and Module Structure

In [1, Chap. 6] we have shown uniqueness of the external and internal product between absolute differential characters. This in particular implies uniqueness of the ring structure on differential cohomology. The proof starts from an axiomatic definition of the cross product. The axioms essentially coincide with the properties in Theorem 26 (for the absolute case).

The methods of proof used in [1, Chap. 6] directly apply to the external product between relative and absolute differential characters defined in the present work. Thus we could have defined the external product axiomatically by the properties in Theorem 26. Then we could have derived the formulae (56) and (57) from this axiomatic decription. This would have proved uniqueness of the external and internal product and hence uniqueness of the right $\widehat{H}^*(X;\mathbb{Z})$-module structure on the relative differential cohomology $\widehat{H}^*(\varphi;\mathbb{Z})$ of a smooth map $\varphi : A \to X$.

Thus we note without explicit proof here:

**Corollary 29.** *The relative differential cohomology $\widehat{H}^*(\varphi;\mathbb{Z})$ of a smooth map $\varphi : A \to X$ carries the structure of a right module over the ring $\widehat{H}^*(X;\mathbb{Z})$. The module structure is uniquely determined by the properties in Theorem 28.*

### 4.2.6 The Module Structure on Parallel Characters

In Theorem 17 we have shown that the graded abelian group $\widehat{H}^*(X,A;\mathbb{Z})$ defined by characters on the groups of relative cycles coincides with the subgroup of parallel characters in $\widehat{H}^*(i_A;\mathbb{Z})$, where $i_A : A \to X$ is the embedding of a smooth submanifold. By Theorems 26 and 28, the external and internal products of relative and absolute differential characters are multiplicative with respect to the covariant derivative. Thus products of flat characters with absolute characters are again flat characters. In other words, we have:

**Corollary 30.** *Let $i_A : A \to X$ be the inclusion of a smooth submanifold. Then there exist unique natural internal and external products*

$$\times : \widehat{H}^k(X,A;\mathbb{Z}) \times \widehat{H}^{k'}(X';\mathbb{Z}) \to \widehat{H}^{k+k'}(X \times X', A \times X';\mathbb{Z}), (h,h') \mapsto h \times h',$$
$$* : \widehat{H}^k(X,A;\mathbb{Z}) \times \widehat{H}^{k'}(X;\mathbb{Z}) \to \widehat{H}^{k+k'}(X,A;\mathbb{Z}), (h,h') \mapsto h * h',$$

*such that curvature, characteristic class and topological trivializations are multiplicative. Moreover, the products are associative with respect to absolute characters.*

*In particular, the graded abelian group $\widehat{H}^*(X,A;\mathbb{Z})$ carries a unique structure of a right module over the ring $\widehat{H}^*(X;\mathbb{Z})$.*

*Remark 31.* The identification of the relative Hopkins-Singer group $\check{H}^*(\varphi;\mathbb{Z})$ as a subquotient of the group $\widehat{H}^*(\varphi;\mathbb{Z})$ of relative differential characters induces external and internal products

$$\times : \check{H}^k(\varphi;\mathbb{Z}) \times \widehat{H}^{k'}(X';\mathbb{Z}) \to \check{H}^{k+k'}(\varphi \times \mathrm{id}_{X'};\mathbb{Z}), \ (h,h') \mapsto h \times h',$$
$$* : \check{H}^k(\varphi;\mathbb{Z}) \times \widehat{H}^{k'}(X;\mathbb{Z}) \to \check{H}^{k+k'}(\varphi;\mathbb{Z}), \ (h,h') \mapsto h * h',$$

This is well-defined, since for $h \in \check{H}^k(\varphi;\mathbb{Z}) \subset \widehat{H}^k(\varphi;\mathbb{Z})$ and $h' \in \widehat{H}^{k'}(X';\mathbb{Z})$ we have $\mathrm{cov}(h \times h') = \mathrm{cov}(h) \times \varphi^*\mathrm{curv}(h') \in \Omega_0^{k+k'-1}(A \times X')$. Hence $h \times h'$ lies in the subgroup $\check{H}^{k+k'}(\varphi;\mathbb{Z}) \subset \widehat{H}^{k+k'}(\varphi;\mathbb{Z})$ of characters with covariant derivative a closed form with integral periods.

This yields a right $\widehat{H}^*(X;\mathbb{Z})$-module structure also on $\check{H}^*(\varphi;\mathbb{Z})$.

## 5 Fiber Integration and Transgression

Let $\pi : E \to X$ be a fiber bundle with closed oriented fibers. There are natural fiber integration maps $\fint_F : \Omega^k(E) \to \Omega^{k-\dim F}(X)$ for differential forms and $\pi_! : H^k(E;\mathbb{Z}) \to H^{k-\dim F}(X;\mathbb{Z})$ for integral cohomology. Thus it is natural to expect that there exists a fiber integration map $\widehat{H}^*(E;\mathbb{Z}) \to \widehat{H}^{k-\dim F}(X;\mathbb{Z})$ that induces the well-known maps on the curvature and characteristic class.

Such fiber integration maps have been constructed for several models of differential cohomology, see [26], [11] for differential cocycles, [17], [29] for simplicial forms, [24] for de Rham-Feder currents and [1] for the original model of differential characters. In [1, Chap. 7] we prove that fiber integration is uniquely determined by the requirements to be compatible with pull-back diagrams and with fiber integration for differential forms (i.e. with curvature and topological trivializations). The proof is constructive in that it yields an explicit formula for the fiber integration map. In particular, the various constructions in the different models for differential cohomology yield the same fiber integration map.

In this section we use the method from [1] to construct fiber integration and transgression maps for relative differential characters. In particular, we make use of the pull-back operation for geometric relative cycles and the transfer maps constructed in Sections 2.6 and 2.7. We show that fiber integration for relative characters is compatible via the homomorphisms $\breve{\imath}$ and $\breve{p}$ with fiber integration for absolute characters. As a corollary, we obtain fiber integration and transgression maps for parallel characters. Moreover, fiber integration and transgression commute with the long exact sequences (34) and (42).

## 5.1 Fiber Integration

Let $\pi : E \to X$ be a fiber bundle and $\varphi : A \to X$ a smooth map. We have the pull-back diagram

$$\begin{array}{ccc} \varphi^*E & \xrightarrow{\Phi} & E \\ {\scriptstyle\pi}\downarrow & & \downarrow{\scriptstyle\pi} \\ A & \xrightarrow[\varphi]{} & X\,. \end{array}$$

In the following we construct fiber integration for relative differential characters. We discuss its compatibility with curvature, covariant derivative, topological trivializations and characteristic class and with fiber integration for absolute differential characters.

### 5.1.1 Construction of the Fiber Integration Map

For convenience of the reader, we recall the formula for fiber integration of (absolute) differential characters obtained in [1, Chap. 7]. For a differential character $h \in \widehat{H}^k(E;\mathbb{Z})$ and a smooth singular cycle $z \in Z_{k-\dim F-1}(X;\mathbb{Z})$, we have:

$$\widehat{\pi}_! h(z) = h(\lambda(z)) \cdot \exp\Big(2\pi i \int_{a(z)} \fint_F \mathrm{curv}(h)\Big)\,. \tag{72}$$

We now adapt this formula to relative differential characters.

Let $h \in \widehat{H}^k(\Phi;\mathbb{Z})$. To evaluate the character $\widehat{\pi}_! h \in \widehat{H}^{k-\dim F}(\varphi;\mathbb{Z})$ on a cycle $(s,t) \in Z_{k-\dim F-1}(\varphi;\mathbb{Z})$, we use the homomorphism $(a,b)_\varphi$ from Sect. 2.6 and the transfer map $\lambda_\varphi$ defined in Sect. 2.7.

**Definition 32.** Let $F \hookrightarrow E \overset{\pi}{\twoheadrightarrow} X$ be a fiber bundle with closed oriented fibers. Let $\varphi : A \to X$ be a smooth map and $\Phi : \varphi^*E \to E$ the induced bundle map.Let $k \geq \dim F + 2$. Fiber integration for relative differential characters is the group homomorphism $\widehat{\pi}_! : \widehat{H}^k(\Phi;\mathbb{Z}) \to \widehat{H}^{k-\dim F}(\varphi;\mathbb{Z})$ defined by

$$(\widehat{\pi}_! h)(s,t) := h(\lambda_\varphi(s,t)) \cdot \exp\Big(2\pi i \int_{(a,b)_\varphi(s,t)} \fint_F (\mathrm{curv},\mathrm{cov})(h)\Big). \tag{73}$$

Here $(s,t) \in Z_{k-\dim F-1}(\varphi;\mathbb{Z})$.

Clearly, the mapping $h \mapsto \widehat{\pi}_! h$ is a additive in $h$, thus $\widehat{\pi}_!$ is a group homomorphism. Moreover, the map $(s,t) \mapsto (\widehat{\pi}_! h)(s,t)$, defined by the right hand side of (73), is a group homomorphism $Z_{k-1-\dim F}(\varphi;\mathbb{Z}) \to \mathrm{U}(1)$, since the maps $\lambda_\varphi$ and $(a,b)_\varphi$ are group homomorphisms. In order to show that this homomorphism is indeed a differential character in $\widehat{H}^{k-\dim F}(\varphi;\mathbb{Z})$, we need to evaluate it on a boundary $\partial_\varphi(v,w)$, where $(v,w) \in C_{k-\dim F}(\varphi;\mathbb{Z})$. This will be done in the proof of Theorem 35 below.

### 5.1.2 Well-Definedness

Before discussing its properties, we show that fiber integration is well-defined, i.e. its definition is independent of the choice of geometric representative $(\zeta,\tau)_\varphi(s,t)$ and chain $(a,b)_\varphi(s,t)$:

**Lemma 33.** *Let $F \hookrightarrow E \overset{\pi}{\twoheadrightarrow} X$ be a fiber bundle with closed oriented fibers. Let $\varphi : A \to X$ be a smooth map and $\Phi : \varphi^*E \to E$ the induced bundle map. Let $k \geq \dim F + 2$. Let $h \in \widehat{H}^k(\Phi;\mathbb{Z})$ and $(s,t) \in Z_{k-1-\dim F}(\varphi;\mathbb{Z})$. Let $(\zeta',\tau') \in \mathcal{Z}_{k-1-\dim F}(\varphi)$ and $(a',b') \in C_{k-\dim F}(\varphi;\mathbb{Z})$ be any geometric cycle and singular chain such that $[\zeta',\tau']_{\partial_\varphi S_{k-\dim F}} = [(s,t) - \partial_\varphi(a',b')]_{\partial_\varphi S_{k-\dim F}}$. Then we have:*

$$(\widehat{\pi}_! h)(s,t) = h([\mathrm{PB}_{E,\varphi^*E}(\zeta',\tau')]_{\partial_\varphi S_k}) \cdot \exp\Big(2\pi i \int_{(a',b')} \fint_F (\mathrm{curv},\mathrm{cov})(h)\Big). \tag{74}$$

*Proof.* Since the geometric cycles $(\zeta,\tau)$ and $(\zeta',\tau')$ represent the same homology class in $H_{k-1-\dim F}(\varphi;\mathbb{Z})$, they are bordant. A bordism $(W,M) \xrightarrow{(F,G)} (X,A)$ from $(\zeta,\tau)$ to $(\zeta',\tau')$ yields a bordism $(F^*E, G^*(\varphi^*E)) \xrightarrow{(\mathbf{F},\mathbf{G})} (E,\varphi^*E)$ from $\mathrm{PB}_{E,\varphi^*E}(\zeta,\tau)$ to $\mathrm{PB}_{E,\varphi^*E}(\zeta',\tau')$. By equation (2) and the assumption, we have

$$\partial_\varphi\Big((F,G)_*[W,M]_{S_{k-\dim F}}\Big) \stackrel{(2)}{=} [\zeta',\tau']_{\partial_\varphi S_{k-\dim F}} - [\zeta,\tau]_{\partial_\varphi S_{k-\dim F}}$$
$$= [\partial_\varphi(a,b) - \partial_\varphi(a',b')]_{\partial_\varphi S_{k-\dim F}}\,.$$

In particular, we find a cycle $(x,y) \in Z_{k-\dim F}(\varphi;\mathbb{Z})$ such that

$$(F,G)_*[W,M]_{S_{k-\dim F}} = [(a,b)-(a',b')-(x,y)]_{S_{k-\dim F}}\,. \tag{75}$$

Using (10), we obtain:

$$\begin{aligned}
&h([\mathrm{PB}_{E,\varphi^*E}(\zeta',\tau')]_{\partial_\varphi S_k})\cdot\big(h([\mathrm{PB}_{E,\varphi^*E}(\zeta,\tau)]_{\partial_\varphi S_k})\big)^{-1}\\
&\stackrel{(10)}{=} h(\partial_\varphi(\mathbf{F},\mathbf{G})_*[F^*E,G^*(\varphi^*E)]_{S_k})\\
&= \exp\Big(2\pi i\int_{[F^*E,G^*(\varphi^*E)]_{S_k}}(\mathbf{F},\mathbf{G})^*(\mathrm{curv},\mathrm{cov})(h)\Big)\\
&= \exp\Big(2\pi i\int_{[W,M]_{S_{k-\dim F}}}(F,G)^*\fint_F(\mathrm{curv},\mathrm{cov})(h)\Big)\\
&\stackrel{(75)}{=} \exp\Big[2\pi i\Big(\int_{(a,b)-(a',b')}\fint_F(\mathrm{curv},\mathrm{cov})(h)+\underbrace{\int_{(x,y)}\fint_F(\mathrm{curv},\mathrm{cov})(h)}_{\in\mathbb{Z}}\Big)\Big]\\
&= \exp\Big(2\pi i\int_{(a,b)-(a',b')}\fint_F(\mathrm{curv},\mathrm{cov})(h)\Big)\,.
\end{aligned}$$

□

*Remark 34.* As a consequence of the preceding lemma, we note that if the cycle $(s,t)$ is a fundamental cycle of a relative geometric cycle $(\zeta',\tau')$, we do not need the chain $(a',b')$. In this case, the formula (73) for fiber integration thus simplifies to

$$(\widehat{\pi}_!h)(s,t) = h(\lambda_\varphi(s,t))\,. \tag{76}$$

### 5.1.3 Naturality and Compatibilities

In order to discuss naturality of fiber integration, we consider pull-back diagrams for smooth maps in the base: Let $\psi : B \to Y$ be a smooth map. Let $(Y,B) \xrightarrow{(f,g)} (X,A)$ be a smooth map, such that the diagram

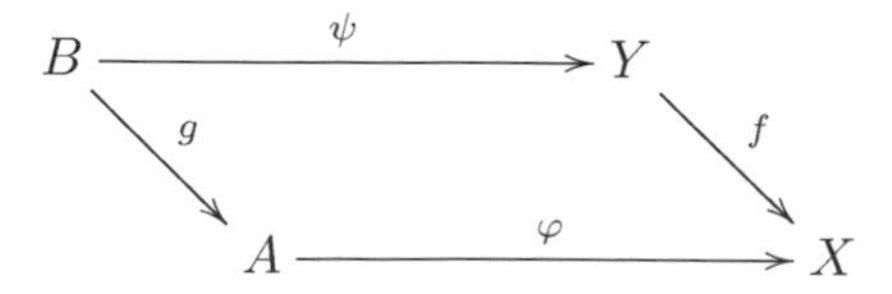


commutes. A fiber bundle $\pi : E \to X$ yields a commutative diagram of pull-back bundles:

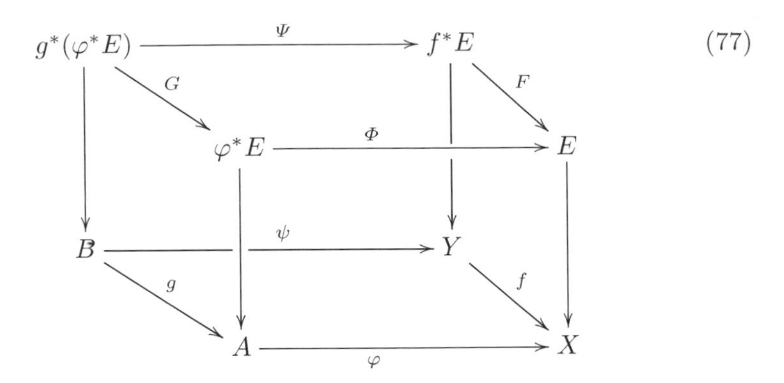

(77)

Here $\Psi : g^*(\varphi^*E) \to f^*E$ is the bundle map induced by $\psi : B \to Y$.

Now we prove the main theorem of this section:

**Theorem 35 (Fiber Integration: Naturality and Compatibilities).** *Let $F \hookrightarrow E \overset{\pi}{\twoheadrightarrow} X$ be a fiber bundle with closed oriented fibers. Let $\varphi : A \to X$ be a smooth map and $\Phi : \varphi^*E \to E$ the induced bundle map.Let $k \geq \dim F+2$. Then the fiber integration map $\widehat{\pi}_! : \widehat{H}^k(\Phi;\mathbb{Z}) \to \widehat{H}^{k-\dim F}(\varphi;\mathbb{Z})$ is natural with respect to pull-back diagrams (77), i.e. for any smooth map $(Y,B) \xrightarrow{(f,g)} (X,A)$ and differential character $h \in \widehat{H}^k(\Phi;\mathbb{Z})$, we have:*

$$\widehat{\pi}_!((F,G)^*h) = (f,g)^*\widehat{\pi}_!(h)\,. \tag{78}$$

*In other words, the following diagram is commutative for all $k \geq \dim F + 2$:*

$$\begin{array}{ccc} \widehat{H}^k(\Phi;\mathbb{Z}) & \xrightarrow{(F,G)^*} & \widehat{H}^k(\Psi;\mathbb{Z}) \\ \downarrow{\scriptstyle \widehat{\pi}_!} & & \downarrow{\scriptstyle \widehat{\pi}_!} \\ \widehat{H}^{k-\dim F}(\varphi;\mathbb{Z}) & \xrightarrow[(f,g)^*]{} & \widehat{H}^{k-\dim F}(\psi;\mathbb{Z})\,. \end{array}$$

*Fiber integration is compatible with curvature and covariant derivative, i.e., the diagram*

$$\begin{array}{ccc} \widehat{H}^k(\Phi;\mathbb{Z}) & \xrightarrow{(\mathrm{curv},\mathrm{cov})} & \Omega^k_0(\Phi) \\ \downarrow{\scriptstyle \widehat{\pi}_!} & & \downarrow{\scriptstyle \int} \\ \widehat{H}^{k-\dim F}(\varphi;\mathbb{Z}) & \xrightarrow[(\mathrm{curv},\mathrm{cov})]{} & \Omega^{k-\dim F}_0(\varphi) \end{array} \tag{79}$$

*commutes.*

*Fiber integration is compatible with topological trivializations, i.e., the diagram*

$$\begin{array}{ccc} \Omega^{k-1}(\Phi) & \xrightarrow{\iota} & \widehat{H}^k(\Phi;\mathbb{Z}) \\ {\scriptstyle f}\downarrow & & \downarrow{\scriptstyle \widehat{\pi}_!} \\ \Omega^{k-1-\dim F}(\varphi) & \xrightarrow[\iota]{} & \widehat{H}^{k-\dim F}(\varphi;\mathbb{Z}) \end{array}$$

*commutes.*

*Fiber integration for relative differential characters commutes with the maps* $\breve{\imath}$ *and* $\breve{p}$*. We thus have the commutative diagram*

$$\begin{array}{ccccc} \widehat{H}^{k-1}(\varphi^*E;\mathbb{Z}) & \xrightarrow{\breve{\imath}_\Phi} & \widehat{H}^k(\Phi;\mathbb{Z}) & \xrightarrow{\breve{p}_\Phi} & \widehat{H}^k(E;\mathbb{Z}) \\ \downarrow{\scriptstyle \widehat{\pi}_!} & & \downarrow{\scriptstyle \widehat{\pi}_!} & & \downarrow{\scriptstyle \widehat{\pi}_!} \\ \widehat{H}^{k-1-\dim F}(A;\mathbb{Z}) & \xrightarrow{\breve{\imath}_\varphi} & \widehat{H}^{k-\dim F}(\varphi;\mathbb{Z}) & \xrightarrow{\breve{p}_\varphi} & \widehat{H}^{k-\dim F}(X;\mathbb{Z}) \,. \end{array} \tag{80}$$

*Proof.* a) We first show naturality: Let $\psi : B \to Y$ and $(Y,B) \xrightarrow{(f,g)} (X,A)$ be smooth maps. Let $h \in \widehat{H}^k(\Phi;\mathbb{Z})$ and $(s,t) \in Z_{k-\dim F-1}(\psi;\mathbb{Z})$. Choose $(\zeta,\tau)_\psi(s,t) \in \mathcal{Z}_{k-\dim F-1}(\psi)$ and $(a,b)_\psi(s,t) \in C_{k-\dim F}(\psi;\mathbb{Z})$ such that $[(s,t)-\partial_\psi(a,b)]_{\partial_\psi S_{k-\dim F}} = [\zeta,\tau]_{\partial_\psi S_{k-\dim F}}$.

Now put

$$\begin{aligned} (\zeta,\tau)_\varphi((f,g)_*(s,t)) &:= (f,g)_*\big((\zeta,\tau)_\psi(s,t)\big) \in \mathcal{Z}_{k-\dim F-1}(\varphi) \\ (a,b)_\varphi((f,g)_*(s,t)) &:= (f,g)_*\big((a,b)_\psi(s,t)\big) \in C_{k-\dim F}(\varphi;\mathbb{Z}) \,. \end{aligned}$$

This allows us to choose the transfer maps such that $\lambda_\varphi((f,g)_*(s,t)) := (F,G)_*\lambda_\psi(s,t)$ holds for any cycle $(s,t) \in Z_{k-1}(\psi;\mathbb{Z})$. We thus obtain:

$$\begin{aligned} &(\widehat{\pi}_!((F,G)^*h))(s,t) \\ &\quad = \big((F,G)^*h\big)(\lambda_\psi(s,t)) \cdot \exp\Big(2\pi i \int_{(a,b)_\psi(s,t)} \fint_F (\mathrm{curv},\mathrm{cov})((F,G)^*h)\Big) \\ &\quad = h((F,G)_*\lambda_\psi(s,t)) \cdot \exp\Big(2\pi i \int_{(a,b)_\psi(s,t)} (f,g)^* \fint_F (\mathrm{curv},\mathrm{cov})(h)\Big) \\ &\quad = h(\lambda_\varphi((f,g)_*(s,t))) \cdot \exp\Big(2\pi i \int_{(a,b)_\varphi((f,g)_*(s,t))} \fint_F (\mathrm{curv},\mathrm{cov})(h)\Big) \\ &\quad = (\widehat{\pi}_!h)((f,g)_*(s,t)) \\ &\quad = \big((f,g)^*\widehat{\pi}_!h\big)(s,t) \,. \end{aligned}$$

b) To compute the curvature and covariant derivative of the character $\widehat{\pi}_!h$, we evaluate it on a boundary $\partial_\varphi(v,w)$, where $(v,w) \in C_{k-\dim F}(\varphi;\mathbb{Z})$:

$$\begin{aligned} &(\widehat{\pi}_!h)(\partial_\varphi(v,w)) \\ &\quad = h(\lambda_\varphi(\partial_\varphi(v,w))) \cdot \exp\Big(\int_{(a,b)_\varphi(\partial_\varphi(v,w))} \fint_F (\mathrm{curv},\mathrm{cov})(h)\Big) \end{aligned}$$

$$\begin{aligned}
&\overset{(18)}{=} h(\partial_\Phi \lambda_\varphi(v,w)) \cdot \exp\Big( \int_{(a,b)_\varphi(\partial_\varphi(v,w))} \fint_F (\mathrm{curv},\mathrm{cov})(h) \Big) \\
&= \exp\Big( 2\pi i \Big( \int_{\lambda_\varphi(v,w)} (\mathrm{curv},\mathrm{cov})(h) + \int_{(a,b)_\varphi(\partial_\varphi(v,w))} \fint_F (\mathrm{curv},\mathrm{cov})(h) \Big)\Big) \\
&\overset{(19)}{=} \exp\Big( 2\pi i \int_{(v,w)} \fint_F (\mathrm{curv},\mathrm{cov})(h) \Big) .
\end{aligned}$$

Thus the homomorphism $Z_{k-\dim F-1}(\varphi;\mathbb{Z}) \to \mathrm{U}(1)$ defined by the right hand side of (73) satisfies condition (22). We conclude that $\widehat{\pi}_! h$ is a differential character in $\widehat{H}^{k-\dim F}(\varphi;\mathbb{Z})$ with curvature $\mathrm{curv}(\widehat{\pi}_! h) = \fint_F \mathrm{curv}(h)$ and covariant derivative $\mathrm{cov}(\widehat{\pi}_! h) = \fint_F \mathrm{cov}(h)$.

c) Now we prove compatibility with topological trivializations: Let $(\omega,\vartheta) \in \Omega^{k-1}(\Phi)$ and $(s,t) \in Z_{k-1-\dim F}(\varphi;\mathbb{Z})$. Then we have:

$$\begin{aligned}
&\widehat{\pi}_!\big(\iota_\Phi(\omega,\vartheta)\big)(s,t) \\
&\quad = \big(\iota_\Phi(\omega,\vartheta)\big)(\lambda_\varphi(s,t)) \cdot \exp\Big( 2\pi i \int_{(a,b)_\varphi(s,t)} \fint_F \underbrace{(\mathrm{curv},\mathrm{cov})(\iota_\varphi(\omega,\vartheta))}_{=d_\Phi(\omega,\vartheta)} \Big) \\
&\quad = \exp\Big( 2\pi i \Big( \int_{\lambda_\varphi(s,t)} (\omega,\vartheta) + \int_{(a,b)_\varphi(s,t)} \fint_F d_\Phi(\omega,\vartheta) \Big)\Big) \\
&\quad \overset{(7)}{=} \exp\Big( 2\pi i \Big( \int_{\lambda_\varphi(s,t)} (\omega,\vartheta) + \int_{(a,b)_\varphi(s,t)} d_\varphi \fint_F (\omega,\vartheta) \Big)\Big) \\
&\quad \overset{(17)}{=} \exp\Big( 2\pi i \int_{(s,t)} \fint_F (\omega,\vartheta) \Big) \\
&\quad = \iota_\varphi\Big( \fint_F (\omega,\vartheta) \Big)(s,t) .
\end{aligned}$$

d) Finally we prove compatibility with the maps $\breve{\imath}$ and $\breve{p}$. It follows from diagram (16), i.e. from compatibility of the transfer maps with the sequence (14). Let $h \in \widehat{H}^{k-1}(\varphi^*E;\mathbb{Z})$ and $(s,t) \in Z_{k-\dim F-1}(\varphi;\mathbb{Z})$. Let $\sigma : Z_{k-\dim F-2}(A;\mathbb{Z}) \to Z_{k-\dim F-1}(\varphi;\mathbb{Z})$ be a splitting as in (14). Write $(s,t) = (z,0) + \sigma(t)$. Then we have:

$$\begin{aligned}
\widehat{\pi}_!(\breve{\imath}_\Phi(h))(s,t) &\overset{(73)}{=} \breve{\imath}_\Phi(h)(\lambda_\varphi(s,t)) \cdot \exp\Big( 2\pi i \int_{(a,b)_\varphi(s,t)} \fint_F \underbrace{(\mathrm{curv},\mathrm{cov})(\breve{\imath}_\Phi(h))}_{=(0,-\mathrm{curv}(h))} \Big) \\
&= h(p(\lambda_\varphi(s,t))) \cdot \exp\Big( 2\pi i \int_{-a(t)} \fint_F -\mathrm{curv}(h) \Big) \\
&\overset{(16)}{=} h(\lambda(t)) \cdot \exp\Big( 2\pi i \int_{a(t)} \mathrm{curv}(h) \Big) \\
&\overset{(72)}{=} (\widehat{\pi}_! h)(t) \\
&= \breve{\imath}_\varphi(\widehat{\pi}_! h)(s,t) .
\end{aligned}$$

Similarly, for a character $h \in \widehat{H}^k(\Phi;\mathbb{Z})$ and a cycle $z \in Z_{k-\dim F-1}(X;\mathbb{Z})$, we have:

$$\begin{aligned}
\breve{p}_\varphi(\widehat{\pi}_!h)(z) &= (\widehat{\pi}_!h)(z,0) \\
&= h(\lambda_\varphi(z,0)) \cdot \exp\Big(2\pi i \int_{(a,b)_\varphi(z,0)} \fint_F (\mathrm{curv},\mathrm{cov})(h)\Big) \\
&\overset{(16)}{=} h((\lambda(z),0)) \cdot \exp\Big(2\pi i \int_{(a(z),0)} \fint_F (\mathrm{curv},\mathrm{cov})(h)\Big) \\
&= \breve{p}_\Phi(h)(\lambda(z)) \cdot \exp\Big(2\pi i \int_{a(z)} \fint_F \mathrm{curv}(\breve{p}_\Phi(h))\Big) \\
&\overset{(72)}{=} \widehat{\pi}_!(\breve{p}_\Phi(h))(z)\,.
\end{aligned}$$

□

As a corollary of Theorem 35, we obtain compatibility of fiber integration with all the maps in the long exact sequence (34) for relative and absolute differential characters groups.

**Corollary 36 (Compatibility with Long Exact Sequence).** *Let $\pi : E \to X$ be a fiber bundle with oriented closed fibers and $\varphi : A \to X$ a smooth map. Then the fiber integration map $\widehat{\pi}_!$ on (relative and absolute) differential characters commutes with all maps in the long exact sequence* (34), *and with the usual fiber integration maps $\pi_!$ on cohomology with* U(1)- *and $\mathbb{Z}$-coefficients, respectively.*

*Proof.* Theorem 28 shows that $\widehat{\pi}_!$ commutes with the maps $\breve{\imath}$ and $\breve{p}$. The rest follows from [1, Prop. 47, 48]. □

### 5.1.4 Compatibility with Characteristic Class

The Leray-Serre spectral sequence of a fiber bundle $\pi : E \to X$ has an obvious modification that converges to the mapping cone cohomology $H^*(\Phi;\mathbb{Z})$. Using this modified spectral sequence, fiber integration on mapping cone cohomology groups $\pi_! : H^k(\Phi;\mathbb{Z}) \to H^{k-\dim F}(\varphi;\mathbb{Z})$ can be defined in the same way as in [5, § 8] for absolute cohomology. Compatibility of fiber integration with the characteristic class is discussed in detail for absolute differential characters in [1, Chap. 7]. The crucial point is that fiber integration for cohomology classes can be realized by pre-composition of cocycles with the transfer map:

$$\pi_! : H^k(E;\mathbb{Z}) \to H^{k-\dim F}(X;\mathbb{Z})\,, \quad [\mu] \mapsto [\mu \circ \lambda]\,.$$

In the same way, we obtain compatibility of fiber integration of relative characters with the characteristic class:

*Remark 37. Compatibility with characteristic class.* Choosing an extension of the transfer map $\lambda_\varphi$ to a homomorphism of chains as in (18), one can show that fiber integration of differential characters is compatible with the characteristic class. Thus for any relative character $h \in \widehat{H}^k(\Phi;\mathbb{Z})$, we have:

$$c(\widehat{\pi}_! h) = \pi_! c(h)\,. \tag{81}$$

### 5.1.5 Fiber Integration of Parallel Characters

By (79) we have $\operatorname{cov}(\widehat{\pi}_! h) = \fint_F \operatorname{cov}(h)$. Thus fiber integrals of parallel characters are again parallel. This way we obtain fiber integration on $\widehat{H}^*(E, E|_A;\mathbb{Z})$:

**Corollary 38 (Fiber Integration of Parallel Characters).** *Let $\pi : E \to X$ be a fiber bundle with closed oriented fibers. Let $i_A : A \to X$ be the embedding of a smooth submanifold. Denote by $I_A : E|_A \to E$ the induced bundle map. Let $k \geq \dim F + 2$. Then the inclusion $\widehat{H}^k(E, E|_A;\mathbb{Z}) \hookrightarrow \widehat{H}^k(I_A;\mathbb{Z})$ induces a natural fiber integration map*

$$\widehat{\pi}_! : \widehat{H}^k(E, E|_A;\mathbb{Z}) \to \widehat{H}^{k-\dim F}(X, A;\mathbb{Z})$$

*that commutes with curvature, characteristic class and topological trivializations. Moreover, it commutes with the long exact sequence* (42) *and fiber integration for cohomology with* U(1)*- and $\mathbb{Z}$-coefficients, respectively.*

By the fiberwise Stokes theorem, the fiber integral of a closed form with integral periods is again closed with integral periods (as long as the fibers are closed). Thus by the idenfication from [7] of the relative Hopkins-Singer group $\check{H}^k(\Phi;\mathbb{Z})$ as the quotient of the subgroup $\widehat{H}^k_0(\Phi;\mathbb{Z}) \subset \widehat{H}^k(\Phi;\mathbb{Z})$ by closed forms with integral periods on $E$, the fiber integration map $\widehat{\pi}_!$ descends to the relative Hopkins-Singer group:

**Corollary 39 (Fiber Integration of Relative Differential Cocycles).** *Let $\pi : E \to X$ be a fiber bundle with closed oriented fibers. Let $\varphi : A \to X$ be a smooth map and $\Phi : \varphi^* E \to E$ the induced bundle map. Let $k \geq \dim F + 2$. Then fiber integration of relative differential characters descends to a fiber integration map*

$$\widehat{\pi}_! : \check{H}^k(\Phi;\mathbb{Z}) \to \check{H}^{k-\dim F}(\varphi;\mathbb{Z})$$

*that commutes with the long exact sequence* (43).

## 5.2 Fibers with Boundary

Let $\pi : E \to X$ be a fiber bundle with compact oriented fibers $F$ with boundary $\partial F$. We have the induced fiber bundle $\pi^{\partial E} : \partial E \to X$ with closed oriented fibers. Fiber integration of differential forms satisfies the fiberwise Stokes theorem:

$$d \fint_F \omega = \fint_F d\omega + (-1)^{k+\dim F} \fint_{\partial F} \omega ,$$

where $\omega \in \Omega^k(E)$. This in particular implies that if a pair of differential forms $(\omega, \vartheta) \in \Omega^k(E) \times \Omega^{k-1}(\partial E)$ is closed in the mapping cone de Rham complex of the inclusion $i_{\partial E} : \partial E \to E$ and has integral periods, then the integrated form

$$\Big((-1)^{k-\dim F} \fint_F \omega - \fint_{\partial F} \vartheta\Big) \in \Omega^{k-\dim F}(X)$$

is also closed and has integral periods.

Likewise, we have the fiberwise Stokes theorem for mapping cone differential forms:

$$d_\varphi \fint_F (\omega, \vartheta) = \fint_F d_\Phi(\omega, \vartheta) + (-1)^{k+\dim F} \fint_{\partial F} (\omega, \vartheta) , \tag{82}$$

where $(\omega, \vartheta) \in \Omega^k(\Phi)$. In particular, fiber integration over the boundary $\partial F$ maps $d_\Phi$-closed forms to $d_\varphi$-exact forms. Likewise, fiber integration in the bundle $\pi : \partial E \to X$ of mapping cone cohomology classes for the bundle map $\Phi : \varphi^* E \to E$ yields the trivial map $\pi^{\partial E} : H^k(\Phi; \mathbb{Z}) \to H^{k-\dim \partial E}(\varphi; \mathbb{Z})$.

In [1, Chap. 7] we show that integration of differential characters on $E$ over the fibers of $\pi : \partial E \to X$ yields topologically trivial characters on $X$. More precisely, for a character $h \in \widehat{H}^k(E; \mathbb{Z})$ with $k \geq \dim F$, we have:

$$\widehat{\pi}_!^{\partial E} h = \iota\Big((-1)^{k-\dim F} \fint_F \operatorname{curv}(h)\Big) . \tag{83}$$

Here we consider two generalizations of this result.

First we consider fiber integration of relative differential characters in the fiber bundle $\pi^{\partial E} : \partial E \to X$.

**Proposition 40.** *Let $E \to X$ be a fiber bundle with compact oriented fibers with boundary. Let $\varphi : A \to X$ be a smooth map and $\Phi : \varphi^* E \to E$ the induced bundle map. Then for any character $h \in \widehat{H}^k(\Phi; \mathbb{Z})$, the integrated character $\widehat{\pi}_!^{\partial E} h \in \widehat{H}^{k-\dim \partial F}(\varphi; \mathbb{Z})$ is topologically trivial, and we have:*

$$\widehat{\pi}_!^{\partial E} h = \iota_\varphi\Big((-1)^{k-\dim F} \fint_F (\operatorname{curv}, \operatorname{cov})(h)\Big) . \tag{84}$$

*Proof.* The integrated character $\widehat{\pi}_!^{\partial E} h$ is topologically trivial since $c(\widehat{\pi}_!^{\partial E} h) = \pi_!^{\partial E}(c(h))$ and $\pi_!^{\partial E} : H^k(\Phi;\mathbb{Z}) \to H^{k-\dim \partial F}(\varphi;\mathbb{Z})$ is the trivial map. We compute the curvature and covariant derivative of $\widehat{\pi}_!^{\partial E} h$ using the fiberwise Stokes theorem (82):

$$\begin{aligned}(\mathrm{curv},\mathrm{cov})(\widehat{\pi}_!^{\partial E} h) &\overset{(79)}{=} \fint_{\partial F} (\mathrm{curv},\mathrm{cov})(h) \\ &\overset{(82)}{=} (-1)^{k-\dim F} d_\varphi\Big(\fint_F (\mathrm{curv},\mathrm{cov})(h)\Big).\end{aligned}$$

Now the exact sequences (24) together with the commutative diagram

$$\begin{array}{ccc} \Omega^{*-1}(\varphi) & \xrightarrow{\;d_\varphi\;} & d_\varphi(\Omega^{*-1}(\varphi)) \\ \downarrow{\scriptstyle \iota} & & \downarrow \\ \widehat{H}^*(\varphi;\mathbb{Z}) & \xrightarrow[(\mathrm{curv},\mathrm{cov})]{} & \Omega_0^k(\varphi) \end{array}$$

yield (84). □

As an application, we obtain the homotopy formula for relative characters:

**Corollary 41 (Homotopy Formula).** *Let $\psi : B \to Y$ and $\varphi : A \to X$ be smooth maps. Let $(f,g) : [0,1] \times (X,A) \to (Y,B)$ be a homotopy between smooth maps $(f_0,g_0),(f_1,g_1) : (X,A) \to (Y,B)$ and $h \in \widehat{H}^k(\psi;\mathbb{Z})$ a relative differential character. Then we have:*

$$(f_1,g_1)^*h - (f_0,g_0)^*h = \iota_\varphi\Big(\int_0^1 f_s^*\mathrm{curv}(h)ds, -\int_0^1 g_s^*\mathrm{cov}(h)ds\Big). \tag{85}$$

*Proof.* As for absolute characters, we obtain the homotopy formula by fiber integration in the trivial bundle[7] $E = X \times [0,1]$:

$$\begin{aligned}(f_1,g_1)^*h - (f_0,g_0)^*h &= \widehat{\pi}_!^{\partial E} h \\ &\overset{(84)}{=} \iota_\varphi\Big((-1)^{k-1}\fint_{[0,1]} (\mathrm{curv},\mathrm{cov})((f,g)^*h)\Big) \\ &= \iota_\varphi\Big(\int_0^1 f_s^*\mathrm{curv}(h)ds, -\int_0^1 g_s^*\mathrm{cov}(h)ds\Big).\end{aligned}$$ □

We obtain another generalization of (83) by weakening the condition on the fiber bundle: instead of a fiber bundle with fibers that bound we consider a fiber bundle $\pi : E \to X$ and a smooth map $\varphi : A \to X$ such that the pull-back bundle $\pi : \varphi^*E \to A$ is the fiberwise boundary of a fiber bundle $\pi' : E' \to A$. For this situation we introduce the following notation:

[7] Note that by the orientation conventions, we have $\fint_{[0,1]} \omega = (-1)^{k-1}\int_0^1 \omega_s ds$ for any $k$-form $\omega$.

**Definition 42.** Let $\pi : E \to X$ be a fiber bundle with closed oriented fibers $F$ and $\varphi : A \to X$ a smooth map. We say that $\pi : E \to X$ *bounds along* $\varphi$ if there exists a fiber bundle $\pi' : (E', \partial E') \to A$ with compact oriented fibers with boundary $(F', \partial F')$ and a fiber bundle isomorphism $\partial E' \to \varphi^* E$ over the identity $\mathrm{id}_A$. For short, we say that $\pi : E \to X$ bounds $\pi' : E' \to A$ along $\varphi : A \to X$.

With this notation, we obtain a generalization of (83) for bundles that bound along a smooth map:

**Proposition 43.** *Let $\pi : E \to X$ be a fiber bundle with closed oriented fibers that bounds a fiber bundle $\pi' : E' \to A$ with compact oriented fibers with boundary $(F', \partial F')$ along a smooth map $\varphi : A \to X$. Then for any differential character $h \in \widehat{H}^k(E;\mathbb{Z})$ the integrated character $\widehat{\pi}_! h \in \widehat{H}^{k-\dim F}(X;\mathbb{Z})$ has a section along $\varphi$ with covariant derivative $(-1)^{k+\dim F'} \cdot \fint_{F'} \mathrm{curv}(\Phi^* h)$.*

*Proof.* The fiber integration map ${\pi'}_!^{\partial E'} : H^k(E';\mathbb{Z}) \to H^{k-\dim \partial F'}(A;\mathbb{Z})$ on integral cohomology is trivial. Thus the integrated character $\widehat{\pi}_! h$ is topologically trivial along $\varphi$, since $\varphi^* c(\widehat{\pi}_! h) = c(\widehat{\pi}'_!(\Phi^* h)) = {\pi'}_!^{\partial E'}(c(\Phi^* h)) = 0$. By the fiberwise Stokes theorem, the curvature satisfies:

$$
\begin{aligned}
\varphi^* \mathrm{curv}(\widehat{\pi}_! h) &= \fint_F \Phi^* \mathrm{curv}(h) \\
&= \fint_{\partial F'} \mathrm{curv}(\Phi^* h) \\
&= (-1)^{k+\dim F'} \cdot d\Big( \fint_{F'} \mathrm{curv}(\Phi^* h) \Big) .
\end{aligned}
$$

Thus the integrated character $\widehat{\pi}_! h$ admits a section along $\varphi$ with covariant derivative $(-1)^{k+\dim F'} \cdot \fint_{F'} \mathrm{curv}(\Phi^* h)$. □

Fiber integration of relative differential cohomology for fibers with boundary has also been discussed recently in [19]. The author uses the model of Deligne cohomology to construct a fiber integration map from the relative differential cohomology of the inclusion $i_{\partial E} : \partial E \to E$ to the absolute differential cohomology of the base. With our methods, we recover his construction and compute the curvature of the integrated character. Moerover, we relate this new fiber integration map to the previously known ones and to the long exact sequences for the maps $i_{\partial E}$ and $\mathrm{id}_X$.

As before, the integrated character will befined by its evaluations on cycles in the base. Choose homomorphisms $\zeta : Z_{k-\dim F-1}(X;\mathbb{Z}) \to \mathcal{Z}_{k-\dim F-1}(X)$ and $a : Z_{k-\dim F-1}(X;\mathbb{Z}) \to C_{k-\dim F}(X;\mathbb{Z})$ such that for any singular cycle $z \in Z_{k-\dim F-1}$ we have $[\zeta(z)]_{\partial S_{k-\dim F}} = [z - \partial a(z)]_{\partial S_{k-\dim F}}$.

**Definition 44.** Let $\pi : (E, \partial E) \to X$ be a smooth fiber bundle with compact oriented fibers with boundary, and let $k > \dim F$. The fiber integration map

$\widehat{\pi}_!^{E,\partial E} : \widehat{H}^k(i_{\partial E};\mathbb{Z}) \to \widehat{H}^{k-\dim F}(X;\mathbb{Z})$ is defined as follows: For a character $h \in \widehat{H}^k(i_{\partial E};\mathbb{Z})$ and a cycle $z \in Z_{k-\dim F-1}(X)$ put

$$(\widehat{\pi}_!^{E,\partial E} h)(z) := h([\mathrm{PB}_{E,\partial E}(\zeta(z))]_{\partial_{i_{\partial E}} S_k}) \cdot \exp\Big(2\pi i \int_{a(z)} \Big((-1)^{k-\dim F} \fint_F \mathrm{curv}(h) - \fint_{\partial F} \mathrm{cov}(h)\Big)\Big). \quad (86)$$

The reason for the choice of sign in the definition will become clear from the comparison of the fiber integration map $\widehat{\pi}_!^{E,\partial E}$ with the previously defined ones as well as with the long exact sequences for the maps $i_{\partial E}$ and $\mathrm{id}_X$, see Theorem 47 below.

Like for the other fiber integration maps we need to show that the above definition is independent of the choice of geometric cycle $\zeta(z)$ and chain $a(z)$. The following lemma and its proof is the analogue of [1, Lem. 42] and Lemma 33 above:

**Lemma 45.** *Let $h \in \widehat{H}^k(i_{\partial E};\mathbb{Z})$ and $z \in Z_{k-\dim F-1}(X;\mathbb{Z})$. Then for any geometric cycle $\zeta'(z) \in \mathcal{Z}_{k-\dim F-1}(X)$ and any chain $a(z) \in C_{k-\dim F}(X;\mathbb{Z})$ satisfying $[\zeta'(z)]_{\partial S_{k-\dim F}} = [z - \partial a'(z)]_{\partial S_{k-\dim F}}$, we have:*

$$(\widehat{\pi}_!^{E,\partial E} h)(z) := h([\mathrm{PB}_{E,\partial E}(\zeta'(z))]_{\partial_{i_{\partial E}} S_k}) \cdot \exp\Big(2\pi i \int_{a'(z)} \Big((-1)^{k-\dim F} \fint_F \mathrm{curv}(h) - \fint_{\partial F} \mathrm{cov}(h)\Big)\Big).$$

*In other words, the character in (86) is independent of the choice of geometric cycle $\zeta$ and chain $a$.*

*Proof.* Let the geometric cycles $\zeta(z), \zeta'(z) \in \mathcal{Z}_{k-\dim F-1}(X)$ be represented by smooth maps $S \xrightarrow{f} X$ and $S' \xrightarrow{f'} X$, respectively. Since $\zeta(z)$ and $\zeta'(z)$ both represent the homology class of the singular cycle $z$, they are bordant. Choose a bordism from $\zeta(z)$ to $\zeta'(z)$, represented by a smooth map $W \xrightarrow{F} X$. The induced bundle map $F^*E \xrightarrow{\mathbf{F}} E$ yields a bordism $((-1)^{k-\dim F+1} F^*E, F^*\partial E) \xrightarrow{\mathbf{F}} (E,\partial E)$ from $\mathrm{PB}_{E,\partial E}(\zeta'(z))$ to $\mathrm{PB}_{E,\partial E}(\zeta(z))$. For the refined fundamental classes of $W$ and its boundary, we find

$$\partial(F_*[W]_{S_{k-\dim F}}) = [\zeta'(z)]_{\partial S_{k-\dim F}} - [\zeta(z)]_{\partial S_{k-\dim F}} = [\partial a(z) - \partial a'(z)]_{\partial S_{k-\dim F}}.$$

Thus there exists a cycle $x \in Z_{k-\dim F}(X;\mathbb{Z})$ such that $F_*[W]_{S_{k-\dim F}} = [a(z) - a'(z) - x]_{S_{k-\dim F}}$. Now we compute:

$$\begin{aligned} &h([\mathrm{PB}_{E,\partial E}(\zeta'(z))]_{\partial_{i_{\partial E}} S_k}) \cdot h([\mathrm{PB}_{E,\partial E}(\zeta(z))]_{\partial_{i_{\partial E}} S_k})^{-1} \\ &= h(-\partial_{i_{\partial E}} \mathbf{F}_*[(-1)^{k-\dim F+1} F^*E, F^*\partial E]_{S_k}) \\ &= \exp\Big(2\pi i \int_{(F^*E, F^*\partial E)} \mathbf{F}^*((-1)^{k-\dim F} \mathrm{curv}(h), -\mathrm{cov}(h))\Big) \end{aligned}$$

$$
\begin{aligned}
&= \exp\Big(2\pi i \int_W F^*\Big((-1)^{k-\dim F} \fint_F \operatorname{curv}(h) - \fint_{\partial F} \operatorname{cov}(h)\Big)\Big) \\
&= \exp\Big(2\pi i \int_{(a(z)-a'(z)-x)} \Big((-1)^{k-\dim F} \fint_F \operatorname{curv}(h) - \fint_{\partial F} \operatorname{cov}(h)\Big)\Big) \\
&= \exp\Big(2\pi i \int_{(a(z)-a'(z))} \Big((-1)^{k-\dim F} \fint_F \operatorname{curv}(h) - \fint_{\partial F} \operatorname{cov}(h)\Big)\Big).
\end{aligned}
$$

The last equality holds since the form $(-1)^{k-\dim F} \fint_F \operatorname{curv}(h) - \fint_{\partial F} \operatorname{cov}(h)$ has integral periods. □

Next we prove that $\widehat{\pi}_!^{E,\partial E}(h)$ is indeed a differential character on $X$ and we compute its curvature.

**Theorem 46.** *Let $\pi : (E, \partial E) \to X$ be a smooth fiber bundle with compact oriented fibers with boundary, and let $k > \dim F$. Let $h \in \widehat{H}^k(i_{\partial E}; \mathbb{Z})$. Then $\widehat{\pi}_!^{E,\partial E}(h)$ is a differential character in $\widehat{H}^{k-\dim F}(X; \mathbb{Z})$. Moreover, we have*

$$
\operatorname{curv}(\widehat{\pi}_!^{E,\partial E}(h)) = (-1)^{k-\dim F} \fint_F \operatorname{curv}(h) - \fint_{\partial F} \operatorname{cov}(h). \tag{87}
$$

*Proof.* By construction, the map $z \mapsto \widehat{\pi}_!^{E,\partial E}(h)$ defines a group homomorphism $Z_{k-\dim F-1}(X; \mathbb{Z}) \to \mathrm{U}(1)$. Thus it suffices to evaluate it on boundaries and thereby compute the curvature. Thus let $z = \partial b$ for some singular chain $b \in C_{k-\dim F}(X; \mathbb{Z})$. Let the geometric chain $\zeta(b) \in \mathcal{C}_{k-\dim F}(X)$ be represented by a smooth map $W \xrightarrow{F} X$. Then $W \xrightarrow{F} X$ is a bordism from the empty cycle to $\zeta(\partial b)$. Then the induced bundle map $F^*E \xrightarrow{\mathbf{F}} E$ yields a bordism $((-1)^{k-\dim F+1} F^*E, F^*\partial E) \xrightarrow{\mathbf{F}} (E, \partial E)$ from $\mathrm{PB}_{E,\partial E}(\zeta(\partial b))$ to the empty cycle. We compute:

$$
\begin{aligned}
&h([\mathrm{PB}_{E,\partial E}(\zeta(\partial c))]_{\partial_{i_{\partial E}} S_k}) \\
&\quad = h(-\partial_{i_{\partial E}} \mathbf{F}_*[(-1)^{k-\dim F+1} F^*E, F^*\partial E]_{S_k}) \\
&\quad = \exp\Big(2\pi i \int_{(F^*E, F^*\partial E)} \mathbf{F}^*((-1)^{k-\dim F} \operatorname{curv}(h), -\operatorname{cov}(h))\Big) \\
&\quad = \exp\Big(2\pi i \int_W F^*\Big((-1)^{k-\dim F} \fint_F \operatorname{curv}(h) - \fint_{\partial F} \operatorname{cov}(h)\Big)\Big) \\
&\quad = \exp\Big(2\pi i \int_{[\zeta(c)]_{S_{k-\dim F}}} \Big((-1)^{k-\dim F} \fint_F \operatorname{curv}(h) - \fint_{\partial F} \operatorname{cov}(h)\Big)\Big).
\end{aligned}
$$

Plugging into the definition of $\widehat{\pi}_!^{E,\partial E}$, we find:

$$
\begin{aligned}
&\widehat{\pi}_!^{E,\partial E}(h)(\partial c) \\
&\qquad = h([\mathrm{PB}_{E,\partial E}(\zeta(\partial c))]_{\partial_{i_{\partial E}} S_k})
\end{aligned}
$$

$$
\begin{aligned}
&\cdot \exp\Big(2\pi i \int_{a(\partial c)} \Big((-1)^{k-\dim F} \fint_F \mathrm{curv}(h) - \fint_{\partial F} \mathrm{cov}(h)\Big)\Big)\\
&= \exp\Big(2\pi i \int_{[\zeta(c)]_{S_{k-\dim F}}} \Big((-1)^{k-\dim F} \fint_F \mathrm{curv}(h) - \fint_{\partial F} \mathrm{cov}(h)\Big)\Big)\\
&\quad\cdot \exp\Big(2\pi i \int_{a(\partial c)} \Big((-1)^{k-\dim F} \fint_F \mathrm{curv}(h) - \fint_{\partial F} \mathrm{cov}(h)\Big)\Big)\\
&= \exp\Big(2\pi i \int_c \Big((-1)^{k-\dim F} \fint_F \mathrm{curv}(h) - \fint_{\partial F} \mathrm{cov}(h)\Big)\Big).
\end{aligned}
$$

The last equality follows from compatibility of fiber integration of differential forms with the pull-back operation on geometric chains [1, Chap. 4] together with the fact that the form $(-1)^{k-\dim F} \fint_F \mathrm{curv}(h) - \fint_{\partial F} \mathrm{cov}(h)$ is closed.

Thus $\widehat{\pi}_!^{E,\partial E}(h)$ is indeed a differential character on $X$ with curvature as claimed. □

To complete the discussion of fiber integration along fibers with boundary, we relate the fiber integration map $\widehat{\pi}_!^{E,\partial E}$ to the long exact sequences for the maps $i_{\partial E}$ and $\mathrm{id}_X$.

**Theorem 47.** *The fiber integration map $\widehat{\pi}_!^{E,\partial E}$ defined by (86) fits into the following diagram obtained from the long exact sequences for the maps $i_{\partial E}$ and $\mathrm{id}_X$. The left and right triangles commute. The parallelogram in the middle commutes (up to sign) on the subgroup $\widehat{H}^k(E,\partial E;\mathbb{Z})$ of parallel characters.*

$$
\begin{array}{ccccc}
\widehat{H}^{k-1}(\partial E;\mathbb{Z}) & \xrightarrow{\breve{\imath}} & \widehat{H}^k(i_{\partial E};\mathbb{Z}) & \xrightarrow{\breve{p}} & \widehat{H}^k(E;\mathbb{Z}) \\
\downarrow{\scriptstyle \widehat{\pi}_!^{\partial E}} & \swarrow{\scriptstyle \widehat{\pi}_!^{E,\partial E}} & & \swarrow{\scriptstyle \widehat{\pi}_!^{E}} & \downarrow{\scriptstyle \widehat{\pi}_!^{\partial E}} \\
\widehat{H}^{k-\dim F}(X;\mathbb{Z}) & \xrightarrow[\breve{\imath}]{} & \widehat{H}^{k-\dim F+1}(\mathrm{id}_X;\mathbb{Z}) & \xrightarrow[\breve{p}]{} & \widehat{H}^{k-\dim \partial F}(X;\mathbb{Z}).
\end{array}
$$

*Proof.* Commutativity of the right triangle has been discussed in [1, Chap. 8]. Commutativity of the left triangle directly follows from the definitions: for any character $h \in \widehat{H}^{k-1}(\partial E;\mathbb{Z})$, and any cycle $z \in Z_{k-\dim F-1}(X;\mathbb{Z})$, we have:

$$
\begin{aligned}
&\big(\widehat{\pi}_!^{E,\partial E}(\breve{\imath}(h))\big)(z)\\
&\quad = \breve{\imath}(h)([\mathrm{PB}_{E,\partial E}(\zeta(z))]_{\partial_{i_{\partial E}} S_k})\\
&\quad\quad \cdot \exp\Big(2\pi i\Big(\int_{a(z)} \Big((-1)^{k-\dim F} \fint_F \underbrace{\mathrm{curv}(\breve{\imath}(h))}_{=0} - \fint_{\partial F} \underbrace{\mathrm{cov}(\breve{\imath}(h))}_{=-\mathrm{curv}(h)}\Big)\Big)\\
&\quad = h([\mathrm{PB}_{\partial E}(\zeta(z))]_{\partial S_{k-1}}) \cdot \exp\Big(2\pi i \int_{a(z)} \fint_{\partial F} \mathrm{curv}(h)\Big)\\
&\quad = \big(\widehat{\pi}_!^{\partial E}(h)\big)(z).
\end{aligned}
$$

The parallelogram in the middle in general does not commute, since for any $h \in \widehat{H}^k(i_{\partial E};\mathbb{Z})$, we have

$$\operatorname{cov}(\breve{\imath}(\widehat{\pi}_!^{E,\partial E}(h))) = -\operatorname{curv}(\widehat{\pi}_!^{E,\partial E}(h)) = -(-1)^{k-\dim F} \fint_F \operatorname{curv}(h) + \fint_{\partial F} \operatorname{cov}(h)$$

whereas

$$\operatorname{cov}(\widehat{\pi}_!^E(\breve{p}(h))) = (-1)^{k-\dim F} \fint_F \operatorname{curv}(h).$$

Since by definition $\widehat{\pi}_!^E(\breve{p}(h)) = \operatorname{cov}^{-1}((-1)^{k-\dim F} \fint_F \operatorname{curv}(h))$, we conclude that the parallelogram commutes (up to sign) precisely on the subset of those characters $h \in \widehat{H}^k(i_{\partial E};\mathbb{Z})$ with $\fint_{\partial F} \operatorname{cov}(h) = 0$. This in particular contains the subgroup of parallel characters. □

## 5.3 The Up-Down Formula

Fiber integration for differential forms satisfies the following up-down formula: Let $(\omega,\vartheta) \in \Omega^k(\varphi)$ and $\omega' \in \Omega^{k'}(E)$. Then we have the equality

$$\fint_F \pi^*(\omega,\vartheta) \wedge \omega' = (\omega,\vartheta) \wedge \fint_F \omega'$$

of differential forms in $\Omega^{k+k'-\dim F}(\varphi)$. Likewise, for cohomology classes $c \in H^k(\varphi)$ and $c' \in H^{k'}(E;\mathbb{Z})$, we have

$$\pi_!(\pi^* c \cup c') = c \cup \pi_! c' .$$

More generally, fiber integration in cross products is compatible with cross products of differential forms and cohomology classes in the following sense: Let $\pi : E \to X$ and $\pi' : E' \to X'$ be fiber bundles with closed oriented fibers $F$ and $F'$. Let $\varphi : A \to X$ be a smooth map and $\Phi : \varphi^* E \to E$ the induced bundle map. Let $(\omega,\vartheta) \in \Omega^k(\Phi)$ and $\omega' \in \Omega^{k'}(E')$ be differential forms and $c \in H^k(\Phi;\mathbb{Z})$ and $c' \in H^{k'}(E';\mathbb{Z})$ cohomology classes. Then we have:

$$\fint_{F\times F'} (\omega,\vartheta) \times \omega' = (-1)^{(k'-\dim F')\cdot \dim F} \cdot \fint_F (\omega,\vartheta) \times \fint_{F'} \omega' \tag{88}$$

$$\pi_!(c \times c') = (-1)^{(k'-\dim F')\cdot \dim F} \cdot \pi_! c \times \pi'_! c' . \tag{89}$$

In [1, Chap. 7] we prove the following up-down formula for absolute differential characters: for $h \in \widehat{H}^k(X;\mathbb{Z})$ and $h' \in \widehat{H}^{k'}(E;\mathbb{Z})$, we have the equality

$$\widehat{\pi}_!(\pi^* h * h') = h * \widehat{\pi}_! h'$$

of differential characters in $\widehat{H}^{k+k'-\dim F}(X;\mathbb{Z})$. Here we prove the relative version of this up-down formula and the relative version of the compatibility of fiber integration with cross products. The method of proof is the same as in [1, Chap. 7].

### 5.3.1 Compatibility with Cross Products

We start with compatibility of cross products with fiber integration in fiber products:

**Theorem 48 (Fiber Integration: Compatibility with Fiber and Cross Products).** *Let $\pi : E \to X$ and $\pi' : E' \to X'$ be fiber bundles with closed oriented fibers and $\varphi : A \to X$ a smooth map. Let $\Phi : \varphi^* E \to E$ be the induced bundle map. Let $\pi \times \pi' : E \times E' \to X \times X'$ denote the fiber product with fiber orientation the product orientation of $F \times F'$. Then fiber integration of differential characters is compatible with the fiber product and the cross product in the sense that the following diagram is graded commutative:*

$$\begin{array}{ccc} \widehat{H}^k(\Phi;\mathbb{Z}) \otimes \widehat{H}^{k'}(E';\mathbb{Z}) & \xrightarrow{\quad\times\quad} & \widehat{H}^{k+k'}(\Phi \times \mathrm{id}_{E'};\mathbb{Z}) \\ \downarrow \widehat{\pi}_! \qquad \downarrow \widehat{\pi}'_! & & \downarrow \widehat{\pi \times \pi'}_! \\ \widehat{H}^{k-\dim F}(\varphi;\mathbb{Z}) \otimes \widehat{H}^{k'-\dim F'}(E';\mathbb{Z}) & \xrightarrow{\quad\times\quad} & \widehat{H}^{k+k'-\dim(F\times F')}(\varphi \times \mathrm{id}_{X'};\mathbb{Z}) \,. \end{array}$$

*More explicitly, for characters $h \in \widehat{H}^k(\Phi;\mathbb{Z})$ and $h' \in \widehat{H}^{k'}(E';\mathbb{Z})$ we have:*

$$\widehat{\pi}_! h \times \widehat{\pi}'_! h' = (-1)^{(k'-\dim F')\cdot \dim F} \cdot \widehat{\pi \times \pi'}_!(h \times h') \,. \tag{90}$$

*Proof.* Let $h \in \widehat{H}^k(\Phi;\mathbb{Z})$ and $h' \in \widehat{H}^{k'}(E';\mathbb{Z})$. We compare the two sides of (90) by evaluating them on cycles in $Z_{k+k'-\dim(F\times F')-1}(\varphi \times \mathrm{id}_{X'};\mathbb{Z})$. By definition of the cross product, we consider the evaluation on cross products of cycles and on cycles in the Künneth complement $T_{k+k-\dim(F\times F')-1}(\varphi \times \mathrm{id}_{X'};\mathbb{Z})$ separately. More specifically, we may assume the cycles $(x,y) \in Z_i(\varphi;\mathbb{Z})$ and $y' \in Z_j(X';\mathbb{Z})$ to be fundamental cycles of appropriately chosen stratifolds. For the correction terms are boundaries which may be added to the torsion cycles in $T_{k+k-\dim(F\times F')-1}(\varphi \times \mathrm{id}_{X'};\mathbb{Z})$.

The transfer map on the fiber product can be chosen multiplicatively as in (20). Let $(x,y) \in Z_i(\varphi;\mathbb{Z})$ and $y' \in Z_j(X';\mathbb{Z})$. If $(i,j)$ neither equals $(k-1-\dim F, k'-\dim F')$ nor $(k-\dim F, k'-1-\dim F')$, then both sides of (90) vanish on $(s,t) = (x,y) \times y'$. For $(i,j) = (k-1-\dim F, k'-\dim F')$, we have:

$$\begin{aligned} &(\widehat{\pi}_! h \times \widehat{\pi}'_! h')((x,y) \times y') \\ &\qquad = (\widehat{\pi}_! h(x,y))^{\langle c(\widehat{\pi}'_! h'), y' \rangle} \end{aligned}$$

$$\stackrel{(76)}{=} h(\lambda_\varphi(x,y))^{\langle h',\lambda'(y')\rangle}$$
$$\stackrel{(21)}{=} (-1)^{(k'-\dim F')\cdot\dim F}\cdot(h\times h')(\lambda_{\varphi\times\mathrm{id}_{X'}}(x,y)\times y')$$
$$= (-1)^{(k'-\dim F')\cdot\dim F}\big(\widehat{\pi\times\pi'}_!(h\times h')\big)((x,y)\times y').$$

Similarly, for $(i,j)=(k-\dim F, k'-1-\dim F')$, we find:

$$\begin{aligned}
&\big(\widehat{\pi}_! h\times\widehat{\pi}'_! h'\big)((x,y)\times y')\\
&= (\widehat{\pi}'_! h'(y'))^{(-1)^{k-\dim F}\cdot\langle c(\widehat{\pi}_! h),(x,y)\rangle}\\
&= h'(\lambda'(y'))^{(-1)^{k-\dim F}\cdot\langle c(h),\lambda_\varphi(x,y)\rangle}\\
&= (h\times h')((-1)^{\dim F}\cdot\lambda_\varphi(x,y)\times\lambda'(y'))\\
&\stackrel{(21)}{=} (-1)^{(k'-1-\dim F')\cdot\dim F}\cdot(-1)^{\dim F}\cdot(h\times h')(\lambda_{\varphi\times\mathrm{id}_{X'}}(x,y)\times y')\\
&= (-1)^{(k'-\dim F')\cdot\dim F}\cdot\big(\widehat{\pi\times\pi'}_!(h\times h')\big)((x,y)\times y').
\end{aligned}$$

Finally, it remains to verify equation (90) on the Künneth complement $T_{k+k'-\dim(F\times F')-1}(\varphi\times\mathrm{id}_{X'};\mathbb{Z})$ and on the correction terms obtained from replacing cycles $(x,y)\in Z_i(\varphi;\mathbb{Z})$ and $y'\in Z_j(X';\mathbb{Z})$ by fundamental cycles of appropriately chosen stratifolds. It suffices that (more generally) the two sides of (90) coincide on all torsion cycles. By Remark 6 this follows from the fact that curvature, covariant derivative and characteristic class of the two sides of (90) coincide. The latter follows from multiplicativity (60), (61), compatibility of fiber integration with curvature, covariant derivative and characteristic class (79), (81) and compatibility of fiber integration in fiber products with cross products of differential forms and cohomology classes (88), (89). □

### 5.3.2 The Up-Down formula

As a corollary of the compatibility of the cross product with fiber integration in fiber products we obtain the following up-down formula:

**Corollary 49 (Up-Down Formula).** *Let $\pi: E\to X$ be fiber bundle with closed oriented fibers. Let $\varphi: A\to X$ be a smooth map. Let $h\in\widehat{H}^k(\varphi;\mathbb{Z})$ and $h'\in\widehat{H}^{k'}(E;\mathbb{Z})$. Then we have the equality*

$$\widehat{\pi}_!(\pi^*h * h') = h * \widehat{\pi}_! h' \tag{91}$$

*of differential characters in $\widehat{H}^{k+k'-\dim F}(\varphi;\mathbb{Z})$.*

*Proof.* The method of proof is the same as for absolute differential characters in [1, Chap. 7]. We first consider the fiber product $E\times E\to X\times X$. Write this as the composite fiber bundle $E\times E\xrightarrow{\pi\times\mathrm{id}_E} X\times E\xrightarrow{\mathrm{id}_X\times\pi} X\times X$. Let

$$\Delta_{(E,\varphi^*E)} := (\Delta_E, (\mathrm{id}_{\varphi^*E} \times \Phi) \circ \Delta_{\varphi^*E}) : (E, \varphi^*E) \to (E, \varphi^*E) \times E$$

and

$$\Delta_{(X,A)} = (\Delta_X, (\mathrm{id}_A \times \varphi) \circ \Delta_A) : (X, A) \to (X, A) \times X$$

be the relative diagonal maps as in Sect. 4.2. Then we have the pull-back diagram

$$\begin{array}{ccc} & & (E,\varphi^*E) \times E \\ & \overset{\Delta_{(E,\varphi^*E)}}{\nearrow} & \downarrow {\scriptstyle \pi \times \mathrm{id}_E} \\ (E,\varphi^*E) & \xrightarrow[(\pi \times \mathrm{id}_E)\circ\Delta_{(E,\varphi^*E)}]{} & (X,A) \times E \\ {\scriptstyle \pi}\downarrow & & \downarrow {\scriptstyle \mathrm{id}_{(X,A)} \times \pi} \\ (X,A) & \xrightarrow[\Delta_{(X,A)}]{} & (X,A) \times X \end{array} \tag{92}$$

The map $(\pi \times \mathrm{id}_E) \circ \Delta_{(E,\varphi^*E)} : (E, \varphi^*E) \to (X, A) \times E$ is the bundle map induced by the diagonal map $\Delta_{(X,A)} : (X, A) \to (X, A) \times X$. Now we compute:

$$\begin{aligned} \widehat{\pi}_!(\pi^*h * h') &= \widehat{\pi}_!(\Delta^*_{(E,\varphi^*E)}(\pi^*h \times h')) \\ &\overset{(59)}{=} \widehat{\pi}_!\big(\Delta^*_{(E,\varphi^*E)}((\pi \times \mathrm{id}_E)^*(h \times h'))\big) \\ &\overset{(78)}{=} \Delta^*_{(X,A)}(\widehat{\mathrm{id} \times \pi})_!(h \times h') \\ &\overset{(90)}{=} \Delta^*_{(X,A)}(h \times \widehat{\pi}_!h') \\ &= h * \widehat{\pi}_!h' \,. \end{aligned}$$

In the second last equation, the sign from (90) drops out since we are considering the fiber product $\mathrm{id}_{(X,A)} \times \pi : (X, A) \times E \to (X, A) \times X$, where the bundle on the first factor has point fibers. □

## *5.4 Transgression*

Transgression of differential characters along oriented closed manifolds $\Sigma$ was constructed in [1, Chap. 9]. Here we adapt this construction to relative differential characters. Transgression of relative characters is used in [2].

### 5.4.1 Transgression along Closed Manifolds

Let $\Sigma$ be a closed oriented manifold. Denote by $\mathrm{ev}_\Sigma : C^\infty(\Sigma, X) \times \Sigma \to X$, $(f, m) \mapsto f(m)$, the evaluation map. Let $\varphi : A \to X$ be a smooth map. Denote by $C^\infty(\Sigma, (X, A))$ the space of pairs of smooth maps $(f, g) : \Sigma \to (X, A)$ such that $f \circ \varphi = g$. Composition of smooth maps $f : \Sigma \to X$ with $\varphi : A \to X$ induces a smooth map $\overline{\varphi} : C^\infty(\Sigma, A) \to C^\infty(\Sigma, X)$, $f \mapsto \varphi \circ f$. Moreover, the evaluation map yields a smooth map

$$\mathrm{ev}_\Sigma : C^\infty(\Sigma, (X, A)) \times \Sigma \to (X, A), ((f, g), m) \mapsto (f(m), g(m))$$

.

Transgression of relative differential characters in $\widehat{H}^k(\varphi; \mathbb{Z})$ is defined by pull-back along the evaluation map $\mathrm{ev}_\Sigma$ followed by integration over the fiber $\Sigma$ of the trivial bundle $\pi : C^\infty(\Sigma, (X, A)) \times \Sigma \to C^\infty(\Sigma, (X, A))$:

**Definition 50 (Transgression along $\Sigma$).** Let $\varphi : A \to X$ be a smooth map. Let $\Sigma$ be a closed oriented manifold. Transgression along $\Sigma$ is the group homomorphism

$$\tau_\Sigma : \widehat{H}^k(\varphi; \mathbb{Z}) \to \widehat{H}^{k-\dim \Sigma}(\overline{\varphi}; \mathbb{Z}), \quad h \mapsto \widehat{\pi}_!(\mathrm{ev}_\Sigma^* h).$$

From the commutative diagram (80) we conclude that transgression for relative characters commutes with the maps $\breve{\imath}$ and $\breve{p}$ and transgression for absolute characters constructed in [1, Chap. 9]:

**Proposition 51.** *Let $\varphi : A \to X$ be a smooth map. Let $\Sigma$ be a closed oriented manifold. Then the transgression maps $\tau_\Sigma$ for absolute and relative characters groups commute with the maps $\breve{\imath}$ and $\breve{p}$. Thus we have the commutative diagram (for $k \geq 2$):*

$$\begin{array}{ccccc}
\widehat{H}^{k+\dim \Sigma - 1}(A; \mathbb{Z}) & \xrightarrow{\breve{\imath}_\varphi} & \widehat{H}^{k+\dim \Sigma}(\varphi; \mathbb{Z}) & \xrightarrow{\breve{p}_\varphi} & \widehat{H}^{k+\dim \Sigma}(X; \mathbb{Z}) \\
\downarrow{\scriptstyle \tau_\Sigma} & & \downarrow{\scriptstyle \tau_\Sigma} & & \downarrow{\scriptstyle \tau_\Sigma} \\
\widehat{H}^{k-1}(C^\infty(\Sigma, A); \mathbb{Z}) & \xrightarrow{\breve{\imath}_{\overline{\varphi}}} & \widehat{H}^k(\overline{\varphi}; \mathbb{Z}) & \xrightarrow{\breve{p}_{\overline{\varphi}}} & \widehat{H}^k(C^\infty(\Sigma, X); \mathbb{Z}) .
\end{array} \tag{93}$$

*Proof.* The claim follows from naturality of the homomorphisms $\breve{\imath}$ and $\breve{p}$ and the commutative diagram (80): Let $h \in \widehat{H}^{k-1}(A; \mathbb{Z})$. For the left square in (93) we have:

$$\tau_\Sigma \breve{\imath}_\varphi(h) = \widehat{\pi}_!(\mathrm{ev}_\Sigma^*(\breve{\imath}_\varphi(h))) \overset{(32)}{=} \widehat{\pi}_!(\breve{\imath}_{\overline{\Phi}}(\mathrm{ev}_\Sigma^* h)) \overset{(80)}{=} \breve{\imath}_{\overline{\varphi}}(\widehat{\pi}_!(\mathrm{ev}_\Sigma^* h)) = \breve{\imath}_{\overline{\varphi}}(\tau_\Sigma h).$$

Let $h \in \widehat{H}^k(\varphi; \mathbb{Z})$. Similarly to the above we find for the right square in (93):

$$\tau_\Sigma \breve{p}_\varphi(h) = \widehat{\pi}_!(\mathrm{ev}_\Sigma^* \breve{p}_\varphi(h)) \overset{(33)}{=} \widehat{\pi}_!(\breve{p}_{\overline{\Phi}}(\mathrm{ev}_\Sigma^* h)) \overset{(80)}{=} \breve{p}_{\overline{\varphi}}(\widehat{\pi}_!(\mathrm{ev}_\Sigma^* h)) = \breve{p}_{\overline{\varphi}}(\tau_\Sigma h). \qquad \square$$

### 5.4.2 Transgression along Manifolds with Boundary

Let $W$ be a compact oriented manifold with boundary. We consider the space of smooth maps $(W, \partial W) \xrightarrow{(f,g)} (X, A)$ and the restriction map

$$r : C^\infty(W, (X, A)) \to C^\infty(\partial W, (X, A)) , \quad (f, g) \mapsto (f, g)|_{\partial W} .$$

The trivial fiber bundle $\pi : C^\infty(\partial W, (X, A)) \times \partial W \to C^\infty(\partial W, (X, A))$ bounds the trivial fiber bundle $\pi : C^\infty(W, (X, A)) \times W \to C^\infty(W, (X, A))$ along the restriction map $r$. Let $h \in \widehat{H}^k(X; \mathbb{Z})$. By [1, Cor. 89] the transgressed character $\tau_{\partial W} h$ admits sections along the restriction map $r$ with prescribed covariant derivative $(-1)^{k-\dim W} \fint_W \mathrm{ev}_W^* \mathrm{curv}(h)$. Similarly, for any relative character $h \in \widehat{H}^k(\varphi; \mathbb{Z})$, the transgressed character $\tau_{\partial W} h$ becomes topologically trivial upon pull-back along the restriction map. A topological trivialization of $r^* \tau_{\partial W} h$ is given by $(-1)^{k-\dim W} \fint_W \mathrm{ev}_W^* (\mathrm{curv}, \mathrm{cov})(h)$.

## Appendix: Künneth Splittings

In the appendix we recall the construction of splittings of the Künneth sequence on the level of cycles by using the classical Alexander-Whitney and Eilenberg-Zilber maps. We use these well-known splittings to construct an analogous splitting of the mapping cone Künneth sequence on the level of cycles. In the main text, we refer to these splittings as *Künneth splittings*.

### *Alexander-Whitney and Eilenberg-Zilber Maps*

We denote the well-known Alexander-Whitney and Eilenberg-Zilber maps by

$$C_*(X \times X'; \mathbb{Z}) \underset{EZ}{\overset{AW}{\rightleftarrows}} C_*(X; \mathbb{Z}) \otimes C_*(X'; \mathbb{Z}) .$$

These are chain homotopy inverses of each other with $EZ \circ AW$ chain homotopic to the identity on $C_*(X \times X'; \mathbb{Z})$ and $AW \circ EZ = \mathrm{id}_{C_*(X;\mathbb{Z}) \otimes C_*(X';\mathbb{Z})}$, see [31, p. 167]. They are used in [1, Chap. 6] to construct a splitting of the Künneth sequence on the level of cycles. Moreover, the Alexander-Whitney map relates the cross product of cochains to the tensor product.

In [1, Chap. 6], we construct a splitting of the Künneth sequence

$$0 \to \big[H_*(X;\mathbb{Z}) \otimes H_*(X';\mathbb{Z})\big]_n \xrightarrow{\times} H_n(X \times X';\mathbb{Z}) \to \mathrm{Tor}(H_*(X;\mathbb{Z}), H_*(X';\mathbb{Z}))_{n-1} \to 0$$

on the level of cycles as follows: Let $s : C_*(X;\mathbb{Z}) \to Z_*(X;\mathbb{Z})$ be a splitting of the sequence $0 \to Z_*(X;\mathbb{Z}) \xrightarrow{i} C_*(X;\mathbb{Z}) \xrightarrow{\partial} B_{*-1}(X;\mathbb{Z}) \to 0$. Similarly, we have the inclusion $i'$ and a splitting $s'$ on $X'$. Set $S := (s \otimes s') \circ AW$ and $K := EZ \circ (i \otimes i')$. Denote the cycles in the tensor product complex by $Z(C_*(X;\mathbb{Z}) \otimes C_*(X';\mathbb{Z}))$. Then we obtain:

$$\begin{array}{ccccc} 0 \longrightarrow Z_*(X;\mathbb{Z}) \otimes Z_*(X';\mathbb{Z}) & \underset{s\otimes s'}{\overset{i\otimes i'}{\rightleftarrows}} & Z(C_*(X;\mathbb{Z}) \otimes C_*(X';\mathbb{Z})) \longrightarrow \dots \\ & {\scriptstyle K}\searrow\nwarrow{\scriptstyle S} & AW \uparrow\downarrow EZ \\ & & Z_*(X \times X';\mathbb{Z}) \end{array}$$

In particular $S \circ K = (s \otimes s') \circ AW \circ EZ \circ (i \otimes i') = \mathrm{id}_{Z_*(X;\mathbb{Z}) \otimes Z_*(X';\mathbb{Z})}$. We refer to the map $S$ as the *Künneth splitting* map.

The splitting allows us to decompose any cycle $z \in Z_{k+k'-1}(X \times X';\mathbb{Z})$ according to $z = K \circ S(z) + (z - K \circ S(z))$. By the Künneth sequence, the latter represents a torsion class in $H_{k+k'-1}(X \times X';\mathbb{Z})$, whereas the former is a sum of cross products of cycles in $X$ and $X'$, respectively. Thus

$$K \circ S(z) = \sum_{\substack{(i,j) \\ i+j=k+k'-1}} \sum_{\alpha \in I} z_i^\alpha \times z'^\alpha_j$$

for appropriate cycles $z_i^\alpha \in Z_i(X;\mathbb{Z})$ and $z'^\alpha_j \in Z_j(X';\mathbb{Z})$.

## *The Mapping Cone Künneth Splitting*

Let $\varphi : A \to X$ be a smooth map. We consider the induced map $\varphi \times \mathrm{id}_{X'} : A \times X' \to X \times X'$. We use the Alexander-Whitney and Eilenberg-Zilber maps above to define Alexander-Whitney and Eilenberg-Zilber maps for the mapping cone complexes such that the following diagram commutes:

$$\begin{array}{ccc} C_*(\varphi;\mathbb{Z}) \otimes C_*(X';\mathbb{Z}) & = & \big(C_*(X;\mathbb{Z}) \otimes C_*(X';\mathbb{Z})\big) \oplus \big(C_{*-1}(A;\mathbb{Z}) \otimes C_*(X';\mathbb{Z})\big) \\ {\scriptstyle AW_{\varphi\times\mathrm{id}_{X'}}} \uparrow\downarrow {\scriptstyle EZ_{\varphi\times\mathrm{id}_{X'}}} & & {\scriptstyle AW_{X\times X'}} \uparrow\downarrow {\scriptstyle EZ_{X\times X'}} \quad {\scriptstyle AW_{A\times X'}} \uparrow\downarrow {\scriptstyle EZ_{A\times X'}} \\ C_*(\varphi \times \mathrm{id}_{X'};\mathbb{Z}) & = & C_*(X \times X';\mathbb{Z}) \oplus C_{*-1}(A \times X';\mathbb{Z}) \end{array}$$

Explicitly, we set $AW_{\varphi\times\mathrm{id}_{X'}} := AW_{X\times X'} \otimes AW_{A\times X'}$ for the mapping cone Alexander-Whitney map and $EZ_{\varphi\times\mathrm{id}_{X'}} := EZ_{X\times X'} \oplus EZ_{A\times X'}$ for the mapping cone Eilenberg-Zilber map. Since the usual Alexander-Whitney and Eilenberg-Zilber maps are natural chain maps, so are the maps on the mapping cone complexes. Moreover $AW_{\varphi\times\mathrm{id}_{X'}} \circ EZ_{\varphi\times\mathrm{id}_{X'}} = \mathrm{id}_{C_*(\varphi\times\mathrm{id}_{X'};\mathbb{Z})}$ and $EZ_{\varphi\times\mathrm{id}_{X'}} \circ AW_{\varphi\times\mathrm{id}_{X'}}$ is chain homotopic to $\mathrm{id}_{C_*(\varphi;\mathbb{Z})\otimes C_*(X';\mathbb{Z})}$. The algebraic

Künneth sequence for the homology of the complexes $C_*(\varphi;\mathbb{Z})$ and $C_*(X';\mathbb{Z})$ now reads:

$$0 \to \big[H_*(\varphi;\mathbb{Z}) \otimes H_*(X';\mathbb{Z})\big]_n \to H_n(C_*(\varphi;\mathbb{Z}) \otimes C_*(X';\mathbb{Z})) \to \mathrm{Tor}(H_*(\varphi;\mathbb{Z}), H_*(X';\mathbb{Z}))_{n-1} \to 0.$$

The mapping cone Alexander-Whitney and Eilenberg-Zilber maps yield isomorphisms:

$$H_*(\varphi \times \mathrm{id}_{X'};\mathbb{Z}) \underset{EZ_{\varphi\times\mathrm{id}_{X'}}}{\overset{AW_{\varphi\times\mathrm{id}_{X'}}}{\rightleftarrows}} H_*(C_*(\varphi;\mathbb{Z}) \otimes C_*(X';\mathbb{Z}))\,.$$

We thus obtain the topological Künneth sequence:

$$0 \to \big[H_*(\varphi;\mathbb{Z}) \otimes H_*(X';\mathbb{Z})\big]_n \to H_n(\varphi \times \mathrm{id}_{X'};\mathbb{Z}) \to \mathrm{Tor}(H_*(\varphi;\mathbb{Z}), H_*(X';\mathbb{Z}))_{n-1} \to 0.$$

We construct a splitting of this Künneth sequence at the level of cycles. Since the group of boundaries $B_*(\varphi;\mathbb{Z})$ of the mapping cone complex is a free $\mathbb{Z}$-module, we have the split exact sequence

$$0 \longrightarrow Z_*(\varphi;\mathbb{Z}) \underset{s_\varphi}{\overset{i_\varphi}{\rightleftarrows}} C_*(\varphi;\mathbb{Z}) \overset{\partial_\varphi}{\longrightarrow} B_{*-1}(\varphi;\mathbb{Z}) \longrightarrow 0\,.$$

Let $i_\varphi : Z_*(\varphi;\mathbb{Z}) \to C_*(\varphi;\mathbb{Z})$ be the inclusion. Fix a splitting $s_\varphi : C_*(\varphi;\mathbb{Z}) \to Z_*(\varphi;\mathbb{Z})$. Similarly, we denote by $i' : Z_*(X';\mathbb{Z}) \to C_*(X';\mathbb{Z})$ the inclusion and by $s' : C_*(X';\mathbb{Z}) \to Z_*(X';\mathbb{Z})$ a splitting for the smooth singular chain complex on $X'$.

Now put $S := (s_\varphi \otimes s') \circ AW_{\varphi\times\mathrm{id}_{X'}}$ and $K := EZ_{\varphi\times\mathrm{id}_{X'}} \circ (i_\varphi \otimes i')$. Then we have

$$S \circ K = (s_\varphi \otimes s') \circ AW_{\varphi\times\mathrm{id}_{X'}} \circ EZ_{\varphi\times\mathrm{id}_{X'}} \circ (i_\varphi \otimes i') = \mathrm{id}_{Z_*(\varphi;\mathbb{Z})\otimes Z_*(X';\mathbb{Z})}\,. \tag{94}$$

For relative and absolute cycles $(x,y) \in Z_*(\varphi;\mathbb{Z})$ and $z' \in Z_*(X';\mathbb{Z})$ we have $(x,y) \times z' = K((x,y) \otimes z')$. Likewise, for chains $(a,b) \in C_*(\varphi;\mathbb{Z})$ and $c \in C_*(X';\mathbb{Z})$ we have $(a,b) \times c' = EZ_{\varphi\times\mathrm{id}_{X'}}((a,b) \otimes c')$.

Denote by $Z(C_*(\varphi;\mathbb{Z}) \otimes C_*(X';\mathbb{Z}))$ the group of cycles in the tensor product complex. By (94), we obtain a splitting of the Künneth sequence on the level of cycles as in [1, Chap. 6]:

$$\begin{array}{ccccc}
0 \longrightarrow Z_*(\varphi;\mathbb{Z}) \otimes Z_*(X';\mathbb{Z}) & \underset{s_\varphi\otimes s'}{\overset{i_\varphi\otimes i'}{\rightleftarrows}} & Z(C_*(\varphi;\mathbb{Z}) \otimes C_*(X';\mathbb{Z})) \longrightarrow \dots \\
 & {\scriptstyle K}\searrow \nwarrow {\scriptstyle S} & AW_{\varphi\times\mathrm{id}_{X'}} \uparrow \downarrow EZ_{\varphi\times\mathrm{id}_{X'}} \\
 & & Z_*(\varphi \times \mathrm{id}_{X'};\mathbb{Z})
\end{array}$$

We refer to this as the *mapping cone Künneth splitting*.

## References

1. C. Bär, C. Becker: *Part I: Differential Characters and Geometric Chains*, this volume
2. C. Becker: *Cheeger-Chern-Simons theory and differential String classes.* arXiv:1404.0716, 2014
3. C. Bohr, B. Hanke, D. Kotschick: *Cycles, submanifolds, and structures on normal bundles.* Manuscripta Math. **108** (2002), 483–494
4. S. Bloch, H. Esnault: *Relative algebraic differential characters.* in: Motives, polylogarithms and Hodge theory, Part I, Int. Press Lect. Ser. **3**, Int. Press, Somerville 2002, 47–73
5. A. Borel, F. Hirzebruch: *Characteristic classes and homogeneous spaces I.* Amer. J. Math. **80** (1958), 458–538
6. R. Bott, L.W. Tu: *Differential Forms in Algebraic Topology.* Springer, New York-Berlin 1982
7. M. Brightwell, P. Turner: *Relative differential characters.* Comm. Anal. Geom. **14** (2006), 269–282
8. J.-L. Brylinski: *Loop Spaces, Characteristic Classes and Geometric Quantization.* Progress in Mathematics, Birkhäuser, Boston 1993
9. U. Bunke: *Differential cohomology.* arXiv:1208.3961, 2012
10. U. Bunke, D. Gepner: *Differential function spectra, the differential Becker-Gottlieb transfer, and applications to differential algebraic K-theory.* arXiv:1306.0247, 2013
11. U. Bunke, M. Kreck, T. Schick: *A geometric description of differential cohomology.* Ann. Math. Blaise Pascal **17** (2010), 1–16
12. U. Bunke, T. Schick: *Smooth K-theory.* Astérisque **328** (2009), 43–135
13. U. Bunke, T. Schick: *Uniqueness of smooth extensions of generalized cohomology theories.* J. Topol. **3** (2010), 110–156
14. A. L. Carey, S. Johnson, M. K. Murray: *Holonomy on D-branes.* J. Geom. Phys. **52** (2004), 186–216
15. J. Cheeger: *Multiplication of differential characters.* Symposia Mathematica, Vol. XI, 441–445. Academic Press, London 1973
16. J. Cheeger, J. Simons: *Differential Characters and Geometric Invariants.* Geometry and topology, Lecture Notes in Math.**1167**, Springer, Berlin 1985, 50–80
17. J.L. Dupont, R. Ljungmann: *Integration of simplicial forms and Deligne cohomology.* Math. Scand. **97** (2005), 11–39
18. C.-O. Ewald: *A de Rham isomorphism in singular cohomology and Stokes theorem for stratifolds.* Int. J. Geom. Methods Mod. Phys. **2** (2005), 63–81
19. F. Ferrari Ruffino: *Relative Deligne cohomology and differential characters.* arXiv:1401.0641v1, 2014
20. D. Freed, J. Lott: *An index theorem in differential K-theory.* Geom. Topol. **14** (2010), 903–966
21. R. Harvey, B. Lawson: *Lefschetz-Pontrjagin duality for differential characters.* An. Acad. Brasil. Ciênc. **73** (2001), 145–159
22. R. Harvey, B. Lawson: *From sparks to grundles – differential characters.* Comm. Anal. Geom. **14** (2006), 25–58
23. R. Harvey, B. Lawson: *D-bar sparks.* Proc. Lond. Math. Soc. (3) **97** (2008), 1–30
24. R. Harvey, B. Lawson, J. Zweck: *The de Rham-Federer theory of differential characters and character duality.* Amer. J. Math. **125** (2003), 791–847
25. N. Hitchin: *Lectures on special Lagrangian submanifolds.* In: Winter School on Mirror Symmetry, Vector Bundles and Lagrangian Submanifolds, Amer. Math. Soc., Providence 2001, 151–182
26. M. J. Hopkins, I. M. Singer: *Quadratic functions in geometry, topology, and M-theory.* J. Diff. Geom. **70** (2005), 329–452

27. J. Kalkkinen: *Holonomies of intersecting branes.* Fortschr. Phys. **53** (2005), 913–918
28. M. Kreck: *Differentiable Algebraic Topology. From Stratifolds to Exotic Spheres.* Amer. Math. Soc., Providence 2010
29. R. Ljungmann: *Secondary invariants for families of bundles.* PhD thesis, Aarhus 2006
30. J. Lott: *$\mathbb{R}/\mathbb{Z}$ index theory* Comm. Anal. Geom. **2** (1994), 279–311
31. J. McCleary: *A user's guide to spectral sequences.* 2. ed., Cambridge University Press, Cambridge 2001
32. M.K. Murray: *Bundle gerbes.* J. London Math. Soc. (2) **54** (1996), 403–416
33. C. Redden: *Trivializations of differential cocycles.* arXiv:1201.2919, 2012, to appear in J. Homotopy Relat. Struct.
34. Z. Shahbazi: *Relative gerbes.* J. Geom. Phys. **56** (2006), 1326–1356
35. Z. Shahbazi: *Prequantization of quasi-Hamiltonian spaces.* Int. Math. Res. Not. 2006, Art. ID 29354, 22p.
36. J. Simons, D. Sullivan: *Axiomatic characterization of ordinary differential cohomology.* J. Topol. **1** (2008), 45–56
37. J. Simons, D. Sullivan: *Structured vector bundles define differential K-theory* in: Quanta of maths, 579–599, Amer. Math. Soc., Providence 2010
38. D. Stevenson: *Bundle 2-gerbes.* Proc. London Math. Soc. (3) **88** (2004), 405–435
39. T. tom Dieck: *Algebraic Topology.* EMS Textbooks in Mathematics, European Mathematical Society, Zürich 2008
40. R. Thom: *Quelques propriétés globales des varits différentiables.* Comment. Math. Helv. **28**, (1954), 17–86
41. A.W. Tucker: *Degenerate cycles bound.* Math. Sbornik **3 (45)**, (1938), 287–288
42. M. Upmeier: *Products in Generalized Differential Cohomology.* arXiv:1112.4173v2, 2011
43. K. Waldorf: *Surface holonomy.* Handbook of pseudo-Riemannian geometry and supersymmetry, IRMA Lect. Math. Theor. Phys. **16**, European Mathematical Society, Zrich 2010, 653–682
44. K. Waldorf: *String connections and Chern-Simons theory* Trans. Amer. Math. Soc. **365** (2013), 4393–4432
45. R. Zuccini *Relative topological integrals and relative Cheeger-Simons differential characters.* J. Geom. Phys. **46** (2003), 355–393

# Index

# *LECTURE NOTES IN MATHEMATICS*

Edited by J.-M. Morel, B. Teissier; P.K. Maini

**Editorial Policy** (for the publication of monographs)

1. Lecture Notes aim to report new developments in all areas of mathematics and their applications - quickly, informally and at a high level. Mathematical texts analysing new developments in modelling and numerical simulation are welcome.
   Monograph manuscripts should be reasonably self-contained and rounded off. Thus they may, and often will, present not only results of the author but also related work by other people. They may be based on specialised lecture courses. Furthermore, the manuscripts should provide sufficient motivation, examples and applications. This clearly distinguishes Lecture Notes from journal articles or technical reports which normally are very concise. Articles intended for a journal but too long to be accepted by most journals, usually do not have this "lecture notes" character. For similar reasons it is unusual for doctoral theses to be accepted for the Lecture Notes series, though habilitation theses may be appropriate.

2. Manuscripts should be submitted either online at www.editorialmanager.com/lnm to Springer's mathematics editorial in Heidelberg, or to one of the series editors. In general, manuscripts will be sent out to 2 external referees for evaluation. If a decision cannot yet be reached on the basis of the first 2 reports, further referees may be contacted: The author will be informed of this. A final decision to publish can be made only on the basis of the complete manuscript, however a refereeing process leading to a preliminary decision can be based on a pre-final or incomplete manuscript. The strict minimum amount of material that will be considered should include a detailed outline describing the planned contents of each chapter, a bibliography and several sample chapters.
   Authors should be aware that incomplete or insufficiently close to final manuscripts almost always result in longer refereeing times and nevertheless unclear referees' recommendations, making further refereeing of a final draft necessary.
   Authors should also be aware that parallel submission of their manuscript to another publisher while under consideration for LNM will in general lead to immediate rejection.

3. Manuscripts should in general be submitted in English. Final manuscripts should contain at least 100 pages of mathematical text and should always include

   – a table of contents;
   – an informative introduction, with adequate motivation and perhaps some historical remarks: it should be accessible to a reader not intimately familiar with the topic treated;
   – a subject index: as a rule this is genuinely helpful for the reader.

   For evaluation purposes, manuscripts may be submitted in print or electronic form (print form is still preferred by most referees), in the latter case preferably as pdf- or zipped ps-files. Lecture Notes volumes are, as a rule, printed digitally from the authors' files. To ensure best results, authors are asked to use the LaTeX2e style files available from Springer's web-server at:

   ftp://ftp.springer.de/pub/tex/latex/svmonot1/ (for monographs) and
   ftp://ftp.springer.de/pub/tex/latex/svmultt1/ (for summer schools/tutorials).

Additional technical instructions, if necessary, are available on request from lnm@springer.com.

4. Careful preparation of the manuscripts will help keep production time short besides ensuring satisfactory appearance of the finished book in print and online. After acceptance of the manuscript authors will be asked to prepare the final LaTeX source files and also the corresponding dvi-, pdf- or zipped ps-file. The LaTeX source files are essential for producing the full-text online version of the book (see http://www.springerlink.com/openurl.asp?genre=journal&issn=0075-8434 for the existing online volumes of LNM). The actual production of a Lecture Notes volume takes approximately 12 weeks.

5. Authors receive a total of 50 free copies of their volume, but no royalties. They are entitled to a discount of 33.3 % on the price of Springer books purchased for their personal use, if ordering directly from Springer.

6. Commitment to publish is made by letter of intent rather than by signing a formal contract. Springer-Verlag secures the copyright for each volume. Authors are free to reuse material contained in their LNM volumes in later publications: a brief written (or e-mail) request for formal permission is sufficient.

**Addresses:**

Professor J.-M. Morel, CMLA,
École Normale Supérieure de Cachan,
61 Avenue du Président Wilson, 94235 Cachan Cedex, France
E-mail: morel@cmla.ens-cachan.fr

Professor B. Teissier, Institut Mathématique de Jussieu,
UMR 7586 du CNRS, Équipe "Géométrie et Dynamique",
175 rue du Chevaleret
75013 Paris, France
E-mail: teissier@math.jussieu.fr

*For the "Mathematical Biosciences Subseries" of LNM:*

Professor P. K. Maini, Center for Mathematical Biology,
Mathematical Institute, 24-29 St Giles,
Oxford OX1 3LP, UK
E-mail: maini@maths.ox.ac.uk

Springer, Mathematics Editorial, Tiergartenstr. 17,
69121 Heidelberg, Germany,
Tel.: +49 (6221) 4876-8259

Fax: +49 (6221) 4876-8259
E-mail: lnm@springer.com